David J. Colliver Co

ound Semiconductor Technology D

d J. Colliver Compound Semiconduct

echnology David J. Colliver Compoun

emiconductor Technology David J. Co

er Compound Semiconductor Techn

gy David J. Colliver Compound Semi

ductor Technology David J. Colliver C

pound Semiconductor Technology

avid J. Colliver Compound Semicond

or Technology David J. Colliver Comp

nd Semiconductor Technology Davi

Colliver Compound Semiconductor

nology David J. Colliver Compound

iconductor Technology David J. Colli

Compound Semiconductor Technology

David J. Colliver Royal Signals and Radar Establishment Great Malvern Worcestershire England

610 Washington Street
Dedham, Massachusetts 02026

Printed and bound in the United States of America.

Library of Congress Card Catalog Number: 75-31377

Standard Book Number: 0-89006-052-5

table of contents

preface

Thirteen years have passed since J.B. Gunn first observed microwave oscillations in gallium arsenide, and it is five years since the group at the Royal Radar Establishment showed that indium phosphide had many potential advantages as an oscillator material. Both of these substances owe the great part of their development to the interest in solid state microwave generation by means of exploiting the transferred electron phenomena. But many other device families also have benefited from these advances — microwave mixers and varactors, field-effect transistors, optoelectric devices, fast logic elements, and avalanche diodes.

The technology of these materials is becoming well established. Many of the early techniques were direct adaptations of those already proven in use of silicon technology; now, there are instances where the experience gained from the development of the compound semiconductors has spurred progress in silicon devices. Though the tendency is for these separate technologies to gradually merge into one which covers all materials and devices, there remain essential differences between silicon and the compounds.

The purpose of this book is to introduce the reader to the basic techniques and concepts required for the production of microwave devices from semiconductor compounds. Those methods currently being investigated have been mentioned only briefly because their importance has not yet been established. Techniques such as ion implantation and molecular beam epitaxy show great promise; I would recommend that progress in these subjects be followed in the literature.

The key papers included in the text have been chosen primarily to provide the best introduction, rather than the most recent application, to a given process or technique. Further developments can be found in the references.

Anybody who has worked practically in this field should appreciate that there are many areas open for innovation and continuing experimentation; thus, there are methods which are peculiar to individual laboratories. I have endeavored to restrict the text to generally accepted techniques; where a specific practice is discussed, an appropriate reference is given.

Finally, I would like to acknowledge the contributions of my friends and colleagues at the Royal Radar Establishment, in industry, and abroad. Without their free and open discussion, this book could not have been written.

acknowledgment

The author is grateful to the publishers and authors of the key papers for permission to reproduce them in this volume.

Cover design is a detail from a photograph of Vapour Phase Epitaxy equipment, supplied courtesy of Phillip Schaffer and Joseph White, Microwave Associates, Burlington, Massachusetts, US.

introduction

The investment in semiconductor technology is, and will continue to be, high. While major advances usually require many years in development before the techniques mature to the stage where they have broad commercial significance, smaller refinements and improvements in the basic technology are being made continually. The latter lead to an improved understanding of the device physics and are manifested in greater yields, better performance, and superior overall reliability.

There is yet another difference between the completely new technique (which often leads to new device families) and the general improvement in technology — the former is well reported in the technical journals whereas the latter, as an individual item, is not generally published. This situation occurs either because the general technological improvement is not considered to be significant enough to merit publication, or because the future technical viability of the company depends on such "secret" processes which give one a lead over one's competitors.

This continual advancement has posed its own problems in the preparation of this book. Unfortunately, the body of scientists and technicians throughout the world working on semiconductor technology are not waiting for the text to catch up with them. Consequently, the specialist reader finds that the position in some areas has changed since the book was first drafted. Throughout the text I have endeavoured to introduce the techniques at as general a level as possible, with references to specific methods. Few of these basic techniques will be superceded in the near future.

At the time of going to press, however, there have been reported several major developments which have not been stressed in the text. These developments have resulted in greatly improved devices; it is relevant to draw the reader's attention to them at this point.

The field effect transistor has been developed to the stage where it is a major contender as an amplifier for low noise receiver front-ends at frequencies up to 20 GHz, due to improvements in all the contributory areas of technology. Substrate materials have much improved crystallographic properties, and can be masked by improved control in the epitaxial deposition of semi-insulating buffer layers between substrate and the active region. Ion implantation is now being used successfully to dope the active region into the 10^{17} cm^{-3} range — a fact of particular significance, as the active region is only about 0.2 μm thickness. The analysis of these layers has led to a better understanding of C-V profiling of material grown on semi-insulating substrates.

The exposed planar surface of the FET can be stabilised either by deposited oxides or nitrides, or by chemically formed native oxides. These new oxidation methods have been applied also to planar varactors; they will, no doubt, extend into other areas. Of particular interest for the future is the formation of native oxides on GaAs either by chemical or anodic deposition.

Another major obstacle in planar technology, efficient heat removal from the active region, has been overcome in FET's by the successful inversion of the device onto a heat sink.

In transferred electron devices, particularly with reference to InP oscillators, a much improved understanding of contact phenomena has resulted from experiments with the doping profile adjacent to the semiconductor boundary. These experiments were possible only because epitaxial growth of the vapour deposited layers has been brought under control; variations in the phosphorus trichloride ratio in the gas stream lead to variation in the instantaneous background doping in the as-grown layer. Whether similar profiles could yield improved GaAs transferred electron performance has yet to be demonstrated.

IMPATT performance and reliability continue to improve, largely through the control of existing technology, again, control of the layer doping profile is critical.

The growing importance of optical devices, particularly for energy conversion is leading to a renewed incentive for the large scale economic production of GaAs, for the formation of junctions, and for forward looking work on other compound materials. This field will be an interesting one to follow in the future.

It can thus be seen how varied and numerous are the advances being made, even though this survey has only covered the major items, hopefully with no serious omissions.

The following chapters describe the techniques, processes, and technology required to produce a working device from raw starting materials. It is on this foundation that all present day compound semiconductor devices are based. The significance of new developments can be judged by the reader in relation to this established foundation.

chapter one

materials growth and impurity control

Any advanced semiconductor device requires bulk material for substrates and epitaxial material for the high quality active region. Both must be doped to be of the correct carrier type and impurity level. In addition, it is often subsequently necessary to produce junction structures and to passivate the surface for isolation or control of junction geometries. The management of these processes is extremely critical. It is the determining factor in the final device performance, particularly for the microwave devices where the operation depends so much on the purity and characteristics of the epitaxial material. Indeed, the transferred electron effect is a bulk property of the substance itself.

Substrate material must be either highly conductive or semi-insulating, depending on the device for which it is used. This substrate is necessary as a mechanical support for the comparatively thin epitaxial layer, and it acts as a nucleation surface during subsequent growth. Substrate thicknesses are in the 100-250 micron range. For the highly conductive case (varactors, transferred electron devices, IMPATT's), n-type doping is from 10^{18}–10^{19} cm^{-3}; for semi-insulation, resistivities are in the area of 10^4–10^8 Ω-cm (planar devices such as FET's and logic elements).

Epitaxial material thicknesses span the range from sub-micron to perhaps 100 microns. Transferred electron devices have an active region thickness dependent on operating frequency, with $\ell = 100/f$ (f in GHz, ℓ in microns). Impurity doping level requirements are from mid 10^{14} to mid 10^{16} cm^{-3} for active regions and somewhat higher for contact layers and junction structures.

1.1 bulk growth

Bulk material is grown in the form of large ingots of single crystal; the major problem to be overcome during growth of the compound semiconductors results from the high vapour pressures encountered at growth temperature. These volatile components must be contained either by growth in a completely sealed system or by suppression of the volatilisation by encapsulation of the melt in an inert liquid. These considerations have led to a number of different growth techniques with respective advantages and disadvantages.

The method of melt growth which has been proved to be suitable for a large number of compounds is that of Czochralski or vertical pulling, in which a seed crystal is slowly rotated and withdrawn from a melt contained in a crucible. It is possible to suppress the volatilisation from the melt by covering it with a layer of inert liquid; Mullin[1] achieved this end for GaAs by using boric oxide and pressurising the growth chamber. This procedure has been extended to InP[2], although the extremely high vapour pressure of phosphorus requires pressurisation of the chamber to 27 atmospheres.

The apparatus and method of growth are described fully in the key paper by Mullin et al. The purity of material is dependent on the contaminants introduced from the growth apparatus, the silica crucible, and the starting materials used.

The technique widely used for production of large quantities of GaAs substrate material is the horizontal Bridgman method.[3] The main advantage of horizontal growth is that the process is readily automated. A boat containing the semiconductor compound and crystal seed is sealed inside a quartz ampoule, the other end of which contains a small arsenic excess. The furnace consists of at least two zones, one held at 612°C in which the arsenic end of the bulb sits. The ingot is melted at 1270°C and moved slowly out of the hot area into the cooler one, and crystallisation takes place[86].

Another variation of this method is to maintain both furnace and ampoule in a fixed state and programme the furnace profile to give a temperature gradient across the melt. This technique is known as the gradient freeze, and is also suitable for automation. Growth rates are on the order of 1 cm/hour.

The higher vapour pressure of InP makes the Czochralski method of growth more advantageous from a safety point of view; however, this material has not yet been produced in as great a quantity as has GaAs.

Crystals of GaAs commercially available have dislocation densities in the range of 10^3-10^4 cm^{-2}, permitting the growth of dislocation free material by the Czochralski method, provided that conditions are carefully monitored. This comment is not true for the Bridgman technique because of the wetting of the boat by the grown ingot.

Doping of the crystal is achieved by adding the required amount of impurity to the melt. The final doping level is determined by the distribution coefficient of the particular impurity:

$$K = \frac{\text{concentration in solid}}{\text{concentration in liquid}}$$

For GaAs, the common n-type impurities are silicon or tellurium, with K of 0.14 and 0.06 respectively. Heavily doped n^+ material is usually doped to approximately the 10^{18} cm^{-3} level. For InP, the n-type dopants commonly used are tin or germanium, with K of $2.1.10^{-2}$ and $2.4.10^{-2}$, respectively. Zinc or cadmium are used for p doping in both materials, and chromium converts both to a semi-insulating form with resistivities in the range 10^4 (InP) to 10^8 (GaAs)Ω-cm.

After completion of its growth, the crystal is cooled slowly and the boric oxide is removed with hot water before cooling is finished. Then the crystal is oriented to the desired plane before it is sliced into individual wafers with a diamond or multiple bladed saw. The choice of cutting procedure varies with individual preference; the criteria are minimum surface damage, minimum kerf loss, and maximum yield of usable wafers.

Typical wafer size is about 2 cm diameter by 0.5 mm thick, cut from crystals of around 100g weight. Each crystal boule is analysed (usually twice and separately) by various techniques which we discuss later.

key paper

Liquid Encapsulation Crystal Pulling At High Pressures

J.B. Mullin, R.J. Heritage, C.H. Holliday, and B.W. Straughan
(Ministry of Technology, Royal Radar Establishment, Great Malvern, Worcs., UK)

Journal of Crystal Growth, Volume 3, 4 (1968)
Pages 281-285

The technique of Liquid Encapsulation that is the use of an inert liquid as a seal in preventing the loss of volatile components from dissociable compounds or their melts is discussed with particular reference to the growth of compounds having dissociation pressures in excess of one atmosphere. A pressure chamber is described in which it is possible to pull crystals in a gas pressure of 200 atm. The practical application of the technique to the growth of good quality InP and GaP is also described; and a comparison of the impurity content of these crystals grown either from silica or vitreous carbon crucibles is made.
The relationship between the impurity content of the crystals and their electrical properties is also briefly considered.

1. Introduction

The technique of Liquid Encapsulation, that is the use of an inert liquid as a seal in preventing the loss of volatile components from dissociable compounds or their melts is a useful and technologically important development particularly in the controlled solidification of semiconductors. It has, for example, been suggested by Mullin, Straughan and Brickell[1]) as having potential advantages in the crystal growth of GaP. It is the purpose of this paper to discuss these advantages and report the results of the practical application of a Liquid Encapsulation technique to the vertical pulling of GaP and InP, both compounds having dissociation pressures at their melting points in excess of atmospheric pressure.

InP and GaP can be grown by conventional techniques in which the melt is maintained in equilibrium with its vapour in a sealed or semi-sealed system. Most of these have already been reviewed[2]). A recent development of the vertical Bridgeman technique by Blum and Chicotka[3]) appears as an improvement over existing techniques and has resulted in the growth of good quality crystals.

In these techniques the continuous loss of volatile components from a dissociable compound or its melt is prevented by the use of a closed or effectively-closed container in which the vapour pressure of the volatile component(s) is allowed to build up to the dissociation pressure of the compound or its melt. The technical difficulties in these processes arise from the necessity to contain the pressure of a reactive gas or gases.

The containment problem can be overcome with the use of an inert liquid such as boric oxide. This was first demonstrated by Metz, Miller and Mazelsky[4]) who were able to contain the volatile compounds PbTe and PbSe during crystal growth.

The use of liquid seals for containing dissociable materials is as a concept of general applicability and a number of potentially useful applications have been discussed[1]). Some of these such as Liquid Encapsulated zone refining[5]) and doping crystals with volatile elements during crystal growth have been demonstrated practically[5,6]). The concept has also been applied to the crystal pulling of device-quality InAs[1]), GaAs[1,7]) $Pb_xSn_{1-x}Te$[8]) and $PtSb_2$[9]). None of these compounds, however, develops appreciable dissociation pressures on melting.

2. Basic principles

The principles involved in the application of the technique to highly dissociable compounds can be considered with reference to fig. 1. This is a diagrammatic representation of the pressure chamber in which the crystals were pulled. A conventional arrangement of crucible, susceptor and internal rf heating coil (l) was adopted. The encapsulant, a pretreated charge of B_2O_3, was added to the starting material.

The main physical requirement of the system is that the pressure of *inert* gas P_g on the surface of the liquid

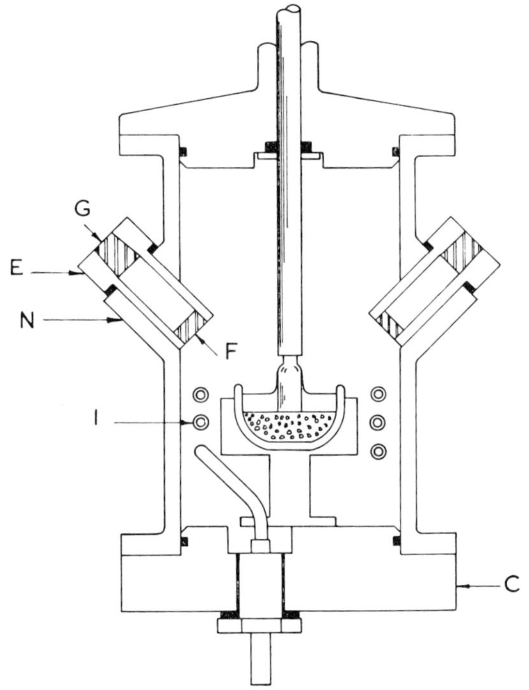

Fig. 1. Diagrammatic representation of 200 atm pressure chamber. Labelled parts noted in text.

encapsulant is greater than the dissociation pressure P_d of the molten compound. The encapsulant can then act as a liquid seal preventing vapour loss. If $P_d > P_g$ volatile constituents simply bubble through the encapsulant.

The principal technical advantage of this arrangement is that the walls of the containing vessel can be kept at room temperature and out of contact with the components of the compound. The walls of the chamber need to be sufficiently strong to withstand a pressure somewhat in excess of the dissociation pressure of the melt.

A conventional sealed or semi-sealed system would require the chamber to be physically and chemically stable at a temperature above the condensation point of the volatile species.

In the case of GaP this temperature would be in excess of 700 °C. Since phosphorus will react with most engineering metals at this temperature the problem of chemical stability is a severe one.

The use of an encapsulant also presents a problem of chemical stability, albeit a less severe one. The basic properties required of an encapsulant – physical and chemical stability with respect to the melt crucible and gaseous environment, transparency in the visible, a density less than that of the melt and a low melting point – have already been stressed[1]). A property of the encapsulant which deserves further consideration is its ability to wet the crystal, the melt and the crucible. Whilst B_2O_3 is particularly effective in this respect, the alkaline earth halides and many other salt-type encapsulants are not so effective. It is the temperature dependence of the viscosity combined with the wetting properties of the encapsulant which determine the thickness of the film which adheres to the hot crystal and prevents vapour loss from the solid. The depth of encapsulant is therefore important since it is an adjustable parameter which can be used to control the temperature at which the crystal emerges above the level of the encapsulant on the melt. For example the most effective depths have been found to be less than 5 mm for InAs about 5–8 mm for InP and 8–10 mm for GaAs and GaP. These figures can only serve as a guide since they will vary with each individual system and hence have to be found empirically.

3. Apparatus

The water cooled pressure chamber has been designed for use up to 200 atm. It can be obtained as a standard attachment to commercial versions* of the crystal puller as shown in fig. 2. The three main sections which are manufactured from austenitic stainless steel comprise a top plate, the main body and a bottom plate. The top plate contains the pull rod seal and also functions as a pull rod housing. The main body has been designed with a shoulder (N) so that the window housings (E) and thermocouple leads-through can readily be incorporated. The window housings contain two silica windows (G and F) so designed that the upper one can withstand the chamber pressure and the lower one can prevent condensible material reaching the upper one. The lower one is readily demountable and replaceable; under operating conditions it is sufficiently hot due to proximity to the crucible that volatile matter does not condense on it.

The crucible is mounted on a support pedestal attached to the base plate C. The leads of the rf induction heating coil pass through the base plate via insulators.

* Metals Research, Cambridge; or Electroheating, Lombard Road, London, S.W. 19.

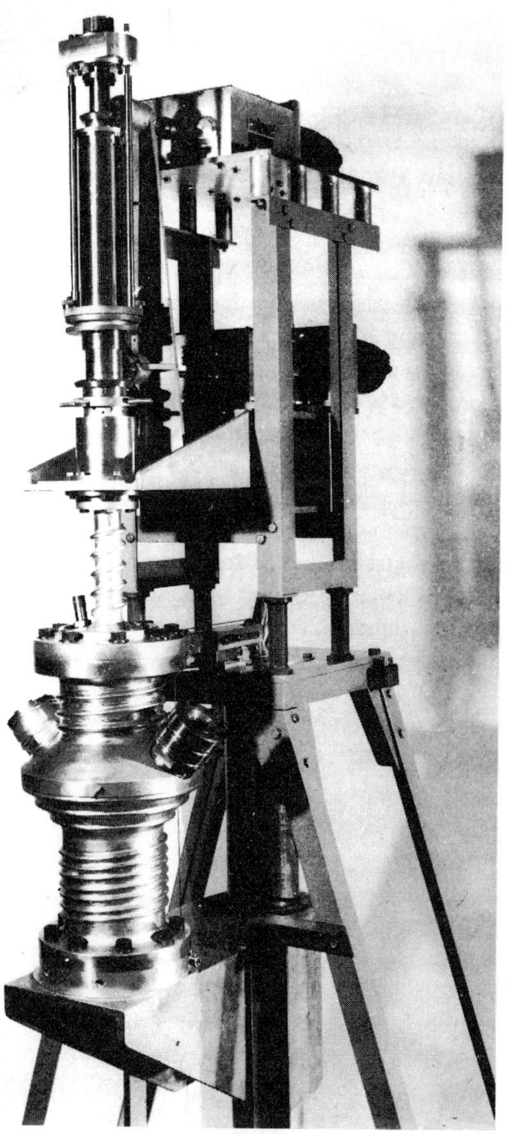

Fig. 2. Crystal puller and pressure chamber used for pulling InP and GaP.

The base plate is attached to a bracket which can be raised or lowered on precision guides; hence the crucible, coil, susceptor and lower windows are all readily accessible as is the seed chuck with the pull rod in the down position. The complete assembly is shown in fig. 2.

The crystal pulling process is viewed by closed circuit television; in the initial stages additional illumination is provided through the second window, above 600 °C the crucible is sufficiently self illuminating.

4. Experimental

The experimental procedure is quite straightforward; it is very similar in practice and ease of operation to the procedure adopted for the growth of GaAs[1]). The starting charge of InP or GaP solution-grown crystallites or occasionally bulk polycrystalline material is loaded into a silica or vitreous carbon* crucible together with a suitable dry lump of B_2O_3. The chamber is sealed and evacuated: the vacuum assists the sealing of the base plate. It is then pressurised with an inert gas, usually helium, to a pressure P_g in excess of the dissociation pressure of the compound. Values of P_g are 25 and 40 atm in the case of InP and GaP respectively.

When the charge is molten the seed is rotated and lowered through the encapsulant until it is wetted by the melt. Crystal pulling then follows standard practice. It is possible to confirm contact between the seed and the melt by the use of an electrical circuit involving the seed and crucible. Its use is not essential however.

5. Results and conclusions

Approximately 15 InP and GaP crystals have been grown. Observation of the crystal during growth either visually using mirrors – there is a psychological objection to direct viewing – or by television camera can be very good both for InP and GaP; but, the transparency of the B_2O_3 can be affected by the quality of the starting material. Small solution-grown crystallites or powder tend to disperse in the encapsulant and this can produce a cloudiness which impairs visibility. Good quality bulk starting material normally gives a clear layer and the conditions of crystal growth enable good diameter control to be achieved. Examples of InP and GaP crystals are shown in fig. 3.

The present study has been aimed at confirming the viability of the technique and assessing the chemical purity of crystals grown in silica and vitreous carbon. Some electrical measurements have also been made. These results are listed in table 1.

The most significant feature of the analytical results is the difference in the boron content between the InP (L series) and the GaP (K series) crystals. The boron

* Vitreous Carbons Ltd., Bishop Auckland, Co. Durham.

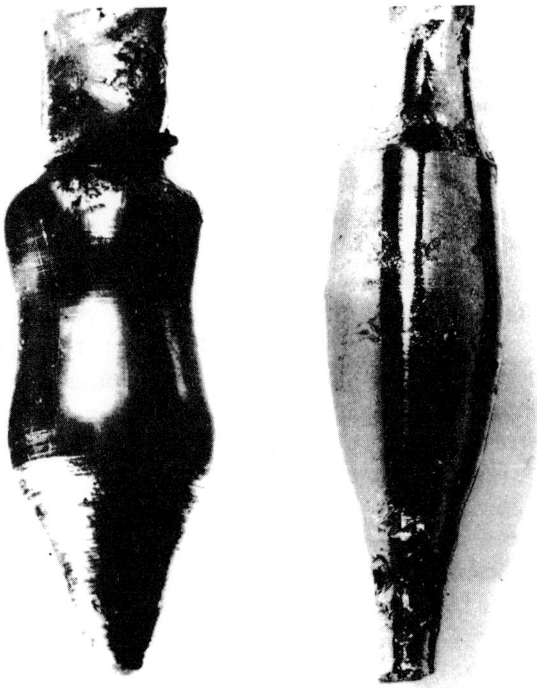

Fig. 3. Crystals of GaP (left) and InP grown by the liquid encapsulation technique. Maximum diameter of GaP is ~ 15 mm and of InP is 11 mm.

content in the InP crystals is less than 1 ppm whereas in the GaP crystals it ranges from 250–2000 ppm. The reason for the high boron content in the GaP crystals is not known. From elementary considerations one might expect the solubility of B_2O_3 in a similar series of materials at their melting points to increase with temperature. The melting points of GaP and InP[2]) are respectively 1470 °C and 1050 °C. However, the melting point of GaAs is 1237 °C and the boron content of GaAs crystals grown by the liquid encapsulation technique using B_2O_3 is usually in the range 0.01–0.1 atomic ppm[7,10]).

The results of the analyses of the silicon content of the crystals are also significant. They show that GaP crystals grown from silica crucibles contain more silicon than the starting material whereas GaP crystals grown from vitreous carbon contain less silicon than the starting material. The distribution coefficient of silicon in GaP is less than 1. This indicates that silica crucibles do contaminate GaP melts, a fact noted as unpublished comment in the discussion of other techniques[2]). The InP results also indicate a tendency for crystals grown in silica to take up silicon; L1 was grown from LSM1. Crystals L3 etc. contained both LSM1 and LSM2 and

Table 1

Analytical results on InP and GaP crystals

	InP								GaP				
Crystal number	LSM1	LSM2	L1	L3	L5	L6	L7	L8	KSM1	K1	K2	K3	K4
Crucible material	—	—	SiO_2	SiO_2	SiO_2	SiO_2	SiO_2	SiO_2	—	VC	VC	SiO_2	SiO_2
Element	ppm atomic								ppm atomic				
B	ND	ND	ND	ND	ND	ND	ND	ND	ND	250	500	700	2000
Si	<2	40	10	15	50	10	20	50	15	5	2	50	100
C	5	20	10	40	50	30	20	10	150	40	25	130	230
As	15	ND	3400	140	2000	20	25	5	—	—	—	—	—
Ga	8	ND	650	2500	1500	25	200	20	M	M	M	M	M
No.* carriers/cm^3 at 20°C	2.9×10^{16}		4.5×10^{16}	2.16×10^{16}	5.5×10^{16}	2.92×10^{16}	2.58×10^{16}			9.5×10^{17}	7.4×10^{17}		7.3×10^{17}
Mobility cm^2/V sec	1570		2920	3460	3070	3260	3060			39	105		108

General: Analyses were made using an AEI-MS702 mass spectrometer. Accuracy ±30%, all results from densitometer measurements.

Notes: No other elements at concentrations >1 ppm were detected, except:
S present in all samples; $^{16}O_2$ interferes with ^{32}S but result confirmed on ^{34}S.
O present in all samples; evidence of very high surface contamination on GaP.
The residual hydrocarbons in the mass spectrometer have enhanced the carbon analysis but this is not greater than a few ppm.

* All specimens n-type except K1 which was p-type.

Symbols: ND = no line detected; hence concentration <1 ppm,
VC = vitreous carbon,
LSM = InP starting material,
KSM = GaP starting material.
M = main constituent.

parts of previous crystals and consequently no positive conclusions about their Si-contents can be drawn.

The carbon content both of the GaP starting material and grown crystals are high; but the carbon content of K1 and K2, crystals grown from vitreous carbon crucibles is nevertheless significantly lower than the rest. K3 and K4 starting charges were initially melted in vitreous carbon. High carbon contents (10^{18}–10^{19} atoms/cm^3) in GaP prepared in carbon and BN have been noted[2]) in other work. Although, volatile CP compounds are known[11]) and could be important the starting material is the most likely potential source of carbon.

The Ga and As in the InP crystals is due to the use of a GaAs seed for L1. Since the generation of crystals usually involved pulling parts of previous crystals the Ga and As contents of crystals remains in high but diminishing concentration in succeeding crystals.

The electrical measurements indicate that the Si and C is probably not fully active i.e., one electron or hole per atom. Boron is probably present as a boron–oxygen complex; even elemental boron would be electrically inactive. A more exhaustive and systemmatic study will be required before significant quantitative relationships between the chemical and electrical properties can be deduced.

The main conclusion from the results is that good quality crystals of InP and GaP can be grown by the liquid encapsulation technique under B_2O_3, in a pressure chamber and that the technique offers considerable potentiality for the crystal growth of other highly dissociable materials. The problem of crucible contamination and the role of B_2O_3 in GaP needs however much more extensive investigation.

Acknowledgements

The authors gratefully acknowledge the assistance of Miss A. Royle for the electrical measurements and Dr. S. E. R. Hiscocks and colleagues for useful discussion and help.

References

1) J. B. Mullin, B. W. Straughan and W. S. Brickell, J. Phys. Chem. Solids **26** (1965) 782.
2) D. Richman, Preparation of Indium Phosphide; J. F. Miller, Preparation of Gallium Phosphide, in: *Compound Semiconductors*, Vol. 1. Preparation of III–V Compounds (Reinhold, New York, 1962) p. 214; p. 194.
3) S. E. Blum and R. J. Chicotka, J. Electrochem. Soc. **115** (1968) 298.
4) E. P. A. Metz, R. C. Miller, and R. Mazelsky, J. Appl. Phys. **33** (1962) 2016.
5) E. M. Swiggard, J. Electrochem. Soc. **114** (1967) 976.
6) H. A. Chedzey and D. T. J. Hurle, Brit. J. Appl. Phys. **17** (1966) 699.
7) S. J. Bass and P. E. Oliver, 1966 *Symposium on GaAs* (Institute of Physics and the Physical Society, London, 1967) p. 41.
8) S. E. R. Hiscocks and P. D. West, J. Mater. Sci. **3** (1968) 76.
9) C. T. Elliott and S. E. R. Hiscocks, J. Mater. Sci. **3** (1968) 174.
10) Unpublished results.
11) C. J. Frosch and L. Derick, J. Electrochem. Soc. **108** (1961) 251.

Discussion

J. S. SHAH, H. H. Wills Phys. Lab., Bristol, England

Does the presence of a B_2O_3 layer change the nature of the interface? If so, which parameters.

B. W. STRAUGHAN

We have made no experimental studies of this.

G. S. MEILING, Corning Glass Co., U.S.A.

(1) What were the pulling rates for GaP, and

(2) Was any provision available for adjustment of the pulling rod to the thermal centre of the melt?

B. W. STRAUGHAN

(1) Typically 2 cm/hr.

(2) No, it is not necessary.

1.2 epitaxial growth

The production of a controlled film known as epitaxy, on a single crystal substrate, is essential to all state-of-the-art devices. The active region of the component can be defined in thickness, carrier type, and impurity level and profile to optimise performance. Growth of the epitaxial layer can be achieved with a liquid or vapour phase reaction; more recently, molecular beam reactions have been used. The current techniques for growth and control of layer properties, and the relative advantages and disadvantages of each, are outlined.

1.2.1 substrate preparation

Substrates are severed from bulk grown crystals, and must be subsequently treated to produce a mirror surface, free from imperfections and work damage. This procedure involves the removal of approximately 100 μm from the cut surface through chemical polishing. The most commonly used polishing solution is either of two – a solution of bromine in methanol (often continuously diluted during the polishing process) or sodium hypochlorite (GaAs only). The technique is carried out on a rotating Hyprocel pad with a continuous supply of the chemical etchant. After chemipolishing (and often a free etch), several rinse steps precede loading of the substrates into the reactor. The extent of work damage introduced during the sawing process is open to controversy; in InP there is evidence that it can extend completely through the cut wafer. The role of this damage in final device behaviour has not been determined, neither is it known how far such impairment can propagate into the grown epitaxial layer. These questions raise the point that the subject of substrate-layer interaction needs a far more thorough investigation than it has attracted so far.

Crystal preparation, impurity incorporation, and growth rate are all dependent on the specific orientation of the substrate; however, subsequent die fabrication is made much easier if $< 100 >$ orientation is chosen – the wafer will cleave into square chips with vertical edges. This orientation is preferred for the majority of devices; the details in this section refer specifically to $< 100 >$.

1.2.2 vapour epitaxy

Vapour epitaxy of GaAs and InP can take a number of forms. Preferred methods have been described by Knight (for GaAs) and Clarke (for InP). The growth apparatus for GaAs is shown in Figure 1-1. It is an open flow system with a silica reaction tube heated with a resistance wire that winds in two temperature zones. In the first is gallium in a silica boat (800°C); in the second, the substrates (750°C) on which growth is to take place.

Hydrogen (purified by diffusion through a palladium diffuser) is bubbled through arsenic trichloride that is being maintained just below room temperature, then passed through the furnace. The initial reaction is between the arsenic trichloride and hydrogen forming arsenic vapour and HCl:

$$4AsCl_3 + 6H_2 \rightleftharpoons 12\ HCl + As_4$$

The arsenic vapour is completely absorbed by the gallium source until saturation occurs at 2.25 atomic % arsenic; a skin of gallium arsenide forms over the melt's surface at that point. Subsequent reactions occur between the GaAs solid skin and the gaseous HCl. But, since HCl is still present after the source reaction, it is

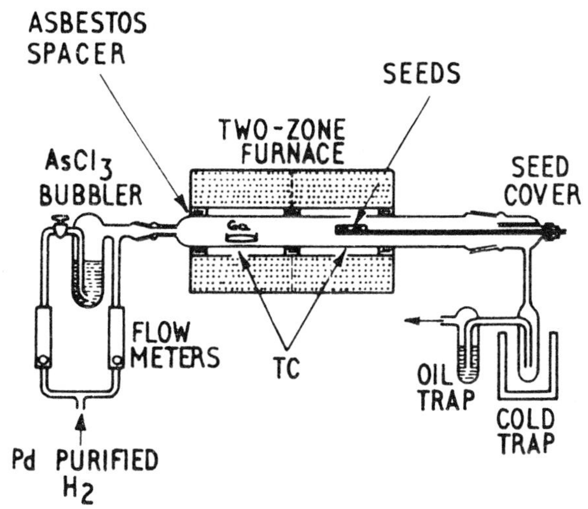

Figure 1-1/GaAs Vapour Epitaxy Kit

possible to carry out substrate etching (by raising the substrate temperature) or to initiate growth (by lowering the temperature). At this stage the substrates are inserted into the reactor and heated to around 900°C at which point vapour etching takes place, ensuring that the growth starts on an extremely clean surface. After the substrates are cooled to 750°C growth takes place according to the reaction:

$$6GaCl + As_4 \rightleftharpoons 4GaAs + 2GaCl_3$$

A special dump tube is used to collect the gallium and arsenic compound deposits so that the main work tube can be reused after each run.

The InP apparatus (Figure 1-2) is very similar, but the source is now In and the phosphorus trichloride is contained in the bubbler. The temperature of the source and seed are 750° and 650°C, respectively. An alternative vapour growth technique is that used predominantly by RCA in America, where the $AsCl_3$ and PCl_3 are replaced by pure gaseous sources of AsH_3 and PH_3 and a liquid source of HCl. The HCl passes directly over the gallium (or indium) source and transports it as the subchloride.

The arsine (or phosphine) is added down stream from the source, so that no source saturation phase is necessary in this process; this process is described in the paper by Tietjen and Amick.

The total impurity concentrations ($N_D + N_A$) are now down in the 10^{13} cm^{-3} range for these vapour grown materials, many orders of magnitude smaller than those achieved by the melt growth procedures described previously. The low level of impurity has made possible many device applications that would not have been considered before the advent of epitaxial methods. Several factors are responsible for the high purity. The most striking difference between epitaxy and growth from the melt is the temperature (750°C as compared to 1240°C); contaminants introduced from the system are reduced considerably.

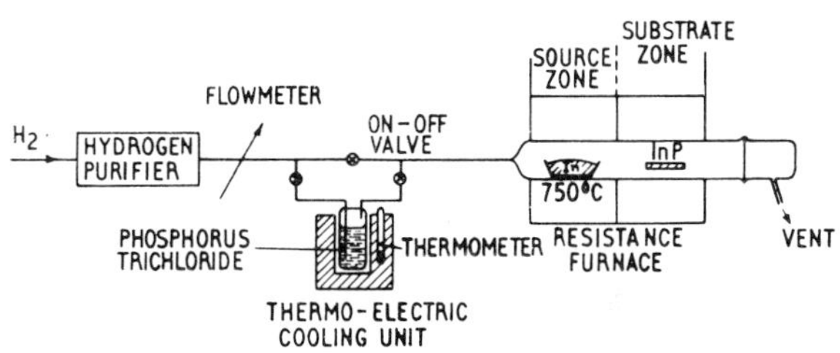

Figure 1-2/InP Vapour Epitaxy Kit

The purity of the starting materials is of paramount importance; significant amounts of research have focused on the refinement of indium, gallium, and the arsenic and phosphorus trichlorides. It is obvious that the apparatus must be leak-free — all the glassware and control taps must be of the highest quality. To reduce this problem it is now common to enclose the whole apparatus in an inert gas blanket. Cleaning of the reaction tube usually involves acid etching with aqua regia followed by water rinsing, wet assembly, and drying *in situ* in a stream of pure hydrogen. "Aristar" grade chemicals and high quality water must be used throughout the process; these practices apply throughout semiconductor device technology.

In the case of GaAs, the mole fraction of $AsCl_3$ in the gas stream can be significant in the determination of the as-grown carrier concentration in the layer, as indicated in Figure 1-3. This effect is often used to produce controlled doping. The InP picture is less clear; Joyce[5] found little

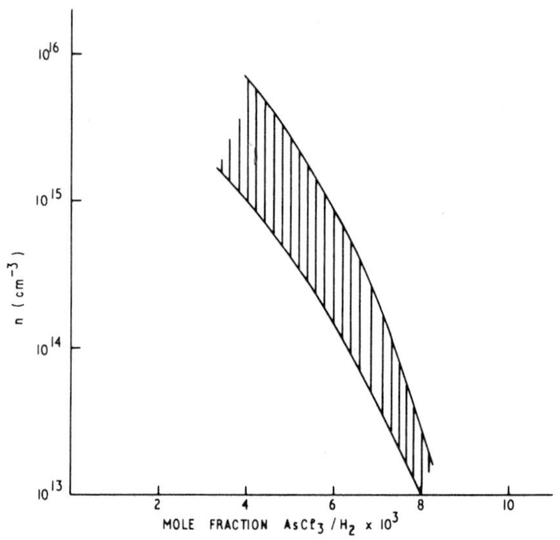

Figure 1-3/Variation of Carrier Concentration with Mole Fraction of $AsCl_3$ in H_2[4] (Shaded region shows range of "n" for a given mole fraction obtained with differing flow rates.)

variation when changing the PCl_3/H_2 mole fraction over the range 4.10^{-2} to 6.10^{-3}, but Clarke[6] discovered considerable change and Knight[7] produced similar results with the controlled addition of oxygen into the gas stream.

Silicon and oxygen contribute to changes in doping level as the mole fraction in the gas stream is varied. There is still controversy, however, over the exact mechanisms involved.

Growth rate is dependent on substrate orientation, but for < 100 > (the most commonly used orientation), it is around 8-10 μm/hr for these vapour growth systems.

Basically, both of the major vapour phase methods are capable of producing high quality material. The initial selection of a particular method is largely governed by the availability of high purity reagents and the simplicity of the growth system.

The arsine/phosphine process uses HCl, which is difficult to purify and handle. The system requires supplies of pure arsine/phosphine gas, as well as hydrogen, which results in a more complex gas-handling facility (which increases the probability of leaks) and, obviously, greater danger for the operator. By comparison, the trichlorides are relatively easy to purify and handle, and the system becomes simpler.

Perhaps the major advantage of the arsine/phosphine system is its flexibility, particularly when the growth of mixed crystals (such as InAsP and GaAsP) is undertaken.

key paper

The Preparation Of High Purity Gallium Arsenide By Vapour Phase Epitaxial Growth

J.R. Knight, D. Effer*, and P.R. Evans
(Allen Clarke Research Centre, Plessey-UK Ltd., Caswell, Towcester, Northants., UK)
Solid State Electronics, Volume 8 (1965)
Pages 178-180

*Now with Ferranti Limited, Chadderton, Lancs., UK

VAPOUR phase epitaxial growth is an attractive method for preparing high purity semiconductors, since low preparation temperatures may be employed and consequently contamination by the container material should be less severe than for melt growth processes.

In this laboratory, epitaxial layers of gallium arsenide have been prepared using hydrogen, arsenious chloride and gallium as reagents. Although other reagents might be used, those mentioned above are advantageous since they are readily available in a state of high purity.

The epitaxial layers are prepared in an open tube flow system as shown in Fig. 1. The three zone furnace is lightly insulated and is constructed so that the gallium (A.I.A.G. 99·999%) source can be observed during the experiment. Hydrogen is bubbled at a rate of 50 ml/min through arsenious chloride (Mining and Chemical Products 99·999%) held at room temperature, and is then passed through the furnace. The initial reaction taking place in zone 1 of the furnace, is as follows:

$$2AsCl_3 + 3H_2 \rightarrow 6HCl + \tfrac{1}{2}As_4$$

This arsenic vapour is completely absorbed by the gallium source until saturation occurs at 2·25 at.% arsenic.[1] During this period no free arsenic condenses beyond the furnace but gallium is transported as a lower chloride. When the melt becomes saturated with arsenic, a skin of gallium arsenide forms over its surface and free arsenic condenses beyond the furnace. At this stage the hydrogen is diverted to by-pass the bubbler, and the substrates are inserted into zone 8 of the furnace and allowed to reach a steady temperature. The arsenious chloride is then re-introduced into the gas stream. We find that the substrates are vapour etched at

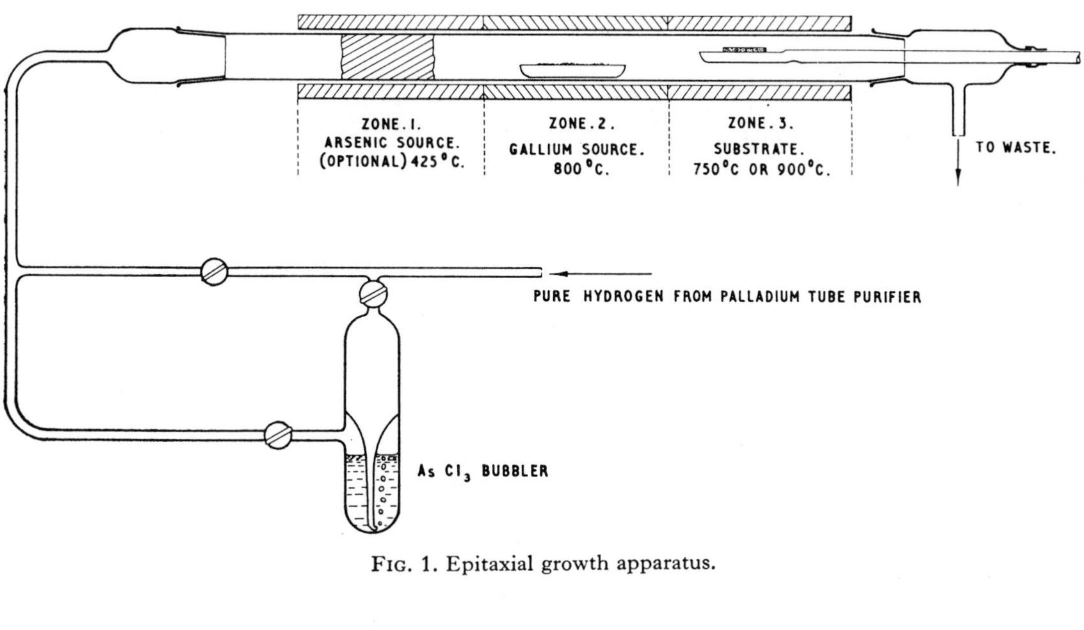

FIG. 1. Epitaxial growth apparatus.

1 μ/min if they are maintained at 900°C, but that epitaxial growth occurs at 0·5 μ/min at 750°C. By a simple adjustment it is therefore possible to switch from vapour etching to epitaxial growth and thereby ensure that the layers grow on a clean surface. The layers described here were 50–100 μ thick and were grown upon {110} slices of semi-insulating gallium arsenide (Mining and Chemical Products Ltd.).

For assessment purposes, a multi-arm specimen was ultrasonically cut from the centre of each slice and indium contacts were alloyed to the epitaxial layer on the current and potential arms. With the geometry employed here, errors in the measurement of resistivity due to the finite width of the potential arms would not exceed 5 per cent.

The epitaxial material appeared to be homogeneous, and of a total of 21 specimens, 11 had room temperature electron mobilities above 7000 cm^2/V-sec. The resistivity, Hall constant and mobility of specimen PM129 over a range of temperature, are shown in Fig. 2. At 60°K the mobility of this specimen reaches a peak value of 40,000 cm^2/V-sec, a higher value than any reported for melt-grown gallium arsenide. In addition several other layers have electron mobilities greater than 30,000 cm^2/V-sec at 77°K.

A series of layers, grown using the same batch of arsenious chloride exhibited an interesting trend in carrier density values. The first two layers had room temperature electron densities of $3{\cdot}5 \times 10^{16}$ and $4{\cdot}4 \times 10^{15}/cm^3$ respectively. Thereafter the carrier densities remained at about $2{\cdot}5 \times 10^{15}$ electrons/cm^3 until the arsenious chloride was nearly consumed. This behaviour suggested the presence of a volatile donor impurity in the arsenious chloride. Accordingly, the next batch of arsenious chloride was purified by repeated gradient freezing, the last fraction being rejected each time. Each layer grown from this batch had a low carrier density suggesting that the unknown donor impurity had segregated during the gradient freezing.

It is reported[2] that the carrier density of epitaxial gallium arsenide depends upon the partial pressure of arsenic above the substrates during growth. To investigate this effect, we formed an arsenic source *in situ* by reducing arsenious chloride with hydrogen. During the epitaxy experiment, this source was placed in zone 1 of the furnace and maintained at 425°C to give a vapour pressure of 7 mm. We could detect no change in carrier density which could be attributed to the increased partial pressure of arsenic.

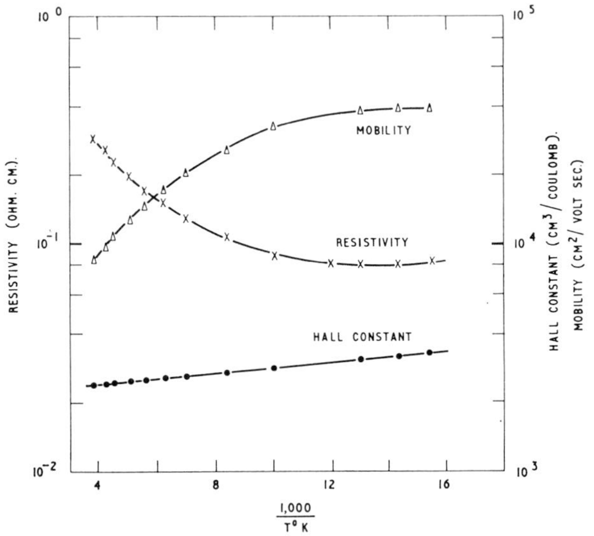

FIG. 2. Electrical properties of specimen. PM 129 as a function of temperature.

This work indicates the possibility of using the open tube epitaxy process to prepare high quality gallium arsenide for devices. If satisfactory *p–n* junctions can be formed by diffusion into this material, it should be possible to prepare a variety of diodes and field effect devices using epitaxial gallium arsenide.

REFERENCES

1. R. N. HALL, *J. Electrochem. Soc.* **110,** 385 (1963).
2. N. GOLDSMITH and W. OSHINSKY, *RCA Rev.* **24,** 546 (1963).

Acknowledgement—The authors wish to thank the Ministry of Defence and Plessey-UK Limited for permission to publish.

The Preparation And Properties Of Vapor-Deposited Epitaxial $GaAs_{1-x}P_x$ Using Arsine And Phosphine

J.J. Tietjen and J.A. Amick
(RCA Laboratories, Princeton, New Jersey, US)

Journal of the Electrochemical Society, Volume 113 (1966)
Pages 724-728

key paper

ABSTRACT

Single crystal $GaAs_{1-x}P_x$ layers have been grown epitaxially from the vapor phase in a novel apparatus, using gaseous arsine and phosphine as sources of arsenic and phosphorus. These layers exhibit the highest quality, and the widest range of controlled resistivity yet reported for solid solutions in this system. Vegard's law is obeyed over the entire composition range. At room temperature, the change from a direct to an indirect transition across the band gap occurs at the value, $x = 0.44$. Electron mobilities in these layers are high; a sample containing 27% GaP, for instance, had a room temperature mobility of 4900 cm^2/v-sec, and a 69% gallium phosphide alloy had a mobility of 700 cm^2/v-sec at room temperature. Both n- and p-type doping have been obtained, with hydrogen selenide gas and zinc vapor as the sources of donor and acceptor impurities, respectively. In this apparatus it is also easy to prepare multilayered structures incorporating layers of different resistivity and/or composition.

The preparation of homogeneous solid solutions of gallium arsenide-gallium phosphide, $GaAs_{1-x}P_x$, is of both fundamental and practical importance. Throughout most of the composition range, the band gap for this system corresponds to energies associated with the visible region of the spectrum. These materials are therefore potentially useful for the fabrication of luminescent devices, such as electroluminescent diodes and injection lasers. This system also affords opportunity to study changes in band structure with composition, especially in the region of the cross-over from a direct to an indirect (phonon-assisted) band-gap transition. Further, it is desirable to control the electrical characteristics of these crystals by the controlled addition of donor or acceptor impurities. Although it is possible to prepare these compositions from the melt (1), vapor deposition techniques have been more commonly employed (2-6). In both sealed tube and continuous gas flow apparatus, epitaxial growth has been obtained by means of a single transporting agent, such as water vapor or a hydrogen halide. Such techniques, however, do not allow for the independent, precise control of the partial pressure of two volatile species, such as arsenic and phosphorus. Furthermore, if one provides liquid or solid sources of arsenic and phosphorus, their vapor pressures are exponentially dependent on the temperature of the reservoirs. While this dependence may not be critical for the growth of the pure compounds, it becomes extremely critical for the preparation of homogeneous solid solutions.

The epitaxial growth of gallium arsenide layers by direct synthesis from arsenic vapor and gallium chloride was recently reported (7). The apparatus has been redesigned to permit the introduction of arsenic and phosphorus independently in the form of their gaseous hydrides, arsine (AsH_3) and phosphine (PH_3). On heating, these hydrides decompose to form the respective elements, with hydrogen as a by-product. As in the earlier apparatus, gallium is transported as its subchloride, prepared by passing hydrogen chloride over gallium. Since the epitaxial growth is carried out in an excess of hydrogen, the substitution of arsine and phosphine for the elements does not substantially alter the chemistry of the growth. This substitution offers the following advantages, however: (*i*) The partial pressure of arsenic and of phosphorus in the growth region is determined by the rate of introduction of arsine and phosphine, respectively. Since both of these compounds are gases at room temperature, their vapor pressure is not a strong function of the temperature. (*ii*) These gases can conveniently be obtained commercially, diluted in hydrogen to any desired concentration, and stored in pressurized cylinders for indefinite periods of time. (*iii*) They can be metered into the apparatus through flowmeters, allowing the introduction rate to be held constant and measured with precision. (*iv*) The rates of introduction of arsine and phosphine can be changed independently, permitting gradual or rapid changes in the composition of the growing layer. (*v*) Since no temperature control of these gases is necessary, the apparatus is simplified; only three critically controlled temperature zones are required.

Still another feature of the new apparatus is that n- and p-type doping over a wide range of resistivities can be obtained easily. Selenium is added as the n-type dopant by introducing gaseous hydrogen selenide; zinc is introduced directly in the vapor form to provide p-type doping. Because of the characteristics of the flow apparatus, doping can be initiated and discontinued at any time, at almost any rate, and during any stage of the crystal growth.

These features have made possible the epitaxial growth of single crystal $GaAs_{1-x}P_x$ solid solution layers of unusually high purity and homogeneity over the entire compositional range. In addition, crystals having a wide variety of resistivities, both p- and n-type, have been prepared. Finally, multilayer structures have been prepared sequentially within the growth apparatus, the samples not being removed from the apparatus until all growth operations are completed.

Experimental

Apparatus.—The apparatus consists principally of a single 2.5 cm ID quartz tube 80 cm long; it is shown schematically in Fig. 1. The three heated zones are surrounded by conventional temperature controlled, resistance heated furnaces. A large bore (35 mm) stopcock separates the growth portion of the apparatus from a forechamber. As described previously (7), substrates mounted at the end of a precision ground (Trubore) rod can be inserted into the forechamber. This chamber can then be purged with hydrogen before the samples are inserted through the stopcock into the growth portion of the apparatus. In this manner, contamination of the growth region during insertion and withdrawal of the samples is minimized. Furthermore, the substrates may be etched or heat-treated in this forechamber prior to their insertion into the growth region. Finally, following

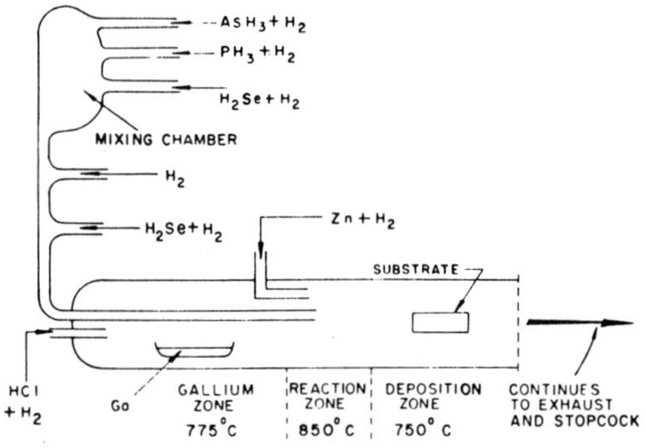

Fig. 1. Schematic representation of vapor-deposition apparatus

deposition, the substrates can cool to room temperature in this chamber in a controlled atmosphere before being withdrawn into the laboratory.

Arsine, phosphine, and hydrogen selenide are introduced through flow meters into a 2-liter ballast vessel which empties into the hydrogen carrier gas stream. This permits these reagents to be thoroughly mixed before they enter the reaction zone, and helps to smooth out small sporadic fluctuations in the individual gas flows. If desired, the hydrogen selenide can be made to by-pass the mixing bulb if abrupt gradients in donor impurity concentrations are required. As shown in Fig. 1, the arsine-phosphine-hydrogen mixture is introduced into the reaction zone through a long quartz tube which extends beyond the gallium. Through a shorter quartz tube a mixture of hydrogen chloride and hydrogen is injected. This mixture, passing over gallium in a quartz boat, provides the gallium subchloride. It is a significant feature of this method of introducing the different vapor phase constituents that at the gas stream velocities used in this investigation only the hydrogen chloride and hydrogen come into contact with the gallium. Since the equilibrium for the reaction between gallium and hydrogen chloride is rapidly attained, and since the contents of the boat are essentially pure gallium throughout the entire deposition, the composition of the gas phase components can be maintained constant during the entire growth cycle. In addition, abrupt changes in the crystal composition can be introduced during the growth period, since the crystal growth at any time does not depend on the previous history of the growth apparatus. In contrast, when arsenic trichloride and phosphorus trichloride are used as transport agents, the contents of the gallium boat change as the run proceeds. As a result, it is difficult, with these transport agents, to start and stop the growth abruptly. P-type doping is obtained from a small amount of metallic zinc held in a quartz "bucket" attached to a Trubore rod. This zinc can be inserted into a heated side arm adjacent to the growth region of the apparatus, and the zinc is transported into the growth region in hydrogen carrier gas. The temperatures of the three zones shown in Fig. 1 are selected so that deposition is emphasized in the vicinity of the substrate and minimized on the long quartz tube through which the arsine-phosphine mixture passes. The concentrations of the reactant species are controlled by adjusting the flow rates of the gaseous species entering the apparatus.

Materials.—The gases employed in this work were obtained commercially and used without further purification. Typically the arsine is obtained diluted in H_2 to a concentration of 10%,[1] the PH_3 as a 5% dilution in H_2,[1] and the H_2Se diluted to 500 ppm in H_2.[1] A typical mass spectrometric analysis of the

Table I. Mass spectrometric analysis of arsine impurities
Average of five analyses

Element detected	Concentration, ppm
Nitrogen	46
Argon	36
Carbon dioxide	15
Oxygen	6

impurities present in the arsine is presented in Table I. It should be noted that, at the levels present here, the three primary impurities apparently do not affect the growth characteristics or electrical properties of the layers. The hydrogen chloride is obtained in standard cylinders.[2] The hydrogen employed as the carrier gas is purified by passage through a standard palladium-alloy diffuser. The manufacturers reported purity for the gallium[3] and zinc[4] is 99.9999 and 99.9998%, respectively.

Both germanium and gallium arsenide substrates were employed in this study. Germanium substrates nominally had {111} surfaces, while gallium arsenide wafers having {111}, {110}, and {100} surfaces were used. Although it was demonstrated that epitaxial layers of $GaAs_{1-x}P_x$ can be successfully deposited on {111} germanium surfaces or on gallium arsenide surfaces oriented normal to the $\langle 110 \rangle$ and $\langle 111 \rangle$ directions, gallium arsenide wafers having {100} surfaces, about 1 cm^2 in area and 0.5 mm thick, were almost exclusively used in this work. These were mechanically polished to a flat, mirror-smooth finish and then chemically polished in a mixture of sulfuric acid, hydrogen peroxide, and water in the ratio, 4:1:1 (Caro's acid). As substrates, a wide variety of n-type, gallium arsenide wafers oriented normal to the $\langle 100 \rangle$ direction were employed, having electron carrier concentrations in the range, 1×10^{16} to $2 \times 10^{18}/cm^3$, with the major impurities including silicon, selenium, tellurium, and tin.

Procedure.—Freshly etched substrates are first mounted on the substrate support. Since growth is usually required on only one surface of a substrate wafer, two wafers are usually mounted in a common slot, back to back. In this manner, two substrates having essentially identical growth history can be obtained in one run. When inserted into the apparatus, the surfaces of these substrates are parallel to the direction of gas flow. With no further treatment, the substrates are introduced into the growth region of the apparatus, the entire apparatus being at room temperature. After purging the system with hydrogen, the furnaces are turned on, and the temperature rises at about 20°C per minute. When the substrate temperature reaches 600°C, the arsine flow is started in order to establish an arsenic atmosphere in the deposition zone and to prevent decomposition of the substrate. When all three zones have reached the operating temperature, the hydrogen chloride flow over the gallium is started; at this time, gallium arsenide begins to deposit epitaxially on the substrate. The phosphine flow is then initiated to produce a gas phase mixture of arsine and phosphine appropriate to the desired gallium arsenide-phosphide composition. Because of the 2-liter ballast vessel in the gas introduction system, the composition of the epitaxial growth on the substrate changes slowly from pure gallium arsenide to the selected alloy composition. It is believed that this gradual transition aids in minimizing strain arising from differences between the lattice constants of the substrate and of the deposited alloy.

Typically, the sum of the flow rates of pure arsine and pure phosphine are about 15 cc/min. The hydrogen chloride flow rate is approximately 5 cc/min. An additional 2.5 liters/min of hydrogen is employed as a

[1] Matheson Company, Rutherford, New Jersey.

[2] Frontier Chemical Company, Wichita, Kansas.

[3] AIAG Metals, New Nork, N. Y.

[4] United Mineral and Chemical Corporation, New York, N.Y.

carrier. With these flow rates, with a substrate temperature in the range 725°-775°C, a reaction zone temperature of 850°-875°C, and a gallium zone temperature of 775°-800°C, growth rates of approximately ½ μ/min are obtained under steady-state conditions. With these growth conditions, the contents of any of the gas cylinders permits at least 60 hr of continuous crystal growth.

To determine the electron carrier concentrations and mobilities in the deposited layers, a technique was developed which permits Hall coefficient and resistivity measurements to be made on layers as thin as 50μ. Indium contacts are alloyed to the grown layer in a hydrogen atmosphere. The temperature required to alloy ohmic contacts ranges from 450°C for GaAs to 850°C for GaP. The sample is then fastened to a quartz lapping plate with glycolphthalate and the substrate is lapped off. Final contacts and measurements are made while the sample remains supported on the quartz lapping plate. In this way the measurements are made on layers that have been completely removed from the substrate but are otherwise unaltered. To test this procedure, several bulk samples of gallium arsenide were measured first by conventional techniques and then, using this procedure, after being lapped to about 100μ thickness. In all cases, the second set of measurements produced identical results to the first.

Results and Discussion

Crystallinity and Homogeneity

X-ray analysis by the Debye-Scherrer technique shows that these layers are single phase, cubic, solid solutions which have a very high degree of homogeneity, as revealed by the sharpness of the diffraction lines. The width of the diffraction lines, for 15 specimens of different alloy compositions, is identical to those for pure gallium arsenide and pure gallium phosphide. Thus, any variation in composition was below the detection limit. This result is in agreement with electroluminescent data obtained from diodes fabricated from selenium-doped alloys containing up to 30% gallium phosphide. For these diodes, the width of the incoherent emission peak at 77°K is equal to that of pure gallium arsenide. Since the maximum experimental error associated with this measurement is approximately 5%, this result indicates that compositional variations are less than 0.1 mole %. Also, the threshold current densities of laser diodes fabricated from specimens containing as much as 20% gallium phosphide is the same as that for pure gallium arsenide (8). In addition, there is no drop in electron mobility with alloying up to 30% gallium phosphide (9). Moreover, the room temperature electron mobilities of undoped layers, presented in Table II, are high for all alloy compositions. Note especially the values of 4900 and 700 cm²/v-sec for alloys containing 27 and 69% gallium phosphide, respectively. These results are also attributed to the high degree of homogeneity of these materials.

X-ray analysis by the Laue back-reflection technique reveals that the layers are epitaxial. In Fig. 2 the lattice constant is presented as a function of alloy composition showing that Vegard's law of solid solutions is obeyed in this system. This result is in agreement with those of Pizzarello (4) and Rubenstein (6). The compositions were determined by chemical analysis (10) with an accuracy of ±0.5%. It was also demonstrated by chemical analysis that compositions can be reproduced in sequential runs to within ±1%.

Table II. Room temperature electron mobility of $GaAs_{1-x}P_x$ alloys having electron carrier concentrations in the range, 5×10^{15} to 1×10^{16} cm^{-3} (9)

x	μ, cm²/v-sec
0	5100
0.06	5000
0.09	5000
0.12	5200
0.22	4800
0.27	4900
0.37	3800
0.42	1400
0.69	700
1	120

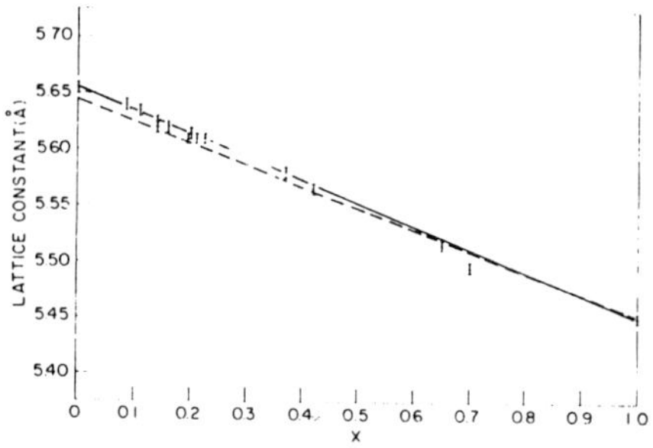

Fig. 2. Dependence of the lattice constant on alloy composition in the system, $GaAs_{1-x}P_x$. The dependence observed by Rubenstein (6) and Pizzarello (4) is also included: ——— Rubenstein; — — — Pizzarello; I this paper.

Growth Morphology

As a result of a novel etchant (11) developed by Abrahams and Buiocchi of the RCA Laboratories, it is now possible to observe dislocation densities on the major low index faces of $GaAs_{1-x}P_x$ alloys. Etch pit densities were obtained (12) using this etchant, which reveal that the density of dislocations in the grown layers are determined almost exclusively by the dislocation densities of the substrates. Practically all dislocations which lie in the substrate, and intersect the plane of the substrate-grown-layer interface, propagate into the grown layer. The additional dislocations which originate in the grown layer are much smaller in number by at least an order of magnitude than those originating in the substrate. Typical dislocation densities for the substrates employed in this study, and therefore for the grown layers, are usually of the order of $10^3/cm^2$, but have ranged up to $10^6/cm^2$.

Electrical Properties

The epitaxial layers were examined by Hall coefficient measurements. Typical electron carrier concentrations for undoped specimens are in the range, 5×10^{15} to $1 \times 10^{16}/cm^3$. The high values of the electron mobility at 77°K of alloys containing up to 20% gallium phosphide, as shown in Fig. 3, attest to the high quality of the undoped samples and demonstrate that this material is noncompensated.

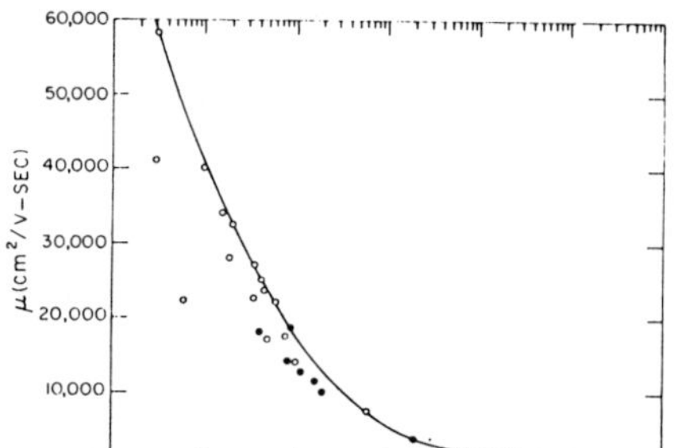

Fig. 3. Dependence of the electron mobility on electron carrier concentration in selenium-doped $GaAs_{1-x}P_x$ at 77°K: ○, X = 0; ●, 0 < X < 0.2.

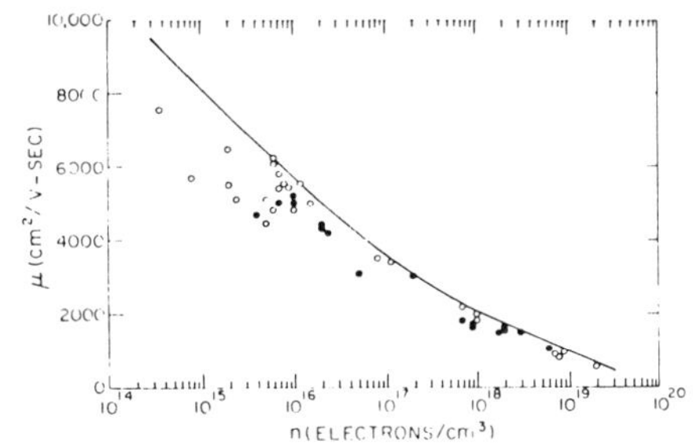

Fig. 4. Dependence of the electron mobility on electron carrier concentration in selenium-doped $GaAs_{1-x}P_x$ at 300°K; symbols are the same as in Fig. 3.

Doping

Attempts were made to dope the layers using hydrides of various donor impurities as the source of the n-type dopant. These included hydrogen sulfide, hydrogen selenide, germane, and silane. However, it was found that the best results were obtained with hydrogen selenide, and it was, therefore, used almost exclusively. The primary advantage of hydrogen selenide is that a lower gas phase concentration of this hydride is required. Apparently, selenium is incorporated into the growing layer much more efficiently than the other doping agents. As compared with the other hydrides, as much as 100 times less hydrogen selenide is needed to produce a given electron carrier concentration in the deposited crystal. This has the inherent advantage of introducing less unknown contamination to the growth system. Using hydrogen selenide diluted in hydrogen in the concentration range 5-5000 ppm allows for doping from 5 x 10^{16} to 2 x 10^{19} electrons/cm^3 with a reproducibility of ±25%.

Selenium-doped samples containing up to 20% gallium phosphide were prepared having electron carrier concentrations as high as 2 x 10^{19}/cm^3. The dependence of the electron mobility on the carrier concentration is shown in Fig. 3 and 4 for 77°K and room temperature, respectively. The relatively high values for the electron mobility of heavily doped samples imply that the crystals are extremely homogeneous and contain few defects. In addition, the dependence of the mobility on electron carrier concentration is the same for the alloy specimens as for pure gallium arsenide. It is concluded that doping to these high donor levels does not interfere with the preparation of homogeneous material.

The investigation of acceptor doping with zinc has been carried out to a lesser extent. However, it has been demonstrated that hole concentrations in the range 2 x 10^{18} to 4 x 10^{19}/cm^3 can easily be attained with mobilities varying from 120 to 60 cm^2/v-sec for this doping range.

In addition to the preparation of single epitaxial layers doped with donor or acceptor impurities, multilayered structures have been fabricated by varying the introduction of the doping agents at different times during crystal growth. Figure 5 shows a section of a gallium arsenide multilayered structure in which an N^+, N, P^+ arrangement has been grown on a {100} surface of a gallium arsenide substrate. The P^+ region has a hole concentration of about 2 x 10^{19}/cm^3 while the N^+ and N regions have electron concentrations of about 4 x 10^{18} and 5 x 10^{15}/cm^3, respectively. Because of the low growth rates that can be realized in this system, the thickness of the individual layers can be controlled to about 1μ. This method of preparing multilayered structures during the growth of the alloy crystal has been used to fabricate several types of semiconductor devices having superior electrical characteristics (13).

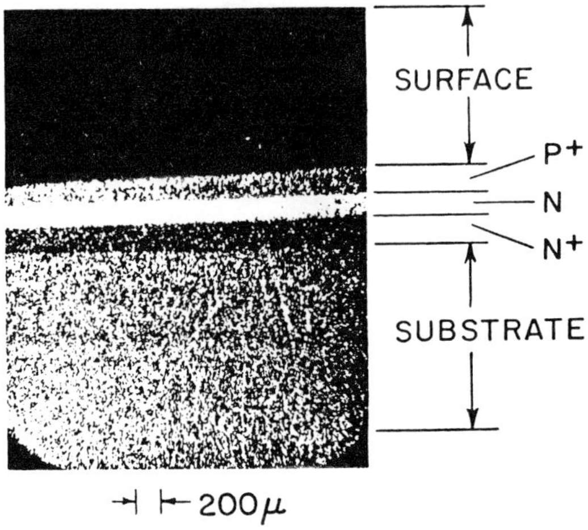

Fig. 5. Cross section of a P^+, N, N^+ GaAs structure, angle-lapped at 3° and chemically stained to reveal the individual regions. The geometric magnification introduced by angle lapping increases the vertical dimension by a factor of approximately 20. Therefore, the actual thickness of each layer is about 10μ.

Optical Band Gaps

Optical absorption measurements were made in a Cary model 14R spectrophotometer on the same epitaxial films used for the Hall coefficient measurements. These measurements revealed that, on the high energy side of the absorption edge the absorption coefficient, α, obeys the relationship

$$\alpha = \alpha_0 (E - E_0)^n$$

where E is the photon energy and E_0 is the energy of the band gap (14). For alloys containing greater than 50% gallium phosphide n is equal to 2, while for alloys containing less than 45% gallium phosphide n is approximately unity. A value of n equal to 2 is indicative of an indirect (phonon assisted) band gap transition while values of n less than unity have been associated with a direct band gap transition (15). The dependence of E_0 on alloy composition at room temperature and 77°K is presented in Fig. 6. It is apparent that a distinct transition occurs at $x = 0.44$ (at room temperature) as the slope of the curve changes abruptly. This is undoubtedly due to the change from a direct to an indirect band gap transition. These data are in excellent agreement with the results of Pilkuhn and Rupprecht (16), Fenner (17) and Cusano, Fenner, and Carlson (18), but are in par-

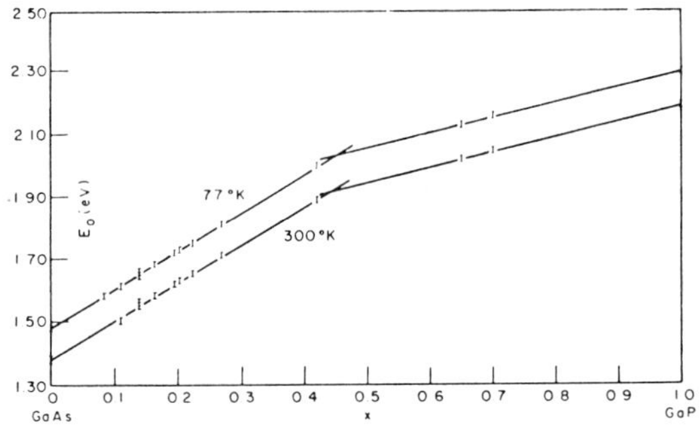

Fig. 6. Dependence of the energy band gap on alloy composition in the system, $GaAs_{1-x}P_x$.

tial disagreement with the results of Pizzarello (4) and Ku (3).

Conclusions

The superiority of this vapor-growth method for the preparation of homogeneous $GaAs_{1-x}P_x$ alloys has been demonstrated. Furthermore, very homogeneous alloys are obtained for a wide range of doping concentrations as both p- and n-type material. In addition, the versatility of this method has been explored in terms of growing a multilayer structure incorporating this high-quality material.

Measurements on the material prepared by this technique reveals that very high mobilities can be attained in this alloy system. It is especially striking that an alloy containing 27% gallium phosphide had a room temperature mobility of 4900 cm^2/v-sec, while a 69% gallium phosphide alloy had a mobility of 700 cm^2/v-sec at room temperature. Vegard's law of solid solutions was found to be obeyed, and optical absorption data showed that the change from a direct to indirect band gap transition occurs at $x = 0.44$ at room temperature.

This technique should also be useful for the preparation of any semiconducting compound or alloy system where volatile halides and/or hydrides occur.

Acknowledgments

The authors wish to express their sincere appreciation to H. F. Gossenberger for his valuable assistance throughout the course of this study, and to D. Richman and L. R. Weisberg for many very helpful suggestions and discussions. Thanks are extended to G. W. Neighbor, R. Paff, and R. Smith for performing the x-ray analysis, to S. Ochs for measuring the electroluminescent spectra, and to I. J. Hegyi and A. Salerno for mobility data. The research reported in this paper was sponsored by the Advanced Research Projects Agency, Materials Sciences Office, under Contract No. SD-182.

Manuscript received Jan. 17, 1966; revised manuscript, received March 21, 1966.

Any discussion of this paper will appear in a Discussion Section to be published in the June 1967 JOURNAL.

REFERENCES

1. H. Flicker and P. G. Herkart, *Bull. Am. Phys. Soc.*, 5, II, 407 (1960).
2. N. Holonyak, Jr., D. C. Jillson, and S. F. Bevacqua, "Metallurgy of Semiconducting Materials," p. 49, Interscience Publishers, Inc., New York (1962).
3. S. Ku, *This Journal*, **110**, 991 (1963).
4. F. A. Pizzarello, *ibid.*, **109**, 226 (1962).
5. G. E. Gottlieb, *ibid.*, **112**, 192 (1965).
6. M. Rubenstein, *ibid.*, **112**, 426 (1965).
7. J. A. Amick, *RCA Review*, **24**, 555 (1963).
8. J. J. Tietjen and S. A. Ochs, *Proc. IEEE (Correspondence)*, **53**, 180 (1965).
9. J. J. Tietjen and L. R. Weisberg, *Appl. Phys. Letters*, **7**, 261 (1965).
10. B. L. Goydish, To be published.
11. M. S. Abrahams and C. J. Buiocchi, *J. Appl. Phys.*, **36**, 2855 (1965).
12. M. S. Abrahams and J. J. Tietjen, To be published.
13. J. J. Tietjen, To be published.
14. R. A. Smith, "Semiconductors," Cambridge University Press, 193 (1959).
15. C. Hilsum and A. C. Rose-Innes, "Semiconducting III-V Compounds," pp. 167-171, Pergamon Press, New York (1961).
16. M. Pilkuhn and H. Rupprecht, *J. Appl. Phys.*, **36**, 684 (1965).
17. G. E. Fenner, *Phys. Rev.*, **134**, A1113 (1964).
18. D. A. Cusano, G. E. Fenner, and R. O. Carlson, *Appl. Phys. Letters*, **5**, 144 (1964).

key paper

The Preparation Of High Purity Epitaxial InP

R.C. Clarke, B.D. Joyce, and W.H.E. Wilgoss
(Royal Radar Establishment, Great Malvern, Worcs., UK)

Solid State Communications, Volume 8 (1970)
Pages 1125-1128

(*Received* 9 *April* 1970 *by* C.W. *McCombie*)

Epitaxial layers of indium phosphide were prepared by the reaction of phospherous trichloride with indium in an atmosphere of hydrogen. The electron mobility of this material was measured as 44,000 cm^2/Vsec at 77°K; higher than any previously reported value for indium phosphide.

INDIUM PHOSPHIDE is a III–V semiconducting compound with an energy gap very close to that of gallium arsenide. Until recently most InP had been prepared by various melt growth techniques[1,2] and the purest material reported to date has been prepared by the gradient freeze method[3] and has an electron mobility of 23,400 cm^2/Vsec. at 77°K. Because of the high equilibrium vapour pressure of phosphorus over InP the preparation of this compound from the vapour at lower temperatures is attractive. Epitaxial layers of InP have been grown on GaAs substrates[4] by the reaction of high purity indium (99.9999%) with PCl_3 (99.999%). This material had carrier concentrations from 1.4×10^{17} to 3.2×10^{18} per cm^3 with mobilities from 2000 to 165 cm^2/Vsec at 77°K. It was postulated that these high carrier concentrations and low mobilities were due to the formation of phosphorus vacancies in the layers which contributed to the ionized impurity scattering of the carriers.

More recently vapour grown layers of InP have been prepared[5] by the reaction of phosphine with indium monochloride, the latter compound being formed by the reaction of hydrogen chloride with indium at elevated temperatures. These layers were grown on InAs substrates and had mobilities of 18,000 cm^2/Vsec. at 77°K.

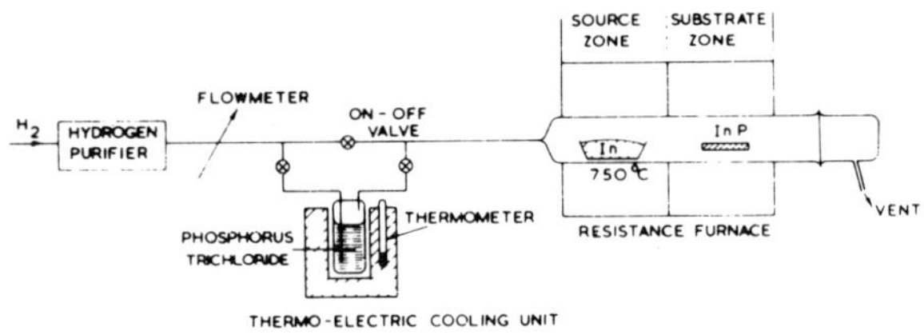

FIG. 1. Epitaxial growth apparatus

In this work the reaction between indium (Johnson Matthey Chemicals Ltd or Mining and Chemical Products, 99.9999%) and phosphorous trichloride (Mining and Chemical Products 99.999%) was used to grow the InP layers and a schematic diagram of the apparatus is shown in Fig. 1. A quartz reaction tube is held in a two-zone restisance furnace and high purity hydrogen from a palladium diffusion cell is metered through the tube. The hydrogen can also be diluted with PCl_3 by passing the gas through a PCl_3 bubbler held at 0°C. In all experiments reported here the hydrogen passing through the reaction tube had a linear velocity of 14 cm/min.

The epitaxial growth process is carried out in two stages, a source saturation and a growth stage. In the first stage a quartz boat containing indium is placed in the source zone and heated in hydrogen to a temperature between 700 and 850°C. PCl_3 is then introduced into the hydrogen and above the indium source it is reduced to phosphorous vapour and hydrogen chloride. Indium monochloride vapour is formed principally by the reaction of hydrogen chloride with the indium and this vapour passes along the reaction tube. The phosphorus dissolves in the indium until saturation is obtained when phosphorus vapour also passes along the reaction tube with the indium monochloride. This completes the first stage and the furnaces are cooled to room temperature in hydrogen.

In the second stage of the reaction, the source is kept in position and a single crystal substrate is placed in the second zone of the furnace and heated to a temperature between 600 and 750°C in hydrogen while the source is not heated. PCl_3 is introduced into the hydrogen for a few minutes to vapour etch and, hence, clean the substrate surface. Hydrogen is again passed through the reaction tube and the source is heated to its previous saturation temperature which in this work was 100°C higher than that of the substrate crystal. The hydrogen is then bubbled through the PCl_3 and the phosphorus vapour and indium monochloride formed at the source zone react to produce a layer of InP on the substrate according to the equation:

$$3\,InCl + P_2 \quad 2\,InP + InCl_3$$

The growth is terminated by passing pure hydrogen through the system and cooling the two zones to room temperature.

The substrate material used in this work was exclusively semi-insulating InP[6] grown by the liquid encapsulation technique.[7] Substrates were cut 2° off the 100 orientation and were chemically polished in a solution consisting of 2% by volume of bromine in methanol.[8] Prior to use the substrates were rinsed for five minutes in a 10% aqueous solution of potassium cyanide to remove metal impurities on the surface[9] and they were finally washed with 18 megohm water. The thicknesses of the epitaxial layers were measured optically on cleaved cross-sections of the samples which had been subsequently etched for periods of up to 5 min at room temperature in a mixture of potassium hydroxide (6 g) + potassium ferricyanide (4 g) in 50 ml. of water[10] to delineate the substrate-layer interface. Electrical characteristics of the layers were measured on clover leaf specimens by the method described by Van der Pauw.[11]

The growth rates achieved in this system with the PCl_3 maintained at 0°C were generally between 8 and 12 μ/hr. Etched cross-sections of the epitaxial layer using the ferricyanide etch show that the interfaces are flat and sharp. The layer surfaces were generally of a highly polished appearance with few, if any, growth defects. The electrical properties of several layers are illustrated in Table 1. Hall measurements were made in a magnetic field of 2.0 kG. It is noticeable that the carrier concentration in these layers varies by four orders of magnitude yet no dopant impurities were added to the growth system deliberately. We interpret the improvement between the first five and the last seven results as indicative of a reduction in the oxygen concentration in the epitaxial equipment. The experimental evidence points to the fact that the high carrier concentration of the InP is a result of traces of oxygen in the gas stream. We are not able to say at this stage whether oxygen is itself a shallow donor of InP or if it indirectly increased the carriers present through the formation of lattice defects. This contrasts to the role of oxygen as an impurity in GaAs and GaP where it is known to form deep levels in the forbidden gap.

The low background impurity levels in the InP prepared by this technique are comparable with those obtained in GaAs[12] formed by the reation of arsenic trichloride with gallium. This result is to be expected as the purity of the starting materials and the reaction mechanisms are quite similar. The 77°K electron mobility of 44,000 cm^2/Vsec. is the highest value yet reported for InP.

Carrier conc. $N_D - N_A$ (cm^{-3})	Electron mobility ($cm^2/Vsec$)	Carrier conc. $N_D - N_A$ (cm^{-3})	Electron mobility ($cm^2/Vsec$)
9.4×10^{17}	1250	6.7×10^{17}	1,770
4.9×10^{17}	2600	5.0×10^{17}	2,150
4.0×10^{17}	2500	3.5×10^{17}	2,550
1.8×10^{17}	2950	1.4×10^{17}	3,400
7.1×10^{16}	2400	4.8×10^{16}	3,720
8.9×10^{15}	3800	6.0×10^{15}	20,000
5.6×10^{15}	3350	3.0×10^{15}	20,000
4.0×10^{15}	3949	3.0×10^{15}	24,500
3.0×10^{15}	3630	1.6×10^{15}	19,000
3.0×10^{14}	4740	2.8×10^{14}	44,000
1.4×10^{14}	3591	1.4×10^{14}	34,700
8.6×10^{13}	4330	8.2×10^{13}	34,000

Table 1. Measured carrier concentrations and mobilities in epitaxial InP *layers.*

Acknowledgements – Valuable discussions with Dr. W. Bardsley and our colleagues are gratefully acknowledged. The authors are indebted to Mrs. K. Fairhurst and Mr. P. Born for assistance in this work, 'to Dr. J.B. Mullin and Mr. B.W. Straughan for the supply of the substrate material and to Miss A. Royle for the electrical assessments.

This paper is published by permission of the Director R.R.E., Copyright Controller, H.B.M.S.O.

REFERENCES

1. RICHMAN D., Preparation of indium phosphide, in: *Compound Semiconductors*, Vol. 1, *Preparation of* III–V *Compounds*, p.214. Reinhold, New York, (1962).
2. WINKLER J.H. *et al.*, Paper 525, Electro chemical Society Meeting, Montreal, October 6–11, (1968).
3. GLISKMAN M. and WEISER K., *J. Electrochem. Soc.*, **105**, 728 (1958).
4. NARITA S., CHOE B. and HARADA H., *Japan. J. appl. Phys.*, **8**, 500 (1968).
5. TIETJEN J.J., MARUSKA H.P. and CLOUGH R.B., *J. Electrochem. Soc.*, **116**, 492 (1969).
6. MULLIN J.B., ROYLE A. and STRAUGHAN B.W., Unpublished work.
7. MULLIN J.B., STRAUGHAN B.W. and BRICKELL W.S., *J. Phys. Chem. Solids*, **26**, 782 (1964).
8. SULLIVAN M.V. and KOLB G.A., *J. Electrochem. Soc.*, **110**, 583 (1963).
9. HALL R.N. and RACETTE J.H., *J. appl. Phys.*, **35**, 379 (1964).
10. JOEL G., *Etching techniques for Gallium-Arsenide Surfaces*, Technical Report of the Research Institute of the Central Communications Agency, Deutsche Bundespost Report FTZ.A.18.T.Br1, August 1968.
11. VAN DER PAUW L.T., *Philips Res. Rep.*, **13**, 1 (1958).
12. KNIGHT J.R., EFFER D. and EVANDS P.P., *Solid-St. Electron.* **8**, 178 (1965).

Epitaktische Indiumphosphid – Schichten wurden durch die Umsetzung von Indium mit Phosphortrichlorid in einem Luftkreis von Wasserstoff vorbereitet. Die Elektronenbeweglichkeit dieses Stoffs wurde als 44,000 $cm^2/Vsec$ auf 77°K gemessen; höher, also, als jeder bisher berichtete Wert für Indiumphosphid.

1.2.3 liquid phase epitaxy

The liquid epitaxial process was first demonstrated for GaAs by Nelson (see key paper) in 1963 for highly doped junctions intended for semiconductor lasers. The mechanics of the apparatus have been modified and improved, but essentially the method has not changed. Figure 1-4 shows the tilt furnace used. The GaAs substrate seed is held at one end of a graphite or silica boat; at the other end is the melt, consisting of a solution of gallium saturated with GaAs. The boat is contained in a constant temperature furnace under an atmosphere of pure palladium-diffused hydrogen. When the melt and source are in equilibrium at the required growth temperature, the furnace is tilted so that the melt covers the seed and the temperature is lowered. The decreasing solubility of arsenic in gallium with decreasing temperature results in supersaturation and epitaxial growth of GaAs on the seed. After the desired growth period, the furnace is tipped back again to remove the melt from the newly-formed layer.

Optimal growth conditions for this system occur with starting temperatures from 720 to 900°C, final temperatures of 400-750°C, and cooling rates of 0.1 to 7°C per minute. The highest purity gallium is used for the melt; it is usually saturated with boat grown (Bridgman method) gallium arsenide with carrier concentrations around 10^{13} cm^{-3}. The purity of the boats, whether graphite or silica, is essential. The low segregation coefficients of many impurities between the solid and the gallium melt help to improve the purity of the grown material.

InP has been grown in an identical manner at around 650°C with cooling rates of 5°C/min, and the distribution coefficients of n-type impurities (Ge, Sn, and Te) determined. E.W. Williams at RRE used a similar system for growth of junctions in InP for electroluminescent diodes, and extended the knowledge of distribution coefficients to the p-type impurites.

The tilting furnace of Nelson is somewhat cumbersome; it has been replaced by accurately machined graphite slider-boat assemblies. The substrate is placed in a shallow machined well inside the slider; after source saturation, the

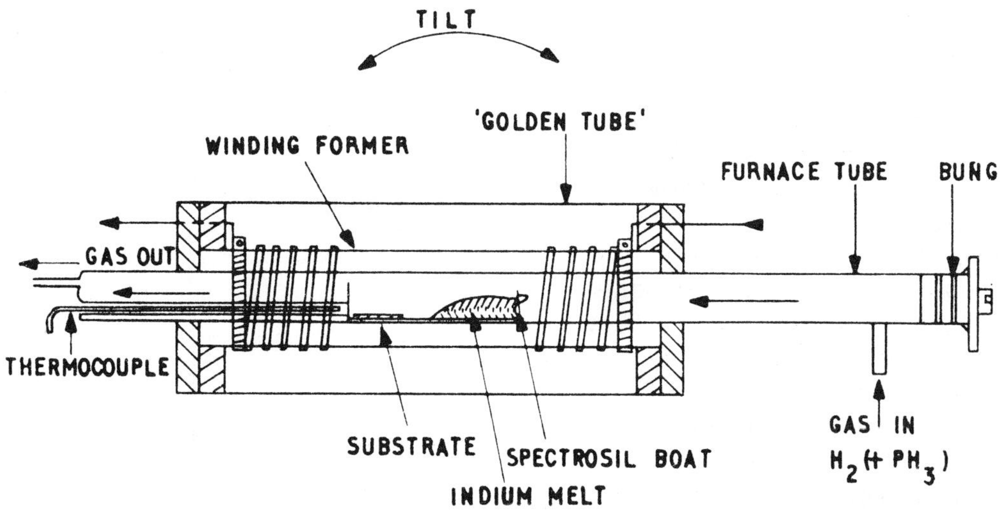

Figure 1-4/Horizontal Tipping Furnace.

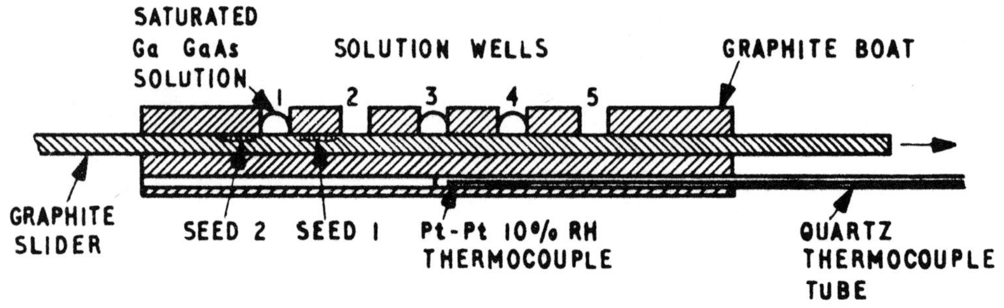

Figure 1-5/Multi-Layer Slider Boat Assembly[8]

slider is moved so as to position the substrate underneath the solution prior to growth. Figure 1-5 illustrates a multi-well slider assembly.

An alternative system to that using the Nelson tilting furnace uses vertical dipping, as in Figure 1-6. The substrate is held in a spring clip or chuck above the saturated melt in a vertical furnace system. At equilibrium the substrate is dipped into the melt, which is then cooled slowly. Upon completion of growth the wafer is withdrawn. Both GaAs and InP crystals have been grown by this method. In lieu of cooling the melt to achieve growth, a temperature gradient can be established.

A further adaptation of the vertical technique is to use the "Transient Growth" of material, first demonstrated by Deitch. The apparatus is similar to that described above, but, prior to dipping, the substrate is maintained at a lower temperature than is the melt (100°C below). When the substrate is immersed, the solution is quenched by the cooler substrate and its holder. This process supersaturates a thin layer of solution adjacent to the substrate, and growth takes place. Because the furnace is held at constant temperature, the whole system eventually returns to equilibrium. Growth occurs very rapidly and then stops; its rate depends on the relative temperatures and thermal masses of the substrate and the melt before dipping takes place. This process is detailed in the key paper by Deitch.

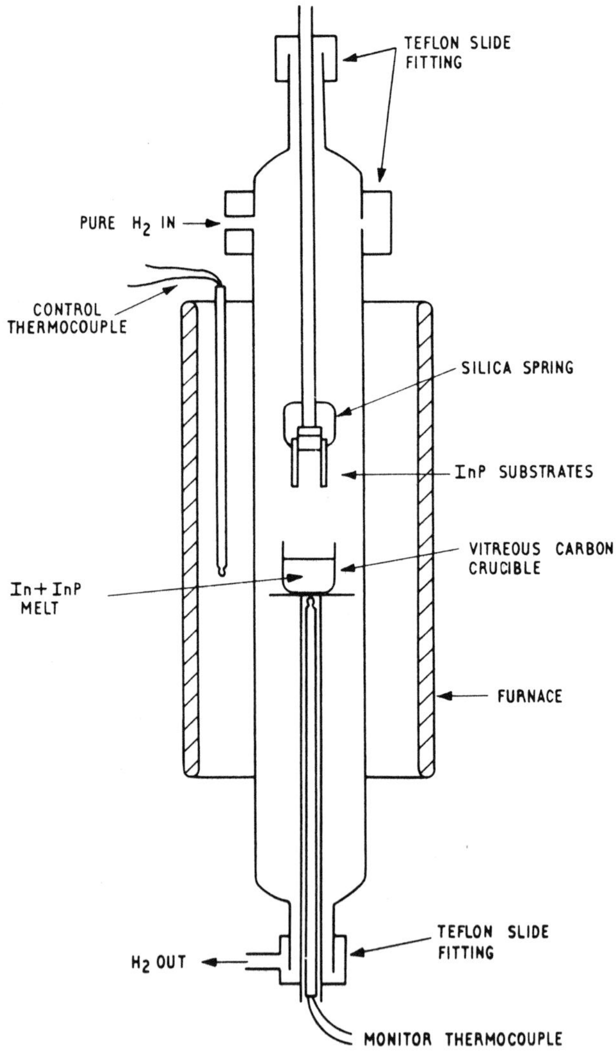

Figure 1-6/InP Vertical Liquid Epitaxy Kit[9]

key paper

Epitaxial Growth From The Liquid State And Its Applications To The Fabrication Of Tunnel And Laser Diodes

H. Nelson
(RCA Laboratories, Princeton, New Jersey, US)

RCA Review, Volume 24 (1963)
Pages 603-615

Summary—*An apparatus and procedures have been developed for the epitaxial growth of GaAs and Ge from the liquid state. The resulting technology has been found to possess advantages over vapor-phase epitaxy in some applications demanding highly doped epitaxial films and high-quality p-n junctions at the substrate–film interface. In this connection, it is an important feature of the liquid-phase process that chemical impurities and mechanical damage of the substrate are removed when material is initially dissolved from the substrate surface prior to epitaxial growth. A clean interface p-n junction is thus obtained. Since liquid-phase epitaxy also favors the achievement of high doping and a steep concentration gradient at the p-n junctions, the process has proved itself eminently suitable for application in the manufacture of Ge tunnel diodes. In its application to the fabrication of the GaAs laser diode, it is an additional advantage of the liquid-phase process that the interface p-n junction is formed on a (100) crystal plane. As a consequence, this p-n junction is perfectly planar and also perpendicular to the (110) cleavage planes of the wafer. An optimum geometry (plane-parallel ends perpendicular to a perfectly flat p-n junction) is thus insured for diodes cleaved from (100) oriented GaAs wafers whose p-n junction has been formed by liquid-phase epitaxy.*

INTRODUCTION

IT IS WELL KNOWN that vapor-phase epitaxy has had a revolutionary impact on semiconductor device technology. It is less well known, however, that for some applications epitaxial growth from the liquid phase can also give excellent results. Investigations carried out in these Laboratories have shown, for instance, that liquid-phase epitaxy possesses advantages over vapor-phase epitaxy in applications demanding highly doped epitaxial films and high-quality p-n junctions at the substrate–film interface. Thus, liquid-phase epitaxy has found important applications in the manufacture of tunnel diodes[1] and in the fabrication of experimental GaAs laser diodes.

In this paper, the apparatus and the processing details involved in

[1] H. Nelson, "Epitaxial Crystal Growth from the Liquid Phase," *Solid State Device Conference*, Stanford University, June 26, 1961.

liquid-phase epitaxy are described. In addition, the physical properties of Ge and GaAs epitaxial material are discussed and device results presented.

Apparatus and Procedure

Liquid-Phase GaAs Epitaxy

Figure 1 illustrates the apparatus used for liquid-phase epitaxy. In the cross section at the top of the figure, the starting condition

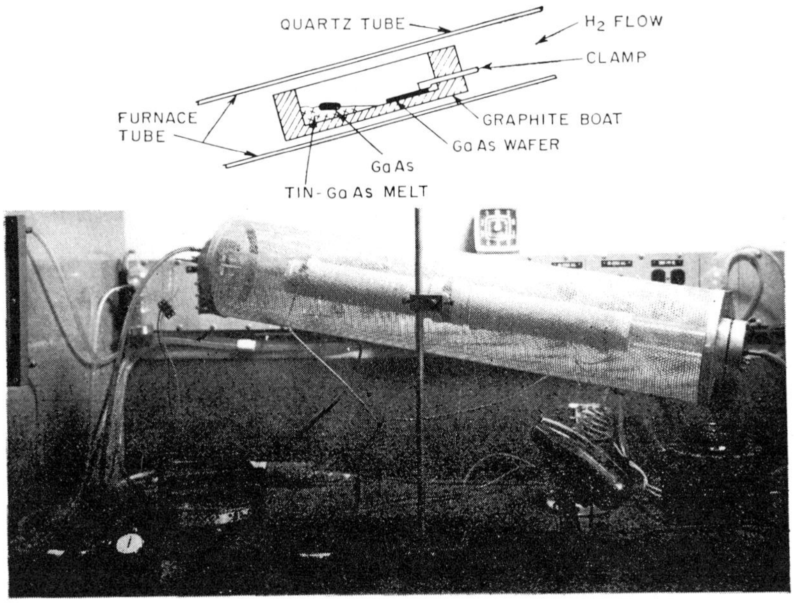

Fig. 1—Apparatus for epitaxial growth from the liquid state.

for the generation of an n-type GaAs film on a p-type GaAs substrate is illustrated. The substrate wafer to be processed is of the (100) orientation and it is held tightly against the flat bottom at the upper end of a graphite boat. A tin–GaAs mixture is placed at the lower end. The graphite boat is fixed in position at the center of a constant-temperature zone of the quartz furnace tube. With the furnace tube tipped as shown in the diagram and with a flow of hydrogen through the tube, the graphite boat is heated to about 640°C. As the temperature rises, GaAs particles dissolve in the tin at the lower end of the boat. When the temperature reaches about 640°C, the heating power is turned off and the furnace is tipped so that the molten tin covers the exposed surface of the GaAs wafer. At this time the tin is nearly saturated with GaAs. As the furnace cools, GaAs initially dissolves from the substrate surface until a solution equilibrium is established. Then upon further cooling, precipitation of GaAs from the solution and epitaxial growth upon the substrate occurs. At about 400°C, the hydrogen flow through the furnace is replaced by nitrogen and the solution is decanted from the wafer surface by tipping the furnace back to its original position. Immediately afterwards the graphite boat is removed from the furnace tube and any remaining tin is wiped off the surface of the epitaxial layer. The same GaAs-Sn solution can be used successively for the processing of several wafers.

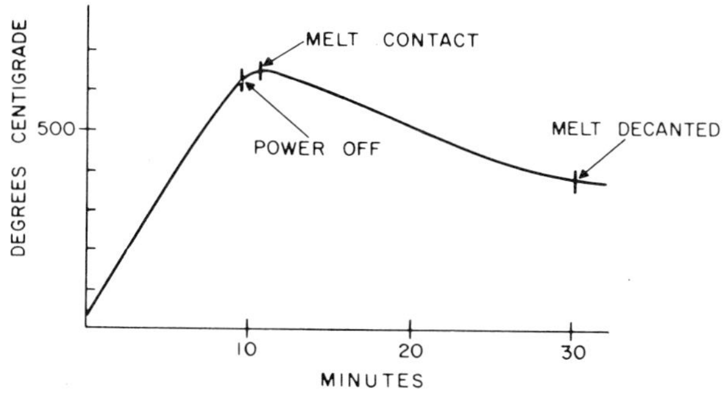

Fig. 2—Heating schedule for epitaxial deposit of GaAs from a tin solution.

More GaAs is occasionally added to the melt to replace that consumed in epitaxial growth. The details of the heating cycle employed are depicted in Figure 2. This heating cycle leads to the formation of an epitaxial layer about 1 mil thick.

It is an important feature of the liquid-phase process described above that during the substrate dissolution, which occurs before epitaxial growth begins, a solution front is formed which coincides with a (100) crystal plane of the wafer. As a consequence, the p-n junction formed is perfectly planar and also perpendicular to the (110) cleavage planes of the wafer. This condition is particularly advantageous when the liquid-phase process is used in the fabrication of laser diodes. An optimum geometry (plane-parallel ends perpendicular to a perfectly flat p-n junction) is insured for the parallelopiped diodes which are cleaved from the processed wafer. It is another desirable feature of the liquid-phase process that chemical impurities and work damage on the substrate surface are removed during dissolution of the surface layer prior to epitaxial growth. Consequently, a clean, interface p-n junction with excellent diode characteristics is insured.

In the example of liquid-phase epitaxy described above, the tin used as a solvent is a donor in GaAs. As a consequence, it can be used for the generation of n-type regions with a net donor concentration of about 5×10^{17} cm^{-3} (determined by the as yet unknown function of the saturation solubility of tin in GaAs at these temperatures). Compensated n-type regions of higher resistivity, as well as p-type regions, can be obtained by adding zinc or cadmium to the tin used as solvent for the GaAs. However, to avoid the need for compensation, the tin

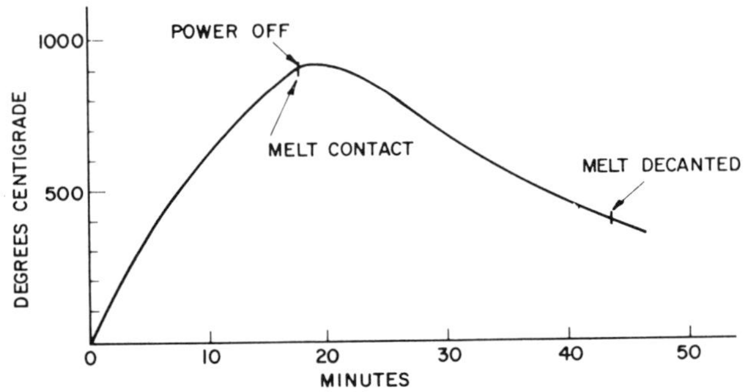

Fig. 3—Heating schedule for epitaxial growth of GaAs from a gallium melt.

can be replaced by gallium as a solvent for the GaAs. Type and degree of doping can then be controlled by adding the requisite amount of donor or acceptor impurity to the solvent melt. Thus, excellent epitaxial material for laser-diode use has been prepared using a solvent melt composed of 4.5 gram Ga and 0.7 gram GaAs. This basic melt has been doped with 0.010 gram of Te for the growth of n-type material or it has been doped with 0.10 gram of Zn for the growth of p-type films. The processing schedule is similar to the one used with tin except that higher temperatures are used. The heating cycle is shown in Figure 3. An epitaxial film about 5 mils thick is formed under these processing conditions.

Liquid-Phase Germanium Epitaxy

In the case of Ge, liquid-phase epitaxy has proved important chiefly in the manufacture of tunnel diodes. Epitaxial material for this application is processed in the same type of apparatus as is used with GaAs and the processing schedules are similar but (111) oriented wafers are employed as substrates. To deposit a degenerate p-type

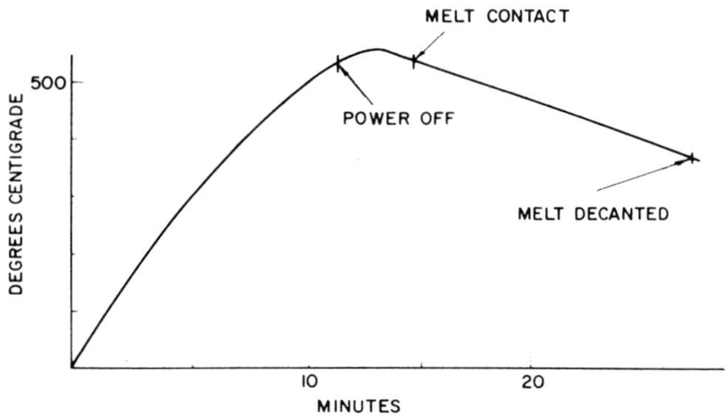

Fig. 4—Heating schedule for epitaxial deposit of Ge from an indium solution.

epitaxial film on an n-type substrate, Ga-doped indium is used as the solvent for the Ge. The n-type substrate material is doped with arsenic to a net donor concentration of, say, 2.5×10^{19} cm^{-3}. When the heating cycle shown in Figure 4 is employed, about 0.3 mil thickness of Ge is dissolved from the substrate after which an epitaxial film about one mil thick is deposited. The net acceptor concentration in the epitaxial layer is a function of the Ga content of the solvent melt. For a melt composed basically of 30 grams In and 3 grams Ge, the net acceptor concentrations varies with the weight of Ga added to the melt as shown in Figure 5.

The interface p-n junctions of epitaxial Ge material processed as above can be perfectly planar as observed by light microscopy. How-

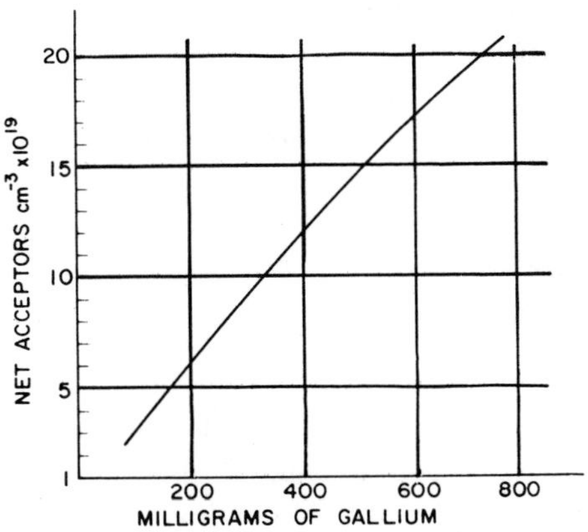

Fig. 5—Acceptor concentration in epitaxial layer versus weight of gallium added to a 30-gram indium solvent melt.

ever, in many instances slight departure from planarity occurs. Thus, for the growth of germanium on a germanium substrate, it is observed that the solution-front formed before regrowth is less likely to coincide with a wafer crystal plane than in the case of GaAs.

Epitaxial Ge material particularly suitable for the manufacture of high-quality tunnel diodes is prepared by the growth of degenerate n-type epitaxial layers on degenerate p-type substrates. A solvent melt composed of 40.0 grams of tin, 25.0 grams of lead, 5.0 grams of germanium, and 1.5 grams of arsenic is used. The heating cycle shown

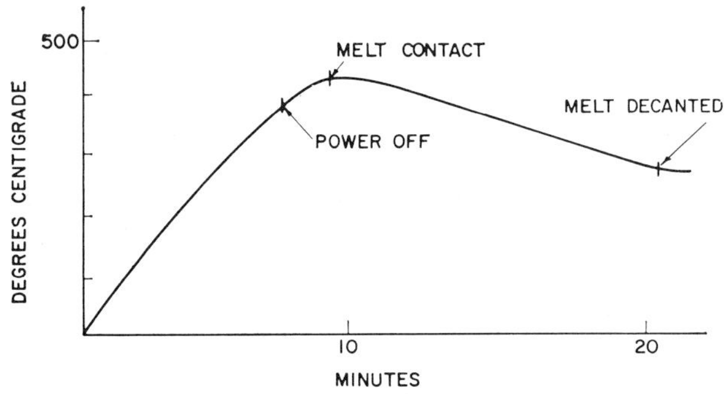

Fig. 6—Heating schedule for epitaxial deposit of Ge from a tin-lead solution.

in Figure 6 results in the growth of an epitaxial film about 0.3 mil thick following substrate dissolution to a depth of about 0.2 mil. The p-n interface region tends to be very jagged.

Additional work has been carried out in which high-purity lead was used as the main constituent of the solvent melt. This has allowed the deposition of high-quality epitaxial films in a range of thickness from 1 to 75 microns and in a range of resistivity from 0.001 to 0.8 cm. However, this work led to the conclusion that the liquid-phase epitaxy is not likely to be competitive with vapor-phase epitaxy in applications where a high degree of control and reproducibility with regard to resistivity and film-thickness is required.

Characteristics of Liquid-Phase Epitaxial Material

In the course of the investigations of liquid-phase epitaxy, device performance has been largely depended upon as a direct measure of the quality of the epitaxial material employed. Additional information has been obtained from x-ray diffraction data and from optical microscope observations on dislocation density and lineage. Except for

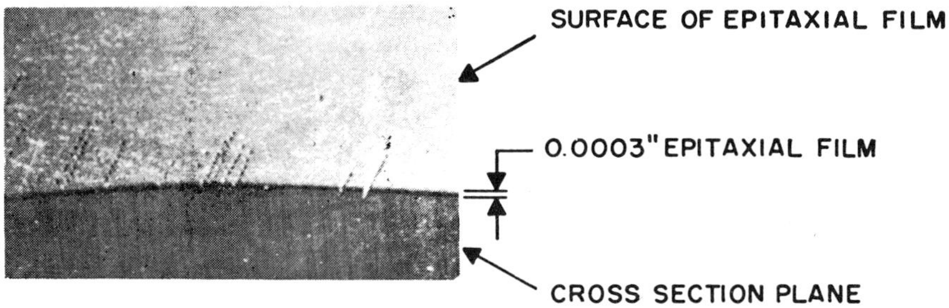

Fig. 7—Angle-lapped cross section of arsenic-saturated epitaxial film on degenerate p-type germanium.

arsenic-doped, highly degenerate germanium films, Laue patterns obtained on the epitaxial films are indistinguishable from those obtained for bulk, single-crystal material. In the case of n-type germanium films grown from a tin–arsenic solution, the crystalline quality deteriorates as the arsenic content is increased to the point where highly degenerate deposits are obtained. The surface texture becomes grainy and inclusions of GeAs are found at the surface of the films. As a result of the high saturation solubility of tin in germanium, the tin content of these films is as high as 0.5 at.%. Since tin is electrically inert in germanium, the semiconductor properties are little affected by its presence. However, it leads to a slight increase in the coefficient of thermal expansion of the germanium as evidenced by a "bimetallic strip effect" exhibited by processed wafers.

An angle-lapped cross section of an arsenic-doped, epitaxial film grown on degenerate p-type Ge is shown in Figure 7. The grainy texture of the epitaxial surface and the jagged p-n junction typical of this epitaxial material is apparent in the photograph. The cross section shown in Figure 8, however, is more typical of liquid-phase epitaxial

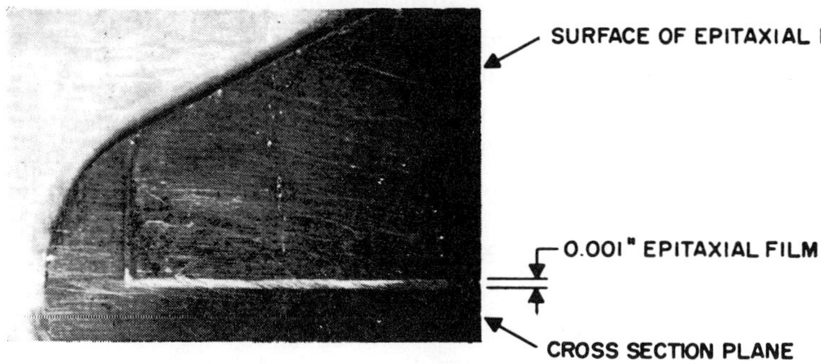

Fig. 8—Angle-lapped cross section of degenerate Ga-doped epitaxial film on degenerate n-type germanium.

material in general. It is an angle-lapped cross section of germanium epitaxial material prepared by growth of degenerate p-type germanium ($N_a \approx 8 \times 10^{19}$ cm^{-3}) on a degenerate n-type substrate ($N_d \approx 2.5 \times 10^{19}$ cm^{-3}). The surface of the epitaxial layer is relatively flat and the p-n junction is seen to be planar.

The microphotographs of Figure 9 show cross sections prepared by etching cleavage planes of typical epitaxial GaAs material prepared by liquid-phase epitaxy. As seen in these micrographs, the interface p-n junctions are planar and perpendicular to the (110) cleavage planes.

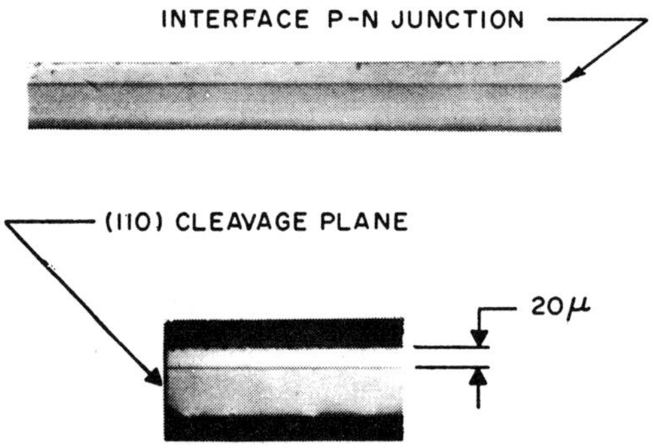

Fig. 9—Cross sections of GaAs epitaxial laser-diode material.

This is evidence of the fact that in GaAs, the pre-growth solution front tends to coincide with a (100) crystal plane.

DEVICE RESULTS

GaAs Laser Diodes

The application of liquid-phase epitaxy to the fabrication of experimental GaAs laser diodes was directed toward two main objectives. One of these was to provide laser diodes for the study of light emission and laser behavior as a function of doping impurities and distribution profiles of dopants, especially of those not readily obtained by diffusion. Another objective was to determine the merits of liquid-phase epitaxy relative to conventional diffusion for the fabrication of high-quality laser diodes.

Although these investigations have not been completed, significant results have been obtained. With regard to the first objective, the effects of different dopants and of unusual concentration gradients at the p-n junction in epitaxial laser diodes have been studied by J. Pankove and by G. Dousmanis.[2] As an example of results obtained, the curves in Figure 10 show a markedly greater frequency shift of the incoherent emission peak with current density for diodes with steep concentration gradients than for diodes with more-shallow p-n junctions. The curves A and B represent data for a diffused and an epitaxial diode, respectively. The doping density on either side of the p-n junction is about the same for both diodes. As determined from

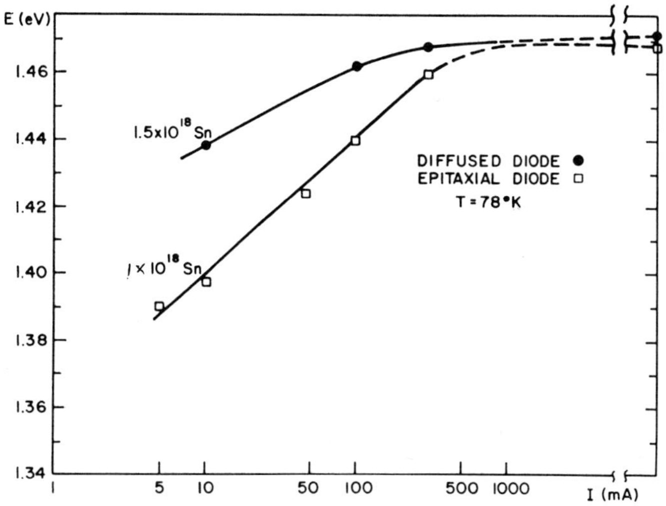

Fig. 10—Energy of emission peak versus current.

capacity measurements, however, the dopant concentration gradient at the p-n junction is 3.5×10^{22} cm^{-4} for the diffused and 1.4×10^{23} cm^{-4} for the epitaxial diode.

With regard to the second objective, tests carried out in the course of our investigations have shown lower laser thresholds for epitaxial than for diffused diodes. However, light emission efficiency and laser threshold for both types of diodes vary greatly from lot to lot of GaAs starting material, apparently as a result of variation in crystal quality. It is possible, therefore, that superiority in crystalline quality of starting material rather than superiority in processing technology is responsible for the improved results obtained with epitaxial laser diodes. It is worth emphasizing, however, that planar p-n junctions which are perpendicular to (110) cleavage planes can be more reliably achieved

[2] G. C. Dousmanis, C. W. Mueller, and H. Nelson, "Effect of Doping on Stimulated and Incoherent Emission in GaAs," to be published in *Jour. Appl. Phys. Letters.*

through liquid-phase epitaxy than through diffusion. Optimum laser-diode geometry should thus be more easily obtainable with epitaxially grown layers.

The laser performance data gathered in the course of this investigation have been obtained for laser diodes cleaved from epitaxy-processed wafers of the (100) orientation. Before cleavage these wafers are heat treated at 900°C for four hours and then lapped on the n-type side to a final thickness of three mils. Also, both of the major surfaces are metallized to provide high-conductivity ohmic contacts to the p- and n-type regions. The parallelopiped diodes cleaved from these wafers are dimensionally nonuniform, but for comparative tests, units are

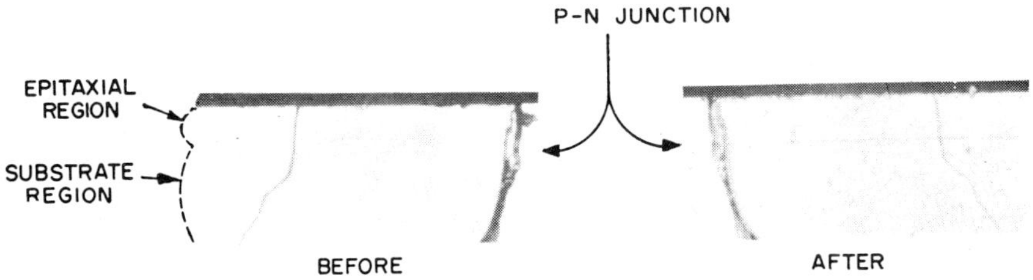

Fig. 11—Cross sections of epitaxial GaAs wafer before and after 4 hours of inter-diffusion at 900°C.

selected which are about 5 mils wide and 20 mils long. The 900°C heating step is essential for good laser performance. The resulting interdiffusion of acceptors and donors that occurs at the p-n junction leads to no perceptible displacement of this junction (see Figure 11). However, the interdiffusion results in an increased depletion-layer width at the p-n junction with a resultant improvement of the reverse and forward characteristics, as shown in Figure 12.

The curves in Figure 12 represent data obtained for a typical diode cleaved from material prepared by epitaxial growth of GaAs from a tellurium-doped gallium solution onto a zinc-doped substrate ($N_a \approx 1 \times 10^{19}$ cm^{-3}). Almost identical diode characteristics are obtained from diodes cleaved from material prepared by GaAs growth from a zinc-doped gallium solution on a tellurium-doped substrate ($N_a \approx 1.1 \times 10^{18}$ cm^{-3}). Table I shows threshold data for three representative diodes from each of the two types of material. For purpose of comparison, data for three diffused diodes are included in this table. These diodes were cleaved from material prepared by the diffusion of zinc into a tellurium-doped wafer ($N_d \approx 1.1 \times 10^{18}$ cm^{-3}). All diodes in-

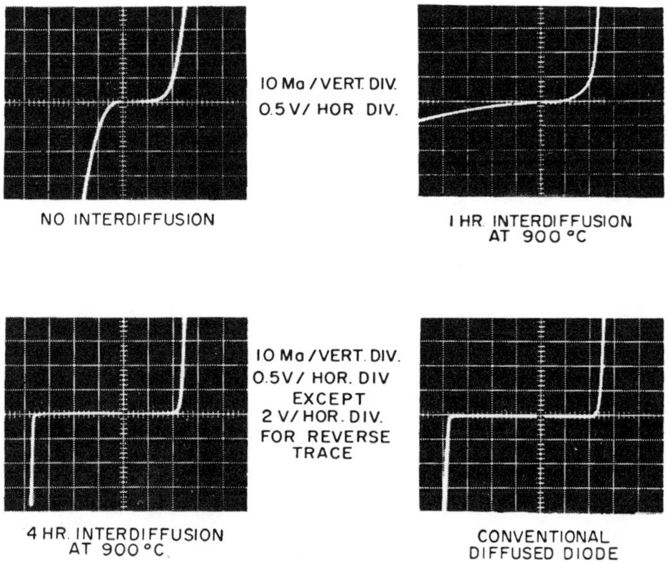

Fig. 12—Epitaxial GaAs diode characteristics.

volved are parallelopiped structures the sides of which have been sawed. They are all approximately 0.003 inch thick, 0.005 inch wide, and 0.020 inch long.

Germanium Tunnel Diodes

In the manufacture of tunnel diodes one of the principle advantages of liquid-phase epitaxy is that it permits the fabrication of numerous

Table I—Threshold Data for Epitaxial and Diffused GaAs Laser Diodes

Diode Type	Laser Threshold ($amp\ cm^{-2}$) Diode #1	Diode #2	Diode #3
Epitaxial n-type film on a p-type substrate	1.8×10^3	1.7×10^3	2.0×10^3
Epitaxial p-type film on an n-type substrate	1.5×10^3	1.6×10^3	1.4×10^3
P-type region diffused into an n-type wafer*	7.0×10^3	8.5×10^3	9.0×10^3

* This wafer and the above n-type substrate wafer were cut from the same GaAs crystal. GaAs laser diodes made from other crystals have yielded thresholds as low as 3×10^3 amp cm^{-2}.

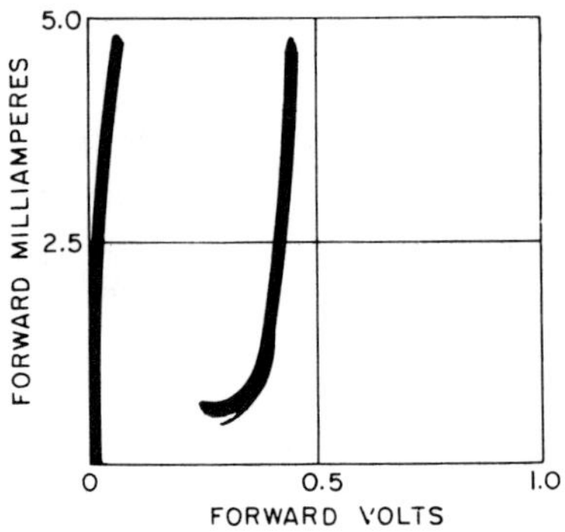

Fig. 13—Forward-bias characteristics of ten tunnel diodes from a 25-unit array.

identical devices from large-area wafers of processed material. The data represented in Figure 13 is testimony to the great uniformity from unit to unit which can be achieved by this technique. The curve shown is an "envelope" of the oscillograph traces for 10 equal-area tunnel diodes of a 25-unit array. This device uniformity is indicative also of the high degree of homogeneity which can be achieved in epitaxial films grown by liquid-phase epitaxy.

For a p-type epitaxial layer on an n-type substrate the forward-bias characteristics shown by the oscillographs traces A and B of Figure 14 are obtained. Traces A and B are typical for low and high doping densities, respectively. Trace A represents a diode with a net donor concentration in the n-region of about 2.2×10^{19} cm^{-3} and a net acceptor concentration in the p-region of about 8×10^{19} cm^{-3}. Trace B represents a diode for which the respective concentrations are 3.2×10^{19} cm^{-3} and 3.0×10^{20} cm^{-3}. The junction area is the same for both diodes. As evidenced by the data represented by Figure 12, it is characteristic of p- and n-type tunnel-diode material that high tunnel-

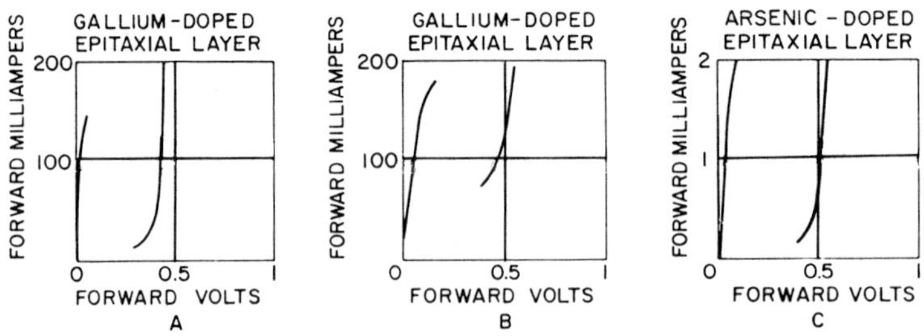

Fig. 14—Forward-bias characteristics of epitaxial Ge tunnel diodes.

current density cannot be achieved by increased doping without simultaneous loss of high peak-to-valley current ratio. In this respect superior results are obtained with n-on-p type material. Trace C of Figure 14 represents a diode of the same junction area as the one represented by traces A and B, but it is fabricated from epitaxial material prepared by growing arsenic and tin-saturated n-type germanium on a p-type substrate with a net acceptor (gallium) concentration of 8×10^{19} cm^{-3}. The high peak-to-valley current ratio evidenced by trace C is characteristic of this type of epitaxial material even at doping densities requisite for peak-currents densities up to 30,000 amp cm^{-2}.

Conclusions

The epitaxial growth of Ge and GaAs from the liquid state has proved a valuable addition to semiconductor technology. This process has found important practical applications in the manufacture of several types of Ge tunnel diodes and in the fabrication of high-quality, experimental GaAs laser diodes.

Compared to vapor-phase epitaxy, the liquid-phase process is characterized by simplicity and speed. Its equipment requirements are very modest and the same apparatus can be used for different semiconductor materials and for a great variety of growth conditions.

Acknowledgments

In the work reported upon, I have had the encouragement and the assistance of numerous coworkers. I am particularly indebted to H. S. Sommers for encouragement in early work on the preparation of epitaxial tunnel diodes and to S. Bragnum for invaluable direct assistance in this work. R. Glicksman, N. Ditrick, and R. M. Minton deserve special credit for major contributions in adapting the liquid-phase process for the successful manufacturing of Ge tunnel diodes.

Laser diode investigations by J. Pankove and G. Dousmanis have provided vital feedback information used in adapting the liquid-phase process for the fabrication of high-quality GaAs laser diodes. Also, the availability of ingenious test equipment developed by F. H. Nicoll has greatly facilitated this work. Finally, it is a pleasure to acknowledge that Frank Hawrylo's participation in the GaAs work has been characterized to an unusual degree by initiative and the contribution of valuable ideas.

key paper

Liquid-Phase Epitaxial Growth Of Gallium Arsenide Under Transient Thermal Conditions*

R.H. Deitch
(Bayside Research Center of General Telephone and Electronics Laboratories Incorporated, Bayside, New York, US)

Journal of Crystal Growth, Volume 7 (1970)
Pages 69-73

*Based in part on presentation having the same title given at the ACCG Conference on Crystal Growth, Gaithersburg, Md., 12 August 1969.

Thin, smooth epitaxial layers of GaAs have been prepared by solution growth using a dipping technique in a vertical system. Unlike previously reported methods for growing semiconductor layers by liquid-phase epitaxy, the transient method does not require extreme temperature control to produce thin layers or the use of very slow cooling rates to maintain interface stability. Layers a few microns thick were generated in a matter of seconds by dipping a cooler substrate into a hotter solution, and, because the extent of growth is necessarily limited, reproducibility in layer thickness was achieved.

1. Introduction

It is well known that epitaxial layers of single-element, binary, and mixed-crystal semiconducting compounds can be grown from metallic solution[1-7]. The epitaxial layers that are grown by the usual liquid-phase techniques for the fabrication of electroluminescent, laser, and Gunn diodes are normally greater than 10 μm thick. For many other device structures such as avalanche diodes, field-effect transistors and special-purpose contacts, thinner epitaxial layers are required. In many cases it also is desirable that the epitaxial layers have very smooth as-grown surfaces.

Recently, a method for growing thin epitaxial layers from solution was reported by Dawson and Whelan[7]. They dipped a GaAs substrate into a Ga solution saturated with As, cooled between 1 °C and 3 °C, and then withdrew the sample. A temperature control of ± 0.01 °C was required to control layer thicknesses within the range of 1 to 15 μm, and, because of the controlled degree of supersaturation obtained as a consequence of their close temperature regulation, excess solution did not adhere to the layer surface. If any saturated solution had remained on the surface of a layer, additional undesirable growth would have occurred as the sample was subsequently cooled to room temperature. Epitaxial layers of GaAs on the order of a few microns thick have also been reported grown from solution by Solomon[8], Kinoshita et al.[9]) and Gorelenok et al.[10]).

This paper describes a novel method of liquid-phase epitaxial growth which has been used to prepare thin layers of GaAs. Growth is initiated by dipping a cooler GaAs substrate into a hotter Ga solution saturated with As. The furnace power is not turned off and no external adjustment of temperature is required to grow a layer. Layers in the range 1 to 10 μm thick have been grown within time intervals of a few seconds to a few minutes at growth rates at least as high as 10 μm/min at the beginning of growth. Epitaxial layers that are grown by this transient* or "quick-dip" method are conveniently limited in thickness by a self-regulating mechanism and can be produced with relatively smooth surfaces.

Briefly, the transient method can be explained in the following way. When the substrate is immersed in the solution, the solution is quenched by the cooler substrate and substrate holder. The quenching supersaturates a thin layer of solution adjacent to the substrate surface and an epitaxial layer grows as the supersaturation is relieved. Since the furnace is not turned off, the solution as well as the substrate will eventually heat up and growth will cease. The extent of growth can be further controlled by removing the substrate from the solution either before the supersaturation has been com-

* We have also referred to this new method as "Transient-Mode Liquid Epitaxy" (TMLE).

pletely relieved or before the temperature has fully recovered.

2. Experimental procedure

The apparatus which is used is shown schematically in fig. 1. The system includes a furnace held in a vertical position; a quartz chamber with provision for keeping a graphite crucible situated in a particular position within the furnace; a graphite substrate holder attached

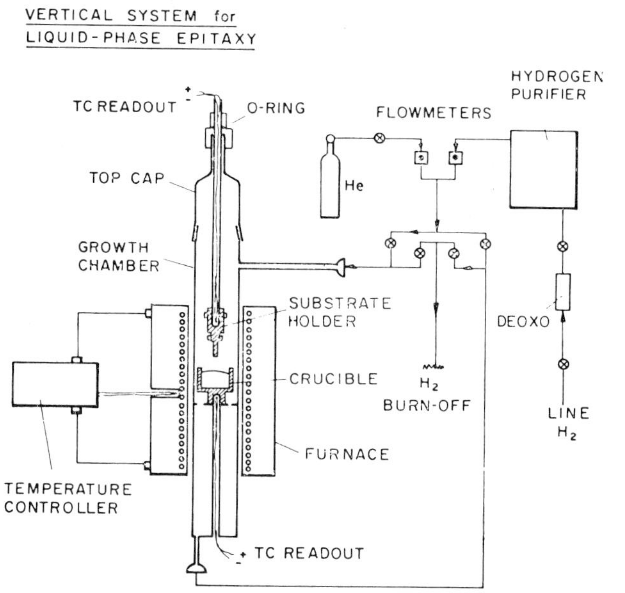

Fig. 1. Sketch of apparatus.

to the end of a quartz tube, which in turn passes through an "O"-ring seal in a removable top cap; a gas train for introducing helium or Pd-purified hydrogen into the chamber; thermocouples to monitor the substrate and solution temperatures, and appropriate temperature controlling equipment.

Up to 70 g of a desired solution of GaAs in Ga is placed in the graphite crucible. Two oriented, chemimechanically polished GaAs substrates approximately 3 cm^2 in area are cleaned and etched, and then clamped in a vertical position onto the substrate holder. The substrates are placed within the growth chamber at a position above the solution with a stream of Pd-purified hydrogen flowing from the top to the bottom of the chamber. The furnace is turned on and the substrates are heated 50 °C to 100 °C below that of the solution. After the solution and substrate temperatures are stabilized, the substrates are dipped into the solution with the heating elements of the furnace maintained at a constant temperature. The substrates, with epitaxial layers, are rapidly withdrawn from the solution to a cooler position in the chamber either after a specified time or when the solution temperature reaches a specified value. The furnace is then turned off and the samples and solution are cooled to room temperature. A substrate can also be held in a horizontal position on a different holder. However, because a droplet of solution adheres to the center of the layer surface when the substrate is removed from the solution, this method of holding a substrate was not used extensively.

3. Results

With a solution temperature of 800 °C, a substrate temperature of 750 °C, and dip times of 3 sec to 7 min, the thickness versus time curve shown in fig. 2 was obtained. It shows that approximately 1 μm grew in 3 sec, 3 μm in 15 sec and 4 μm in 60 sec. By 90 sec the thickness was reduced to about 3 μm and remained near that value for up to 7 min. These thicknesses were obtained primarily from microscopic measurement of cleaved and stained {110} cross-sections of the (100) oriented wafers. Thicknesses were also obtained by measurement of fringes in infrared reflectance spectra. Reproducibility of layer thickness was readily achieved if the same initial conditions and dip times were used.

The surface structures that were obtained are shown in fig. 3. These are typical of layers grown on vertically held substrates for dip times ranging from 6 to 300 sec.

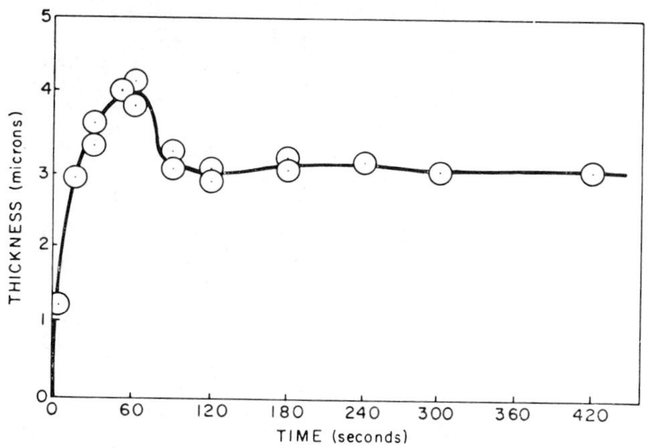

Fig. 2. Effect of immersion time on the thickness of epitaxial layers grown by dipping a 750 °C substrate into an 800 °C saturated solution.

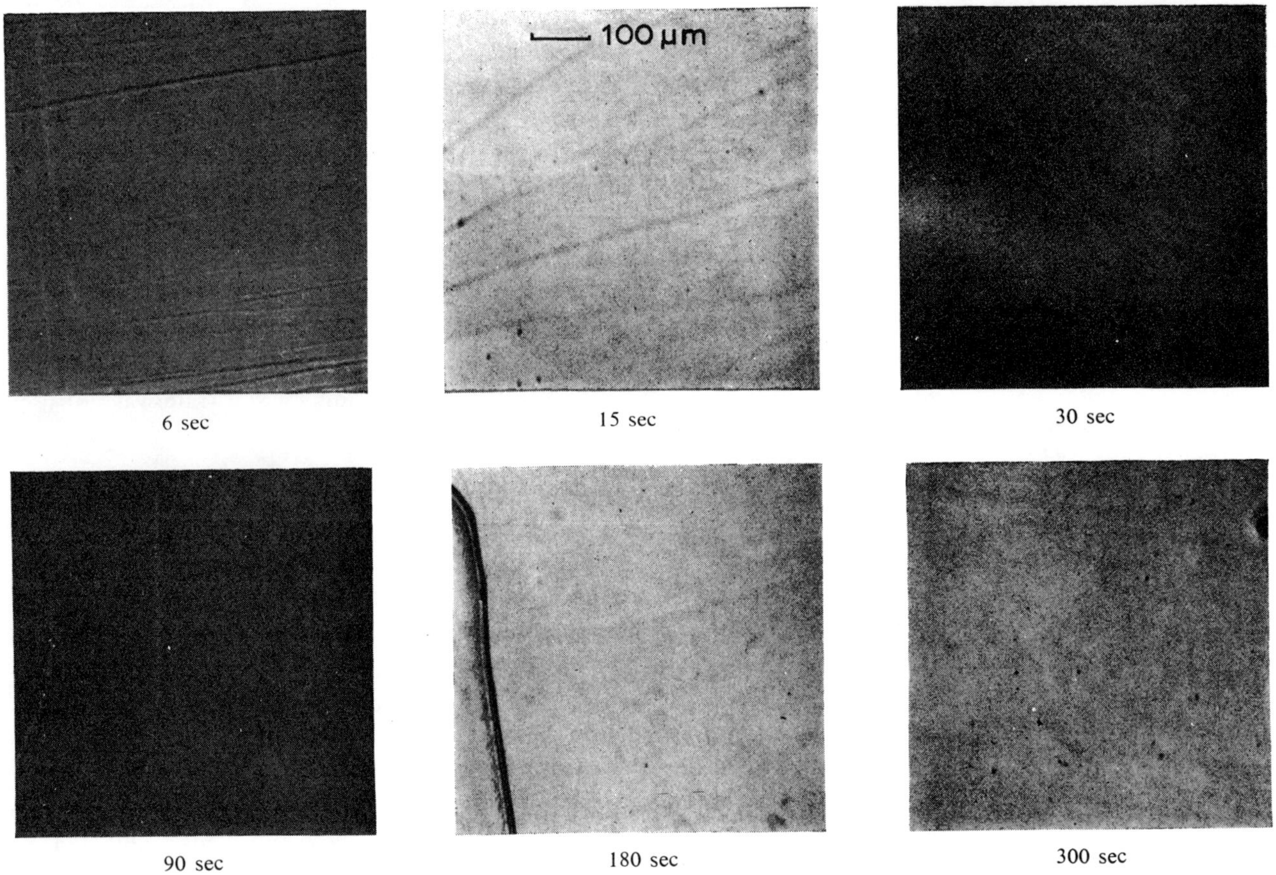

Fig. 3. Surface structures of epitaxial layers grown for dip times of 6, 15, 30, 90, 180 and 300 sec.

Fig. 4. Cross-section and corresponding surface of epitaxial layer grown with a dip time of 7 min. Structure seen at left side of the surface view was used for focusing.

The lines seen on the surfaces of samples withdrawn after 6, 15 and 30 sec are furrows of reduced growth which are curved and generally parallel to the solution surface. These furrows, some of which can be as deep as half the layer thickness for the shorter dip times, are gradually eliminated with increasing dip time. They are not as prominent on layers grown under comparable conditions on horizontally held substrates. The flatness of interface, uniformity of layer thickness and smoothness of surface that can be obtained using the transient method are apparent from fig. 4 in which is shown a typical {110} cross-section and a view of the corresponding (100) surface of a layer grown with a dip time of 7 min.

4. Discussion

It is believed that the experimental results can be explained on the basis of the quenching and subsequent relief of induced supersaturation of the solution as briefly outlined in the Introduction. Fig. 5 is used to explain these results in more detail. The thermal history of a solution in the vicinity of a substrate which has been dipped using the transient method is shown by curve A. Corresponding plots of saturation (curve B), and layer thickness (curve C) as a function of time are also shown.

The state of saturation of the solution can be given

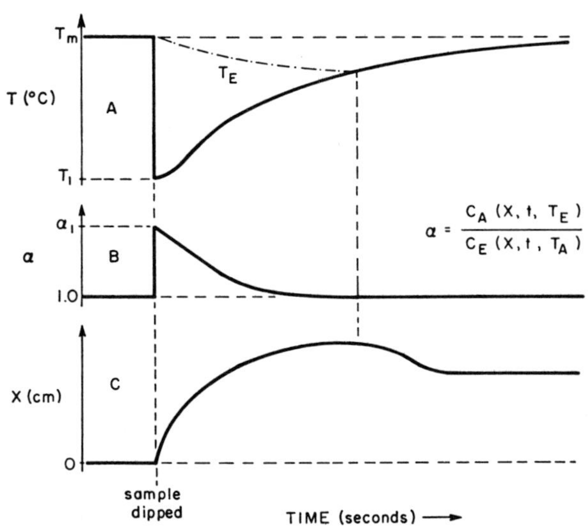

Fig. 5. Idealized curves showing the time dependence of (A) solution temperature, (B) saturation parameter, and (C) layer thickness.

by α, which is defined as the ratio of the actual concentration, C_A, to the equilibrium concentration, C_E, of GaAs in the solution. C_A determines the equilibrium temperature, T_E; while C_E is the equilibrium concentration for the actual temperature, T_A. Thus, as similarly described by Tiller and Kang[11]), the solution is undersaturated if $\alpha < 1$, exactly at saturation if $\alpha = 1$, and supersaturated if $\alpha > 1$.

In curve A, T_m is the starting solution temperature. When the cooler substrate is immersed, a steep temperature gradient is set up across the solution/substrate interface and the temperature of the solution just ahead of the substrate decreases rapidly. This sudden temperature drop causes a sharp increase in α from its initial value of unity. Since heat is continually being furnished to the solution from the crucible walls to bring the solution back to its starting temperature, a minimum temperature, T_1, is reached. Thereafter, the solution gradually heats up and its temperature stabilizes. Although not shown in fig. 5, the temperature of the substrate, which had increased sharply at first, rises in harmony with the temperature of the solution. As GaAs grows onto the substrate, α continually decreases towards unity. Although the rate of growth is initially quite high, the temperature gradient at the growing interface is sufficiently high to prevent constitutional supercooling.

For example, for a growth rate of about 10 μm/min at a temperature of 800 °C, a gradient of about 600 °C/cm is required to prevent constitutional supercooling[12]). A calculation of heat flow considering only thermal conduction with a thermal diffusivity of 10^{-1} cm^2/sec indicates that a temperature gradient of such magnitude is obtained across the growing interface when a 750 °C substrate is dipped into an 800 °C solution. As this temperature gradient decreases, the growth rate also decreases. The relationship between growth rate and temperature gradient is determined by the heat flow, which is governed by the thermal properties and geometry of both the solution and the crucible/substrate-holder system. For a particular system, control over the growth rate and temperature gradient can be obtained by adjusting the absolute temperature of the solution, the relative temperatures of solution and substrate, and the amount of excess GaAs that exists in the solution when the substrate is dipped. For any particular solution temperature there probably is a minimum substrate temperature below which single-crystal growth ceases to occur on the substrate when it is dipped. Such an effect has been observed, for example, when layers of Ge are grown by lowering a cooler single-crystal substrate onto a hotter melt[13]).

When α returns to unity, which occurs when T_E and T_A coincide, equilibrium is reestablished and growth due to supersaturation ceases. If the solution temperature continues to rise, as it did in our experiments, and the sample is left in the solution, material will dissolve from the epitaxial layer until the temperature levels out. If, however, the solution temperature decreases after equilibrium is reestablished, which would be the case if the furnace temperature was being lowered, normal liquid-phase epitaxial growth will occur.

In liquid-phase epitaxy, as it is usually carried out, the growth rate increases with time to a maximum and then decreases asymptotically to zero[11]). Thus, a layer cannot be grown with only a decreasing growth rate, and for thin layers all the growth may occur while the growth rate is increasing. This is not the case when a layer is grown using the transient method. With the transient method the increase in growth rate takes place over a very short interval of time during which a negligible amount of growth is obtained. Such a non-growth period is comparable to the instantaneous increase in growth rate described by Tiller and Kang[11]) for an initially supersaturated condition ($\alpha > 1$). Because an epitaxial layer is grown relatively fast using the transient method, effective impurity distribution coefficients closer to unity than is normally the case in solution growth can be obtained. If growth by the transient method is followed by normal liquid-phase growth, it should be possible to control impurity distributions for junction formation in a way similar to that described by Pankove[14]). By selection of specific impurities, it may be possible to form a junction during only the transient growth cycle.

5. Conclusions

A novel method for growing thin, smooth epitaxial layers from metallic solution has been described. Growth is initiated by dipping a cooler substrate into a hotter solution with no adjustment in temperature control required to obtain a self-regulated growth. The extent and morphology of an epitaxial layer grown by this transient or "quick-dip" method depends upon initial thermal settings and subsequent conditions of heat flow. The sample can be withdrawn before a temperature cycle is completed in order to gain further control over layer thickness. The use of this method for growing thin GaAs layers has been demonstrated. Layers on the order of a few microns thick were grown in less than one minute under the conditions used. It should be possible to apply this method in general to the epitaxial growth of other materials from metallic solution.

Acknowledgments

The author thanks J. Wagner for his valuable computations and discussion of the theoretical aspects of this work, F. J. Reid and J. F. Black for their helpful discussions, and P. Sierminski and W. Powazinik for technical assistance.

References

1) H. Nelson, RCA Rev. **24** (1963) 603.
2) Ivars Melngailis, Appl. Phys. Letters **6** (1965) 59.
3) M. R. Lorenz and M. Pilkuhn, J. Appl. Phys. **37** (1966) 4094.
4) H. Rupprecht, in: *Proc. 1966 Symp. on GaAs, Reading* (Inst. Phys. and Phys. Soc., London, 1967) p. 57.
5) M. S. Panish, H. J. Queisser, L. Derick and S. Sumski, Solid-State Electron. **9** (1966) 311.
6) H. Rupprecht, J. M. Woodall and G. D. Pettit, Appl. Phys. Letters **11** (1967) 81.
7) L. R. Dawson and J. M. Whelan, Bull. Am. Phys. Soc. **13** (1968) 375.
8) R. Solomon, in: *Proc. 1968 Symp. on GaAs, Dallas* (Inst. Phys. and Phys. Soc., London, 1969) p. 11.
9) J. Kinoshita, W. W. Stein, G. F. Day and J. B. Mooney, in: *Proc. 1968 Symp. on GaAs, Dallas* (Inst. Phys. and Phys. Soc., London, 1969) p. 22.
10) A. T. Gorelenok, A. N. Imenkov and B. V. Tsarenkov, Soviet Phys.-Semicond. **2** (1969) 1287.
11) W. A. Tiller and C. Kang, J. Crystal Growth **2** (1968) 345.
12) W. A. Tiller, J. Crystal Growth **2** (1968) 69.
13) L. N. Aleksandrov, E. A. Klimenko and A. G. Klimenko, Soviet Phys.-Crystal. **13** (1969) 962.
14) J. I. Pankove, Proc. IRE **44** (1956) 185.

Received 15 January 1970

1.2.4 comparison of lpe systems

Although the first liquid epitaxy systems were designed for heavily doped materials, all such techniques are capable of producing extremely pure material; each has been refined to a high degree of sophistication.

For the growth of high purity layers, both horizontal and vertical systems are used; however, the current feeling is that the vertical reactor is capable of better temperature control, thereby producing more uniform epitaxy. Uniformity is of particular importance in the growth of thick layers of 100 μm. The transient growth procedure is essentially restricted to thin layers of only a few microns, and thus has been used primarily for heavily doped contact layers or junctions with abrupt interface regions. The horizontal system is generally preferred for good quality layers of 1-10 μm thickness.

1.2.5 present trends in epitaxial systems

Until just recently, the main emphasis in materials preparation has been towards the growth of uniformly doped, high purity, and high quality epitaxial layers. This has led to considerable debate on the comparative merits of liquid or vapour epitaxy. The present trend is towards more complicated layer structuring, demanding considerable dynamic response of the growth system — sophisticated junctions, sub-micron buffer regions, and selective epitaxy. These requirements have emphasised the different capabilities of the two main growth methods, and have led to the investigation of further ways to achieve the layer structures demanded by devices. Vapour phase and liquid phase epitaxy are still the most viable commercial processes.

High Purity Systems

With careful operation, both liquid and vapour epitaxy should produce high purity material, with differences that are primarily mechanical and economic.

Because of the faster growth rates available with LPE, thick layers can be grown in much shorter times and with greater uniformity — the reactor does not have to be stable for an excessively long period of time. The VPE systems, however, lend themselves more readily to the simultaneous growth of many layers, thereby, exploiting all the economic advantages of batch processing, as well as encouraging good reproducibility from wafer to wafer.

Surface finish and thickness control are more difficult to achieve through LPE, with the residual impurities likely to be different from those of the VPE process.

Doping Systems

Thus far, there has been no discussion of the doping of the epitaxial layer with added impurities, because it is during the growth of multilayered structures that the differences between liquid and vapour phase epitaxy are most striking.

With VPE, doping can be achieved by the addition of impurity in solid form to the source. Obviously, though, a new source is required for each impurity concentration and type. A more flexible method is to use gaseous doping introduced into the gas down stream of the source.

The impurity level can thus be controlled independently. Sulphur and zinc are the most common n- and p-type dopants for both GaAs and InP. Both are readily transported by hydrogen passing over the element in a heated tube. For transferred electron devices, a thin layer of sulphur doped material (of a few microns in thickness and at a carrier concentration of around 10^{17}) is often grown directly on the substrate, followed by the high impurity active layer. This intermediate, or buffer, stratum effectively isolates the high purity epitaxy from the substrate; the whole structure thus is grown continuously by varying the gas flow in the doping line. A thin, heavily doped contact layer is often included on the surface to overcome the problems associated with metal semiconductor contacts.

However, some problems do occur in vapour epitaxial layered structures:

1./The reaction tube can have a long-term memory to impurities.
2./The response time of the reaction tube can be significant, so that abrupt changes in impurity profile are difficult to execute.
3./Growth rates are relatively slow; diffusion of the impurity in the semiconductor further "grades" the junction.

Often these effects can be desirable, but, when p-n junctions or abrupt changes in doping profile are required, such is rarely the case.

Liquid phase material is almost always doped from the melt — new melt is required for each layer of different doping. The space restriction in the reaction tube has led to extremely ingenious design of melt boats for multilayer epitaxy (see Figure 1-6).

This approach is extremely flexible; it not only allows for different types and levels of doping to be grown successfully, but it also permits heterojunctions. In this method, the response and growth times are short. A summary of the relative merits of each epitaxy method is shown in Table 1-1.

Device structures utilizing compound semiconductors rely on high purity, high mobility n-type materials. All epitaxial growth modes are enhanced by this fact. The vapour phase systems can achieve purities to around 10^{14} cm^{-3}. The majority of devices require higher doping levels, so the layers are doped additionally. Sulphur is the most convenient n-type dopant for the vapour phase process. It is reduced in a stream of hydrogen and passed into the reaction tube downstream of the source. Selenium can also be transported in a similar fashion; for liquid phase epitaxy, tin is added to the melt. The same procedures are used to dope buffer and contact regions into the 10^{18} cm^{-3} region to form low resistivity n^+ areas.

	System	Purity	Thickness (μm)	Doping Method	Structures Possible	Cost
VPE	$AsCl_3/PCl_3$	Good		Gaseous	Multi-layers of one type only	Suitable for production
	and AsH_3/PH_3	Good	0.5-20 Optimum	or Solid	Multi-layers difficult	
LPE	Horizontal	Good	0.5-20	Solid	Multi-layers Heteroepitaxy	Relatively expensive
	Vertical	Good	10- > 100 Optimum	Solid	Multi-layers difficult	Expensive
	Transient growth	Fair	1-5	Solid	Single doped layers	Cheap

Table 1-1/Comparison of Epitaxy Systems

P-type doping is more difficult to achieve without contamination of the reaction tube. Diffusion is therefore generally preferred, although multi-well slider liquid epitaxy can be used to include p-type solution wells. Zinc is the preferred dopant in this circumstance.

Recently it has been demonstrated that IMPATT's with greatly improved performance can be made if the impurity profile of the n-layer is tailored to a high-low or low-high-low structure. For this process, the ability to grow material with rapid changes in doping level (two orders of magnitude in fractions of a micron of layer width) is required. Such materials have been grown by both liquid and vapour phase epitaxy, although the latter method is more commonly used. Again, though, the vapour phase process would be preferable for production if and when it is perfected.

One new development yet to be discussed is that of selective epitaxy, which was used in the manufacture of planar GaAs devices[10] to create a localised region of heavily doped n^+ material which forms the anode contact. The technology is complex. The surface is masked with pyrolytically grown silica, and holes are etched in unmasked regions. Vapour phase material can then be grown in the latter areas, but very careful control is needed — first, to chemically etch the holes so that they have vertical edges; then, to arrange the growth so that it comes to just the original epitaxial layer surface.

The same technique has been applied to the liquid phase growth of GaP[11]; although no results have been reported, there is no fundamental reason why selective GaAs and InP cannot be grown in the same way.

A completely different procedure is that known as Molecular Beam Epitaxy[17, 18], described in the key paper by Cho et al. The molecular beam sources are Knudsen effusion cells mounted in an ultra high vacuum system capable of working at 10^{-8} Torr; the effusion cells and sample are at operating temperatures. The surfaces are simultaneously monitored with HEED and the system is coupled to a mass spectrometer. The sample (GaAs) is then heated to around 580°C as the molecular beam impinges on the surface with a beam energy of 10-30 keV.

High quality doped sub-micron layers have been grown using this method. Its claimed advantages are a controlled growth environment in UHV and the availability of various methods of cleaning the substrate prior to growth (e.g., high temperature treatment in a controlled atmosphere and ion bombardment).

It is apparent that molecular beam epitaxy is not a commercially viable production process for standard epitaxy; the equipment is expensive and the advantages are relatively minor. However, this epitaxial technique may find application in the growth of very thin structures, where control of the more conventional growth processes becomes increasingly difficult.

key paper

Molecular Beam Epitaxy Of GaAs, $Al_xGa_{1-x}As$, And GaP

A.Y. Cho, M.B. Parish, and I. Hayashi
(Bell Telephone Laboratories, Inc., Murray Hill, New Jersey, US)

Institute of Physics and Physical Society Symposium on GaAs (1970)
Pages 18-29

Abstract. Single crystal epitaxial layers of GaAs and $Al_xGa_{1-x}As$ (doped and undoped) on GaAs, and GaP on GaP were grown by molecular beam epitaxy utilizing Knudsen effusion cells as molecular beam sources. High energy electron diffraction (HEED) studies during growth of the layers indicate that under certain conditions single crystal layers are obtained. The HEED observations are confirmed by phase contrast and electron microscopy. Photoluminescence measurements on the GaAs layers indicate that layers with luminescent efficiencies comparable to bulk and liquid phase epitaxially grown materials are obtained. Doping in the range approximately 10^{16} to at least 10^{18} is achieved with Si, Ge (n-type), and Mg (p-type).

1. Introduction

Arthur (1969) has demonstrated that the sticking coefficient of As from As_2 beams projected onto heated GaAs surfaces is dependent upon the surface coverage of Ga, being large (approximately 1) for high coverage and negligible (much less than 10^{-4}) for zero coverage. The sticking coefficient of Ga is about unity. Arthur and LePore (1969) demonstrated that GaAs can be grown epitaxially on GaAs by projecting simultaneously, beams of Ga and As_2 onto the heated GaAs surface, provided that sufficient extra As_2 beam intensity is provided, to compensate for the low As_2 sticking coefficient. We report here, a continuation of a series of studies of molecular beam epitaxial growth (MBE) of GaAs (Cho 1969, Cho and Chen 1970, Cho and Hayashi 1970) on (111), $(\bar{1}\bar{1}\bar{1})$ and (100) GaAs surfaces, $Al_xGa_{1-x}As$ on $(\bar{1}\bar{1}\bar{1})$ and (100) GaAs surfaces and GaP on $(\bar{1}\bar{1}\bar{1})$ GaP surfaces which have yielded smooth single crystal homo-epitaxial layers with, in the case of GaAs, high photoluminescence efficiency. The attainment of such layers depends upon proper preparation of the substrate surface and upon careful determination of the growth parameters, the latter being accomplished by the use of a combination of Knudsen effusion–mass spectrometric techniques and of high energy electron diffraction (HEED) studies of the crystal structure during growth of the layers.

We have observed, during the course of these studies, that one of the criteria which appear to be necessary for the homo-epitaxial layer growth to be smooth, single crystal having relatively high quality as determined by photoluminescence is that it be possible to observe the presence of characteristic two-dimensional surface structures, frequently described as reconstructed surface structures, by HEED during the beam epitaxial growth. These surface structures are thought to arise from the slight change in the position of surface atoms as the result of interaction of the dangling bonds of the surface atoms with each other. Thus, for instance when a properly prepared wafer of GaP with the $(\bar{1}\bar{1}\bar{1})$ surface exposed is examined with HEED on the [110] azimuth, the typical spotty $(\bar{1}\bar{1}\bar{1})$ pattern resulting from some roughness of the surface is seen as in figure 1(a). During growth of GaP by MBE on this $(\bar{1}\bar{1}\bar{1})$ surface the spots elongate showing that the third-dimensional Laue condition is being relaxed as the result of smoothing of the surface, and eventually a new pattern which represents a structure that has a lattice $\sqrt{19}$ times the size of the underlying bulk lattice is observed (figure 1(b)). This is designated as the GaP $(\bar{1}\bar{1}\bar{1})$–$\sqrt{19}$ surface structure. Similar examination of the (111) surface during growth reveals a GaP (111)–2 structure (figure 1(c)).

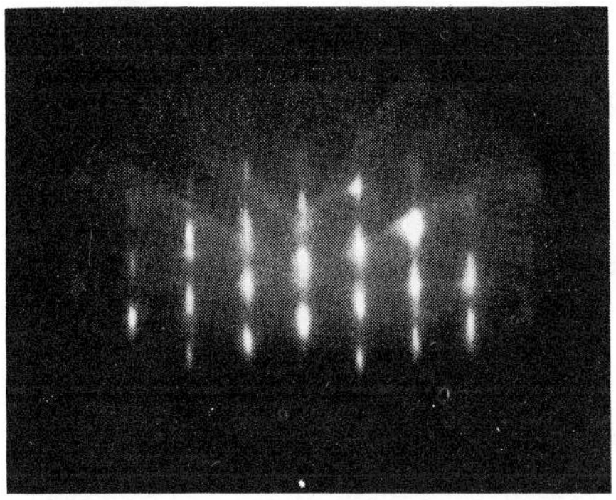

(a)

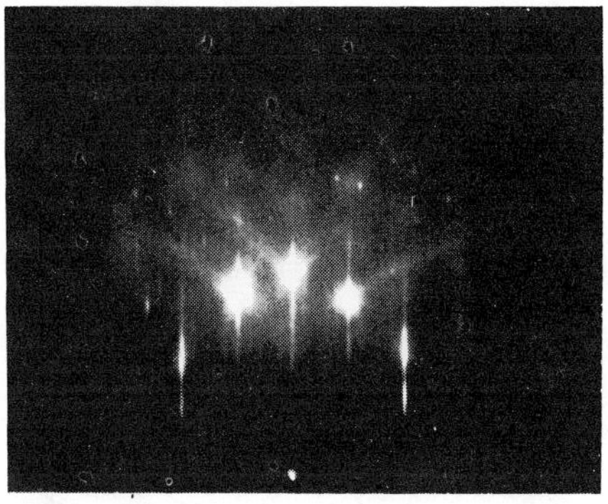

(b)

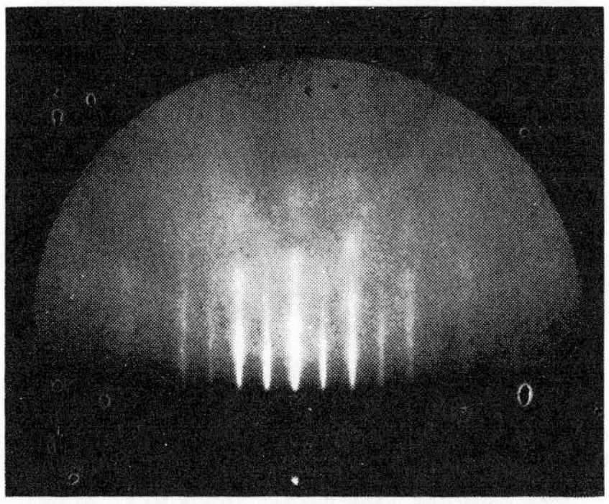

(c)

Figure 1. Reflection electron diffraction photographs showing $(\bar{1}\bar{1}\bar{1})$–$\sqrt{19}$ surface structure and (111)–2 surface structures developed on epitaxial GaP films grown on Br_2–methanol-etched GaP substrates. (a) etched substrate after heating to 545 °C, (b) after depositing 600 Å GaP on the $(\bar{1}\bar{1}\bar{1})$ phosphorus face, (c) after depositing 600 Å GaP on the (111) gallium face. 40 keV [110] azimuth.

Such structures were observed also during the growth of GaAs (Cho 1970, Cho and Hayashi 1970) and it has been demonstrated that on the $(\bar{1}\bar{1}\bar{1})$ surface the $(\bar{1}\bar{1}\bar{1})$–$\sqrt{19}$ structure results from relatively Ga-rich beams and the $(\bar{1}\bar{1}\bar{1})$–2 structure from relatively As-rich beams. On the (111) surface only the (111)–2 structure has been observed. A simple tentative explanation of these two different structures is that the $(\bar{1}\bar{1}\bar{1})$–$\sqrt{19}$ HEED pattern results from a two-dimensional structure resulting from interactions of the three dangling bonds of Ga atoms on the $(\bar{1}\bar{1}\bar{1})$ face (figure 2) and that the $(\bar{1}\bar{1}\bar{1})$–2 and (111)–2 structures result from interactions of the single dangling bonds of As atoms on the $(\bar{1}\bar{1}\bar{1})$ face and Ga atoms on the (111) face (figure 2). The failure to observe (111)–$\sqrt{19}$ is taken to be the result of a very low sticking coefficient for As on GaAs (111). Such structures [(111)–$\sqrt{19}$, (111)–2, (111)–5, (111)–7] have been observed on surfaces cleaned by ion bombardment of Si (Schlier and Farnsworth 1959, Van Bommel and Meyer 1967, Lander 1965) and on Ge. Haneman (1960) observed the (111)–2 surface structure on InSb and GaSb. Both the $(\bar{1}\bar{1}\bar{1})$–$\sqrt{19}$ and $(\bar{1}\bar{1}\bar{1})$–2 structures have been observed during homoepitaxial-MBE growth of $(\bar{1}\bar{1}\bar{1})$ GaAs (Cho 1970, Cho and Hayashi 1970), and as we report here, during the MBE growth of GaP on $(\bar{1}\bar{1}\bar{1})$ GaP and $Al_xGa_{1-x}As$ on $(\bar{1}\bar{1}\bar{1})$ GaAs. Furthermore, as will be described below, surface structures have also been observed during MBE growth of

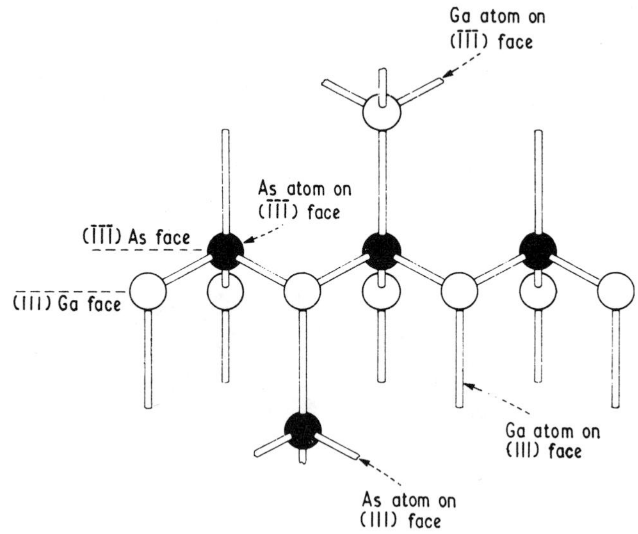

Figure 2. Sectional schematic showing that Group III atoms on the $(\bar{1}\bar{1}\bar{1})$ face or Group V atoms on the (111) face have three dangling bonds per surface atom and Group V atoms on the $(\bar{1}\bar{1}\bar{1})$ face or Group III atoms on the (111) face have one dangling bond.

GaAs and $Al_xGa_{1-x}As$ on (100) GaAs. All of the reconstructed structures observed may be related to beam intensity parameters for Ga and As_2, and to substrate temperatures. Photoluminescence measurements on GaAs layers grown with these surface structures present reveal that high efficiency near gap luminescence is obtained from layers grown from the relatively Ga-rich structures, and low near gap efficiency and an additional peak at about 0·1 eV below the bandgap is obtained with layers grown from the relatively As-rich structure. In this paper we present the experimental procedures and some growth parameters for the growth and of the doping during growth of MBE layers of $(\bar{1}\bar{1}\bar{1})$ GaAs, (100) GaAs, $(\bar{1}\bar{1}\bar{1})$ GaP, $(\bar{1}\bar{1}\bar{1})$ $Al_xGa_{1-x}As$ and (100) $Al_xGa_{1-x}As$ and some description of the surface morphology of the MBE layers. We also compare the photoluminescence efficiency of the near gap radiation from some of these layers with GaAs grown by other techniques with high photoluminescence efficiency.

2. Experimental

2.1. Substrate preparation

Ingots were oriented to either (111) or (100), sliced to wafers of 50 mils in thickness, These wafers were then Syton-polished to remove the surface damage due to cutting and then diced to the final substrate size of $\frac{1}{2} \times \frac{1}{4}$ in. Thermocouple holes were then drilled on the side of the substrates. A final etch was done just before the crystal was placed into the vacuum system by lapping it on a bromine-methanol (1 drop of bromine to 15 cm^3 of methanol) soaked lens paper, followed by pouring this bromine-methanol over the surface and then repeatedly rinsing in distilled water. The substrate was then dried by blowing dry nitrogen over the surface.

2.2. Molecular beam epitaxy

The experiments were conducted in an ultra high vacuum system designed so that the heated sample could be studied by HEED during growth of layers and so that the molecular beams themselves could be examined mass spectrometrically. The Vac-Ion pumping system is capable of maintaining a background pressure of 10^{-8} torr with effusion cells and sample at operating temperatures. The mass spectrometer shows that the background is completely dominated by arsenic vapour during growth of GaAs. The system is shown schematically in figure 3.

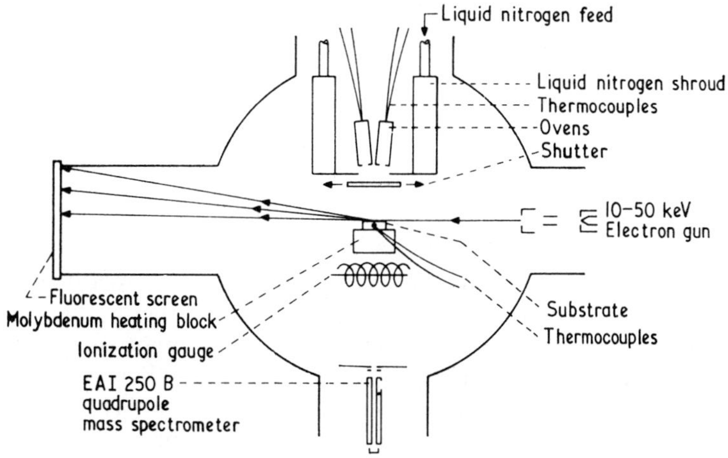

Figure 3. Schematic diagram of the experimental arrangement.

The substrate crystal was mounted on a molybdenum heating block. The temperature of the crystal was measured with a Chromel–alumel thermocouple imbedded in a 10 mm hole drilled 100 mm into the crystal. The Knudson effusion cells were alumina jacketed graphite, wire wound for resistance heating. The temperature of the cells was measured with 5% against 26% W–Re thermocouples.

In a typical run the etched substrate crystal was placed into the vacuum system on the Mo heating block. After evacuation to 10^{-10} torr the sample was annealed while the surface was examined with HEED (850 K for GaAs and 820 K for GaP). Once a clean (111) or (100) pattern was obtained the molecular beams were allowed to impinge upon the heated sample surface. The beam intensities were either calculated from the known vapour pressures in the As or P (Honig 1957), Ga–As (Arthur 1967), or Ga–P (Thurmond 1965, Arthur and Panish 1970) systems or checked mass spectrometrically ourselves in the case where the Al/Ga ratios were needed. If the surface had been properly prepared, the HEED spots elongated (figure 1(*a*)) as the etched surface apparently became smoother, eventually the two-dimensional surface structures were observed (figures 1(*b*), (*c*)). During layer growth the ratio of Group III to Group V atoms on the surface could be shifted by changing the beam intensity ratio of the Group III atoms to the Group V molecules (V_2) when the varying temperatures of the two effusion cells containing either the Group III element in one cell and the III + V elements in another cell or the Group V element in one cell and the III + V elements in another cell. This ratio in turn could be used to grow layers with surface structures characteristic of the relatively III or V rich condition for each crystal surface studied. The surface structure could also be changed reversibly by altering the substrate temperature, since this changes the relative sticking coefficients, or by shutting the beam off and allowing the relatively III rich structure to become predominant.

2.3. Doping

The MBE layers can be doped during growth by allowing molecular beams of dopants to impinge on the growing surface. We have done this so far with Si (Cho and Hayashi 1970), Ge and Mg. J R Arthur (private communication) has demonstrated that the surface lifetime of such atoms as Cd, Zn and Te is too short, at least on GaAs, to permit significant doping with those elements with elemental beams. Low vapour pressure materials such as Si and Ge have sticking coefficients of unity while the high vapour pressure materials such as Zn and Cd are almost instantaneously re-evaporated from the surface at the epitaxial temperature.

2.4. Photoluminescence

Photoluminescence was measured for these MBE GaAs layers at 300 K and 80 K. Layers were illuminated by a focused Argon or He–Ne laser beam, whose power was several milliwatts. Luminescent emission from MBE layers was analysed by a Perkin–Elmer type 112 spectrometer with a cooled (S-1) cathode photomultiplier. A wafer of Te doped boat grown GaAs with an electron concentration of about 2×10^{18} cm^{-3} was always mounted next to the MBE crystals as a reference. Thereby, not only the shape of the spectra but also the photoluminescent efficiency of the MBE layers relative to the reference were measured. This reference crystal was one which had been shown in our earlier study of liquid phase epitaxy and laser growth to be of relatively high photoluminescence efficiency. The MBE layers studied were generally thin, 0·2–1·0 μm. Since this is less than the diffusion length of excited carriers, the measured intensities are probably somewhat different from what would have been obtained with thicker layers.

3. Results and discussion

3.1. Molecular beam epitaxy on {111} faces of GaAs and GaP

As described in the experimental part above, experiments were done with two effusion cells operating. One effusion cell contained Ga + GaP or GaAs, and the other contained either Ga, or As or P. Our variables thus consisted of the arrival rate of Ga, the arrival rate of As or P and the temperature of the substrate. For several of the crystal surfaces we have observed typical surface structures, the presence of which depend upon Ga/As_2 or P_2 ratio in the incident beams. For our experimental system the transitions between structures were observed in GaAs for a given geometrical setup for an effusion cell-sample combination while initially only the effusion cell containing Ga+GaAs was used. For these studies the cell orifice diameter was 4·6 mm and orifice to sample distance was 100 mm for the studies of {111} faces and 55 mm for the studies of {100} faces. The transition was obtained by varying the effusion cell and substrate temperatures. The beam intensities of Ga and the Group V element from the oven containing Ga+GaAs or GaP were then a function of T only. For some of the crystal surfaces on which MBE layers have been grown we find that with this effusion cell alone, we observe a transition from one two-dimensional structure to another which is a function of the cell temperature (and thus Ga and V_2 arrival rate and simultaneously V_2/Ga ratio). This is illustrated by the curves for MBE on the $(\bar{1}\bar{1}\bar{1})$ surface of GaAs (figure 4). The lower curve in figure 4 shows the transition induced by variation of substrate temperatures when only the Ga–As cell is used. The upper curve shows the corresponding As_2/Ga ratio from the first effusion cell at the transition and how the transition is shifted by the addition of As_2 molecules or Ga atoms from an additional effusion cell. When similar experiments were done with GaP $(\bar{1}\bar{1}\bar{1})$ and the same effusion cell-sample geometry, no transition could be observed with only a Ga–P cell. The surface structure observed was always $(\bar{1}\bar{1}\bar{1})$–$\sqrt{19}$. The $(\bar{1}\bar{1}\bar{1})$–$\sqrt{19}$ to $(\bar{1}\bar{1}\bar{1})$–2 transition could, however be induced by a very high additional concentration of P_2 arriving at the growth surface. This is illustrated in figure 4 for a single point.

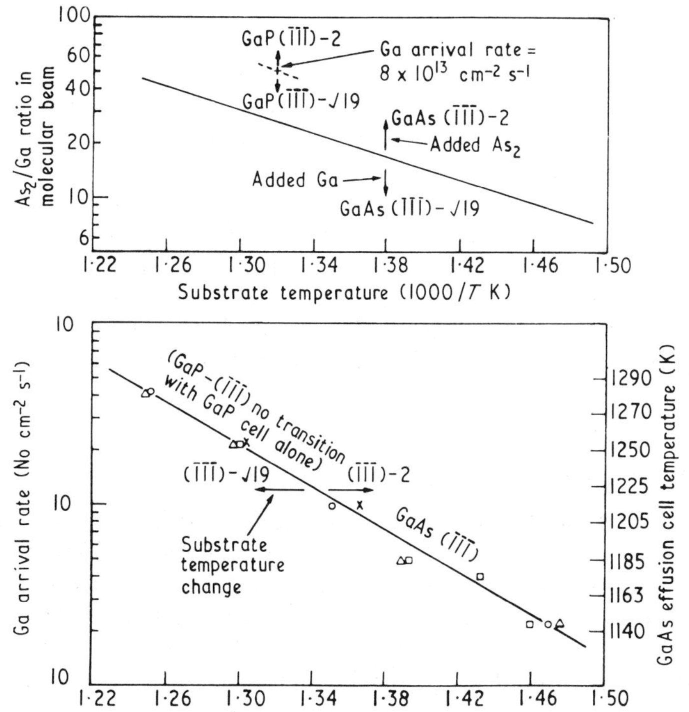

Figure 4. The transition temperature from a $(\bar{1}\bar{1}\bar{1})$–$\sqrt{19}$ surface structure to $(\bar{1}\bar{1}\bar{1})$–2 as a function of the deposition rate and the V_2/III ratio in the molecular beam. The symbols are for runs on several different substrates.

For MBE on (111) surfaces of GaAs and GaP, similar stages of smoothing of the surface and then appearance of the two-dimensional structure were observed, in this case the (111)–2. In neither case, however, was it possible with the available beam intensities, to observe a transition from (111)–2 to (111)–$\sqrt{19}$. We presume that this results from a very low sticking coefficient for the Group V element on the (111) surface, but have not obtained experimental confirmation of this conjecture.

We have found that the macroscopic surface morphology of growth by MBE can best be illustrated by Nemarski interference photomicrography. In the course of this study, fifty-two samples of the ($\bar{1}\bar{1}\bar{1}$) face were used. Surface textures ranging from mirror-appearance to matt-finish to the eye were observed. However, the matt-appearance surfaces were often related to contamination or defects on the substrate surface (contamination from a new oven, heating block, substrate clamp heating up for the first time and defects from a poor quality substrate or insufficient chemical etching or annealing). When there was no obvious contamination during growth, two kinds of texture were observed on the ($\bar{1}\bar{1}\bar{1}$) face by means of the interference phase-contrast microscope: pyramidal-growth at high substrate temperature ($T_S \gtrsim 520\,°C$) and low deposition rate ($\lesssim 2 \times 10^{14}$ Ga cm^{-2} s^{-1}) and a very much smoother step-growth at low substrate temperature ($T_S \gtrsim 520\,°C$) and high deposition rate ($\gtrsim 2 \times 10^{14}$ Ga cm^{-2} s^{-1}). These are illustrated in figure 5(*a*–*c*) with a magnification of $\times 500$. Figures 5(*a*) and (*b*) show decreasing severity of pyramidal growth, figure 5(*c*) shows almost featureless step-growth. The triangular based pyramid occurs because of the three-fold symmetry of the ($\bar{1}\bar{1}\bar{1}$) plane on which the film was grown. These pyramids appeared to be extremely shallow. Several attempts to examine the actual heights of these pyramids by investigating the cleavage cross sections of the film along the (110) planes under scanning electron microscope revealed only structureless layers. The phase-contrast microscope was much more sensitive to the surface texture. The pyramids shown in figure 5(*b*) by the phase-contrast microscope could only barely be seen in the scanning electron microscope when the sample was tilted to 75° with respect to the incident electron beam.

The microscopic growth morphology of GaAs was studied *in situ* in the HEED system where the surface topology can be qualitatively obtained from the diffraction pattern, and subsequently examined in greater detail by the use of the electron microscopy of platinum-carbon replicas. Figure 5(*d*) is an electron micrograph ($\times$ 80 000) of a Pt–C replica with the shadowing angle of 25° of the film shown by phase contrast photomicrography in figure 5(*c*). The steps in the ($2\bar{1}\bar{1}$) direction (approximately 25 Å high) can be seen clearly.

3.1. Molecular beam epitaxy on the (001) face of GaAs

We have studied MBE on the (001) surface of GaAs. Again the typical change of the electron diffraction pattern at the initial epitaxy showed smoothing of the substrate surface, with subsequent appearance of a two-dimensional surface structure. Since the (001) surface is not polar we do not necessarily expect different HEED patterns for the Ga or As stabilized conditions, and indeed, only a pattern corresponding to GaAs (001)–C (2×8) was observed during growth. This pattern, a complete description of the origin of which is beyond the scope of this paper and will be reported in more detail later, results from two different structures, one grown with a relatively As rich beam and one with a relatively Ga rich beam, which are rotated 90° from each other. The transition corresponding to the rearrangement of the surface atoms from one to the

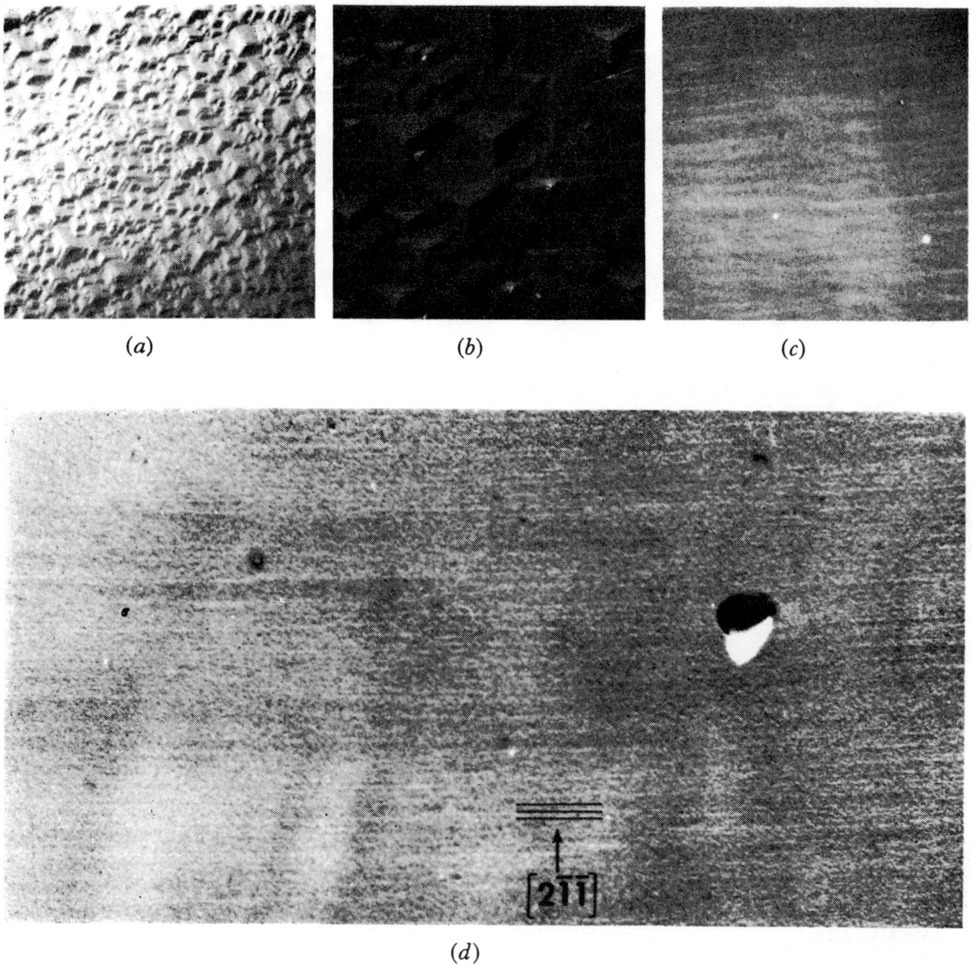

Figure 5. Photomicrographs of epitaxial GaAs layers on $(\bar{1}\bar{1}\bar{1})$ GaAs substrates. (*a*)–(*c*) are phase-contrast micrographs ($\times 500$) for films grown at 520 °C and deposition rates of 5×10^{13} Ga cm^{-2} s^{-1}, 2×10^{14} Ga cm^{-2} s^{-1}, and 8×10^{14} Ga cm^{-2} s^{-1} respectively. Ga and As beams are from a single effusion cell with As_2/Ga ratio of figure 4. ⊢⊣, 10 μm. (*d*) An electron photomicrograph ($\times 80\,000$) of a Pt–C replica of the surface of a film 1 μm thick. ⊢⊣, 0·1 μm.

other could be varied, as with $(\bar{1}\bar{1}\bar{1})$ surfaces, by varying substrate temperature and beam intensities and ratios in a manner eventually identical with that described for the $(\bar{1}\bar{1}\bar{1})$ surface, except that in this case a transition region and some degree of hysteresis upon heating and cooling are obtained. The transitions on the (001) face were observed at slightly higher substrate temperatures and lower beam intensities than those on the $(\bar{1}\bar{1}\bar{1})$ face.

Platinum shadowed carbon replicas figures 6(*a*–*c*) of the etched substrate and the initial and final stages of the deposited films were used to study the growth morphology. These Pt–C replicas were shadowed with an angle of 20° and the electron micrographs were recorded at $\times 80\,000$. The decrease in roughness of the film texture as the film

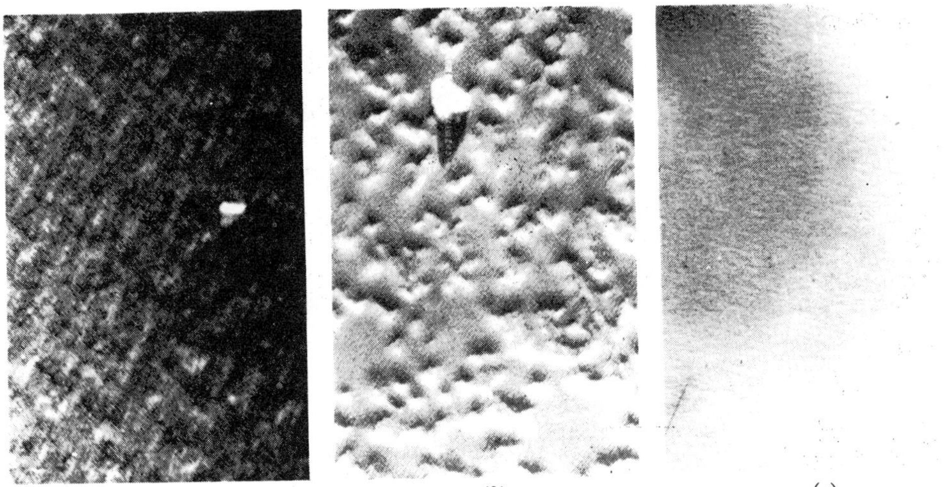

(*a*) (*b*) (*c*)

Figure 6. Electron micrographs (× 80 000) of Pt shadowed carbon replicas of the surface of (*a*) Br_2–methanol polish-etched (001) GaAs substrate heated in vacuum to 580 °C for five minutes. (*b*) The same substrate with 150 Å deposited GaAs at 520 °C. (*c*) The same substrate with 1 μm deposited GaAs at 520 °C. ⊢⊣, 0·1 μm.

grows thicker is consistent with the electron diffraction studies. Phase contrast photomicrographs of layers grown on 001 faces are essentially featureless. Scanning electron photomicrographs of the sheared cross sections show a featureless surface and growth interface. One such layer is shown in figure 7. The layer here is p-type (Mg doped as described below) and reverse biased to enhance contrast between the substrate and the grown layer.

3.3. *Molecular beam epitaxy of $Al_xGa_{1-x}As$ on GaAs*

AlAs and GaAs have very closely matching lattice parameters, and it has been shown with layers of $Al_xGa_{1-x}As$ grown on GaAs by liquid phase epitaxy (Panish *et al.* 1969, Hayashi and Panish 1970) that the resulting heterojunctions are essentially free of states which cause nonradiative recombination. The growth by MBE of $Al_xGa_{1-x}As$ on GaAs is therefore an ideal case for the study of the growth of multilayer and variable bandgap structures.

J R Arthur (private communication) has demonstrated that the sticking coefficient of Al, like that of Ga, is unity for beams on heated GaAs substrates. Thus the ratio of Al/Ga in the beams impinging upon the substrate surface will be the same as that in the crystalline solid solution produced. $Al_xGa_{1-x}As$ layers with x varying from 0·1 to 0·5 have been grown by the molecular beam method. The growth temperature was 550 °C. Reflectivity studies of these layers showed an increase of band gap energy as a function of increasing aluminium content. The composition-band gap variation observed agrees with the results reported by Casey and Panish (1969) for $Al_xGa_{1-x}As$ grown by the liquid phase epitaxy.

3.4. *Doping by molecular beam method*

It is possible to incorporate impurities in MBE films by depositing the film material simultaneously with the dopant from separate sources. In general, low vapour

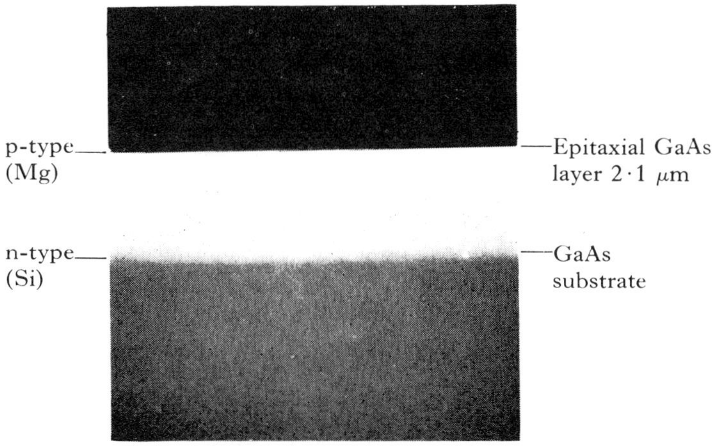

Figure 7. Scanning electron micrograph of a cleaved cross section of a p layer of GaAs grown by MBE on a (100) GaAs surface. The resulting p–n function is reverse biased to enhance contrast.

pressure materials are more favourable for dopants because they are expected to have a relatively high sticking coefficient at the epitaxy temperature (approximately 500 °C). If the sticking coefficient is near unity the doping concentration can be predicted from the beam intensities of Ga and the dopant. Silicon and germanium are examples of dopants with low vapour pressures. Although they are amphoteric for III–V compounds, the results from Schottky barrier diode measurements showed only n-type conductivity for both Si and Ge in GaAs layers grown by the MBE with concentrations from 1×10^{16} cm^{-3} to 5×10^{18} cm^{-3}. The $1/C^2$ against V plot of the Schottky barrier diode, where C is the space charge capacitance and V is the reverse bias voltage, showed that a uniform doping profile could be achieved by the MBE method.

Magnesium has a higher vapour pressure than Si and Ge. The sticking coefficient of Mg on GaAs at the growth temperatures, approximately 500 °C, is less than unity and is a function of temperature. Thus a higher arrival rate has to be used to compensate for this effect. Nevertheless, p-type GaAs with concentrations from 2×10^{16} cm^{-3} to 5×10^{18} cm^{-3} as measured by the Schottky barrier diode method, has been achieved. An Mg arrival rate ranging from two to three orders of magnitude higher than the rate which would have been required if every atom impinging on the surface were incorporated into the film, was used to achieve the above doping levels. These studies are being continued and will be reported later in greater detail. It is interesting to note that zinc has a vapour pressure higher than that of Mg and when Zn beams are used, no Zn can be detected by either photoluminescence or the Schottky barrier diode method in films grown at 500 °C.

3.5. *Photoluminescence*

As mentioned in §1 a correlation has been established between the spectrum obtained for a given layer and the surface structure present during growth of that layer, the near gap efficiency being higher from layers grown with a surface structure characteristic of the Ga-rich growth conditions. This is consistent with observations (Queisser and Panish 1967, Panish 1968) that GaAs grown from Ga solution generally has higher luminescent efficiency than that grown under more As-rich conditions.

Table 1. Relative photoluminescence intensities of the near gap peak in several GaAs crystals†

Orientation and growth	Doping elements	Concentration (net carriers)	Layer thickness (Å)	Relative intensity (80 K)	(300 K)
Boat Grown (standard)	Te	2×10^{18} (n)	—	1·0	1·0
MBE ($\bar{1}\bar{1}\bar{1}$)	None	$\approx 5 \times 10^{16}$ (n)	5 000	0·1	0·04
MBE ($\bar{1}\bar{1}\bar{1}$)	Zn (diffused)	$\approx 10^{18}$ (p)	5 000	0·8	0·25
MBE ($\bar{1}\bar{1}\bar{1}$)	Si	2×10^{17} (n)	7 000	0·4	0·2
MBE (100)	Ge	2×10^{18} (n)	10 000	0·5	0·32
MBE (100)	None	$\approx 2 \times 10^{16}$ (n)	2 000	0·015	0·002
LPE	Zn	2×10^{18} (p)	> 25 000	1·0	0·2
LPE	Si	$\approx 2 \times 10^{18}$ (p)	> 25 000	—	1·0
LPE	Si	$1–4 \times 10^{18}$ (n)	> 25 000	—	1·0–1·5
LPE	Ge	4×10^{18} (p)	> 25 000	0·7	0·25
Boat Grown‡	Si	$3–4 \times 10^{18}$	—	0·3	0·2

† All ($\bar{1}\bar{1}\bar{1}$) layers were grown with Ga-rich growth conditions. (100) layers were grown with As-rich growth conditions.

‡ This is the material used for low threshold single heterostructure laser diodes (Hayashi and Panish 1970).

Table 1 shows normalized 300 K and 80 K near gap photoluminescence intensities of MBE and liquid phase epitaxy layers relative to a reference crystal. No absolute efficiencies are known, but the reference crystal used had a photoluminescence efficiency higher than that of most bulk GaAs crystals studied in this laboratory and was slightly more efficient than the Si doped substrate crystals used by Hayashi and Panish (1970) to obtain low threshold single heterostructure lasers. No correction for the fact that the MBE layers are very thin has been made.

4. Conclusion

A newly developed technique for molecular beam epitaxial growth of single crystal layers of GaAs and $Al_xGa_{1-x}As$ on GaAs, and GaP on GaP has been described. Flattening of the surface during the initial stages of growth and the formation of two dimensional surface structures are observed with HEED. In the case of GaAs, successful doping during growth with Si and Ge (n-type) and Mg (p-type) is reported along with the growth of doped layers of GaAs with photoluminescence efficiencies comparable with the efficiencies obtained with high quality bulk and solution grown material.

References

Arthur J R 1967 *J. Phys. Chem. Solids* **28** 2257
Arthur J R 1969 *Proc. Conf. on Structure and Chemistry of Solid Surfaces* ed G A Somorjai (New York: Wiley) p46
Arthur J R and Le Pore J J 1969 *J. Vac. Sci. Tech.* **6** 545
Arthur J R and Panish M B to be published
Casey H C and Panish M B 1969 *J. appl. Phys.* **40** 4910
Cho A Y 1969 *Surface Sci.* **17** 494
Cho A Y 1970 *J. appl. Phys.* **41** 782, 2780

Cho A Y GaAs Epitaxy by a Molecular Beam Method: Observations of Surface Structures on the (001) Face, to be published
Cho A Y and Chen Y S 1970 *Solid State Commun.* **8** 377
Cho A Y and Hayashi I 1970 *Solid State Elect.*
Haneman D 1961 *Phys. Rev.* **121** 1093
Haneman D 1969 *J. Phys. Chem. Solids* **14** 162
Hayashi I and Panish M B 1970 *J. appl. Phys.* **41** 150
Honig R E 1957 *RCA Rev.* **18** 195
Lander J J 1965 *Progress in Solid State Chemistry*, Vol. II ed H Reiss (Pergamon Press) p26
Panish M B 1968 *J. Phys. Chem. Solids* **29** 409
Panish M B, Hayashi I and Sumski S 1969 *IEEE J. Quantum Elect.* **QE-5** 210
Queisser H J and Panish M B 1967 *J. Phys. Chem. Solids* **28** 1177
Schlier R E and Farnsworth H E 1957 *Semiconductor Surface Physics* ed R H Kingston (Philadelphia: University of Pennsylvania Press) p3
Schlier R E and Farnsworth H E 1959 *J. chem. Phys.* **30** 917
Thurmond C D 1965 *J. Phys. Chem. Solids* **26** 785
Van Bommel A J and Meyer F 1967 *Surf. Sci.* **8** 467

1.3 diffusion

The growth of selective epitaxial regions can be difficult; impurity diffusion is an obvious alternative. Diffusion processes must be understood and regulated wherever junction structures with mandatory controlled impurity profiles are produced. Junction devices utilizing diffused p-n junctions include varactor and paramp diodes, junction IMPATT's, and light emitting diodes.

1.3.1 simple theory

A net transport of impurities occurs in the presence of an impurity gradient. As a result the impurity flux is proportional to the gradient. This process is exactly analogous to thermal diffusion in a solid, described by Fick's Law:

$$\frac{\partial N}{\partial t} = D \frac{\partial^2 N}{\partial x^2}$$

where D is the diffusion coefficient.

D is dependent on both temperature and the properties of the host lattice and diffusion species so that:

$$D = Do \exp\,(-E_A/kT)\ \mathrm{cm^2/sec}$$

where E_A is the "activation energy" of the diffusant in the host lattice.

Essentially, two practical diffusion methods exist — the semiconductor surface is exposed to a volume of gas of uniform impurity concentration or a set number of impurity atoms per unit area are deposited on the surface before diffusion takes place.

These two situations establish different boundary criteria, thereby changing the solution of Fick's Law.

For the constant gas concentration:

$$N(x) = \frac{N_o}{2}\ \mathrm{erfc}\ \frac{x}{2\sqrt{Dt}}$$

where N(x) is the concentration of impurity at x, N_o is the concentration of impurity in the gas, and t is the diffusion time.

The complementary error function term is derived from the distribution below:

$$\operatorname{erfc} \frac{x}{2\sqrt{Dt}} = \frac{1}{2}\left[1 - \frac{2}{\pi}\int_{0}^{x/2\sqrt{Dt}} e^{-\lambda^2}\, d\lambda\right]$$

where λ is the integration variable.

This equation, in turn, leads to the well-known error function distribution of impurity after diffusion.

The distribution becomes Gaussian for the limited source:

$$N(x) = \frac{Q}{\sqrt{\pi Dt}}\, e^{-x^2/4Dt}$$

where Q is the number of atoms/cm^2 at the surface.

The two preceding equations describe the results of diffusion in silicon, but there are variations in the case of the compound semiconductors. These differences become clearer as one observes the ways in which impurities travel through the lattice.

Interstitial Diffusion

The activation energy for interstitial diffusion is approximately 1 eV; impurity atoms move from one interstitial site to the next by squeezing through the constrictions in the voids.

Substitutional Diffusion

The activation energy for substitutional diffusion is approximately 3-4 eV. Impurity atoms move from one lattice site to the next through adjacent vacancies.

The mechanism that dominates in a specific system is dependent upon the defect concentrations in the host crystal. Much research has been published on the interstitial vacancy and antistructure defects in III-V compounds, but there is no intention to consider those arguments here. Both interstitial and substitutional diffusion occur; the situation is far more complicated than for an elemental semiconductor such as Si or Ge, as reflected in all of the work reported on diffusion into GaAs. It is not unusual for measured diffusion coefficients to differ by several orders of magnitude, depending on the precise experimental conditions.

A further complication is introduced by the fact that one component of the compound is volatile, introducing a further variable — the vapour pressure of this volatile component maintained in the particular experiment.

1.3.2 experimental data

The status of diffusion in the compound semiconductor is reviewed in a paper by Yarbrough[87] Table 1-2 illustrates accepted diffusion data for impurities in GaAs. The notable exception in the table is zinc, which is an important acceptor impurity. The reason for this omission is that the diffusion of zinc is concentration dependent in both GaAs and InP. The effect of zinc concentration on the diffusion coefficient for GaAs is seen in Figure 1-7(a). Figure 1-7(b) shows the data for Zn in InP[12]. General data does not exist for other impurities in InP; however, studies

Impurity	Do(cm^2/sec)	E_a(eV)	D at temperature T	
Ag	4.10^{-4}	0.8	10^{-7}	800°C
Au	1.10^{-3}	1.0	10^{-7}	800°C
Be	$7.3\ 10^{-6}$	1.2	$4.1\ 10^{-11}$	880°C
Cd	5.10^{-2}	2.6	$1.1\ 10^{-13}$	850°C
Cu	3.10^{-2}	0.53	10^{-4}	800°C
Mn	$6.5\ 10^{-1}$	2.5	$4.4\ 10^{-12}$	850°C
S	$2.6\ 10^{-5}$	1.86	$2.65\ 10^{-13}$	900°C
Se	3.10^{3}	4.16	10^{-15}	850°C
Sn	6.10^{-4}	2.5	10^{-13}	900°C

Table 1-2/Impurities in GaAs

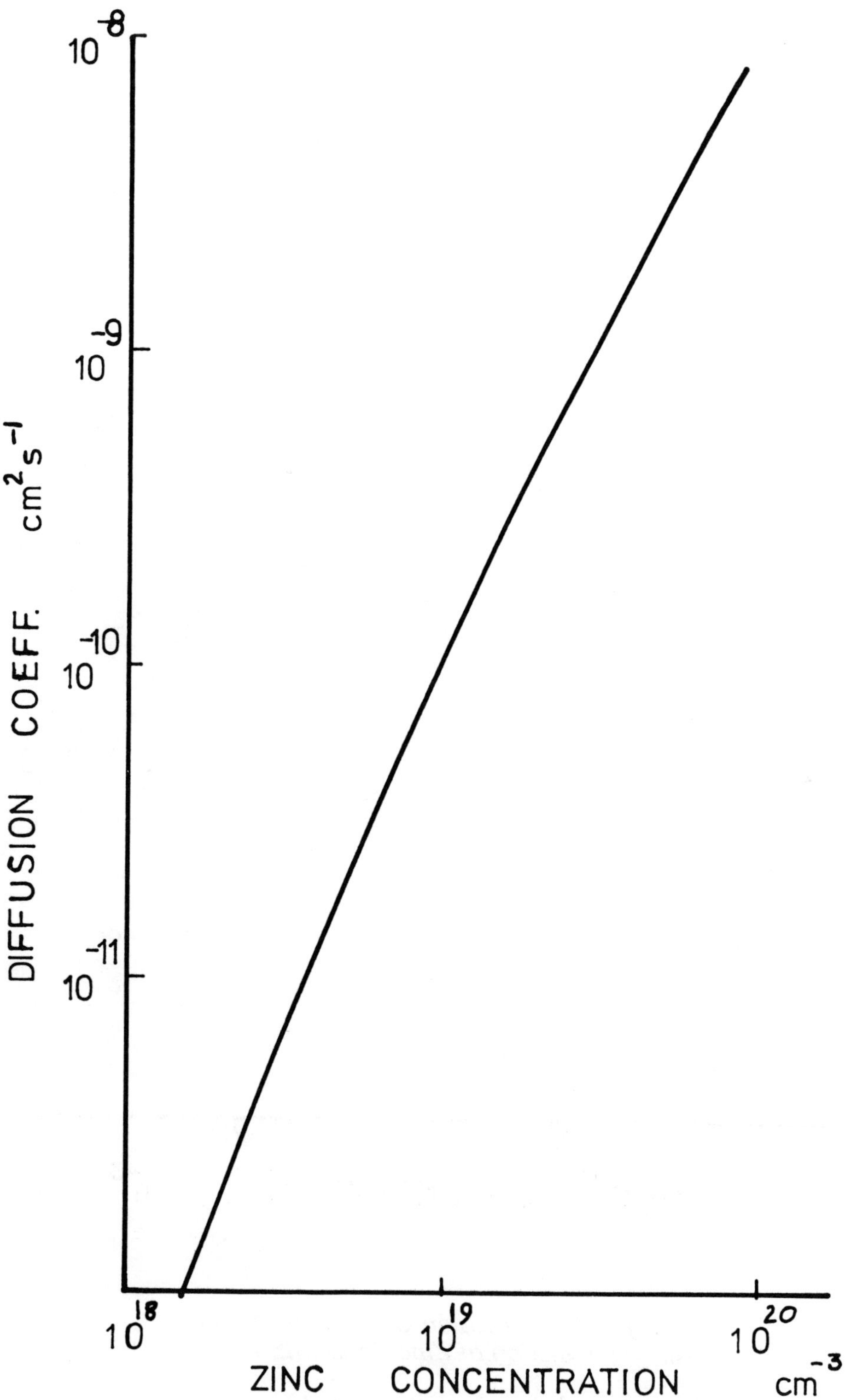

Figure 1-7(a)/Diffusion Coefficient of Zn in GaAs as a Function of Zinc Concentration.

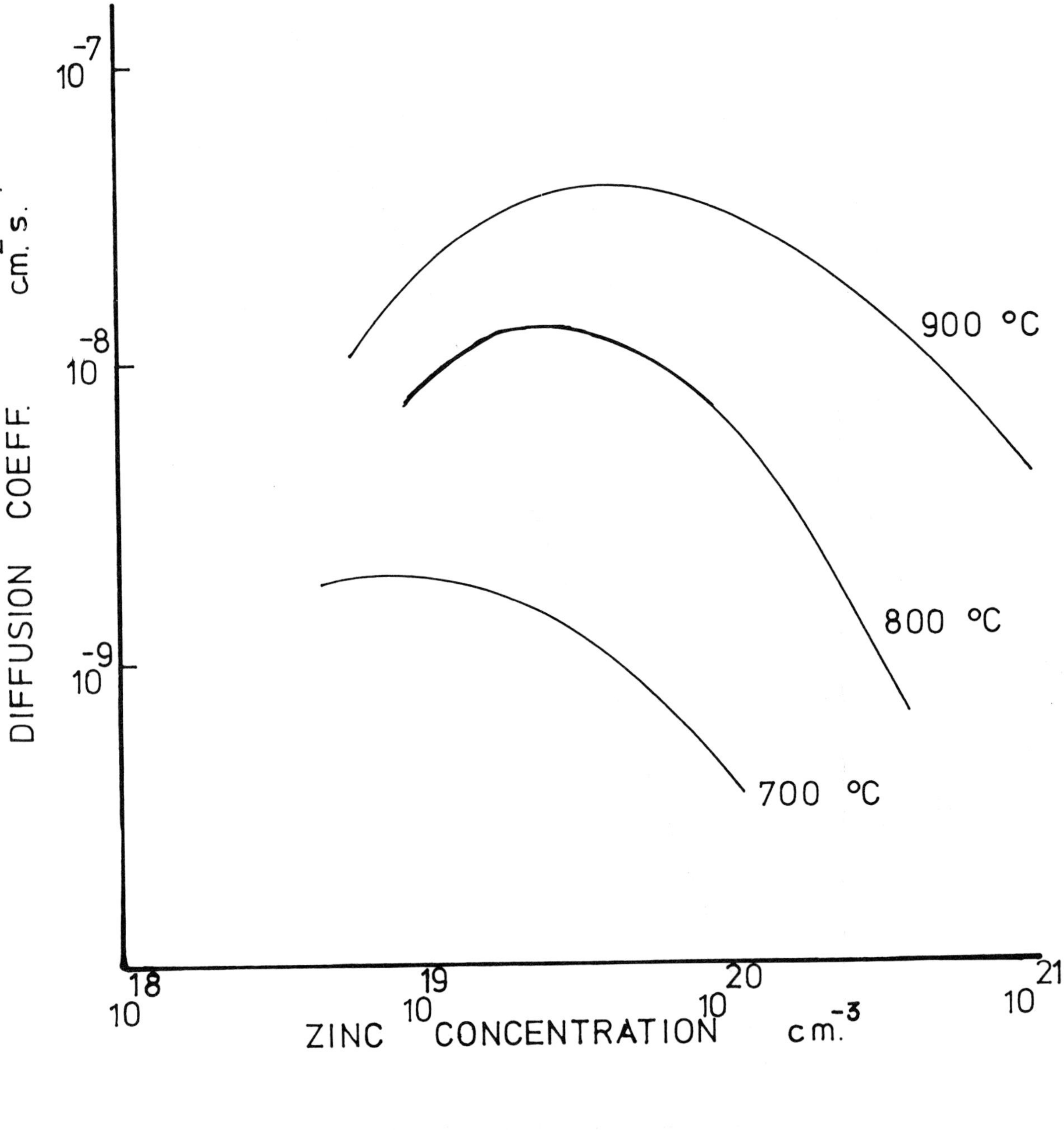

Figure 1-7(b)/Diffusion Coefficient of Zinc in InP as a Function of Zinc Concentration[12].

of diffusion profiles (from substrates during epitaxy[13]) manifest the fact that the behaviours of Sn, Ge, and S are much like those in GaAs but with a higher diffusion coefficient.

The normal technique used for diffusion into these compounds is the closed-tube method. The wafer, together with the diffusion source and some means of providing a vapour pressure of the volatile component at the diffusion temperature (such as pure arsenic or phosphorus, crushed GaAs or InP, or a compound of the diffusant and volatile element (e.g., ZnP_2)) are sealed under vacuum into a quartz ampoule. The ampoule is placed in the diffusion furnace for a specific time at the diffusion temperature, then it is removed and chilled at one end to condense the enclosed gases before cooling the whole unit. In these experiments on diffusion of Zn into InP[14], elemental zinc and phosphorus are used as the source, with varying phosphorus/zinc ratios. The junction depth, surface finish, and diffusion coefficient are dependent on this ratio.

The majority of diffusion work has been using gaseous sources; work, however, has been carried out with solid to solid diffusion[15]. The wafer is coated with the diffusant source which is further covered with SiO_2 to prevent loss. This method provides good sulphur diffusions into GaAs although it cannot be described by a simple complementary error function.

Presently, diffusion into GaAs and InP is limited in that Zn has been well investigated for varactors, IMPATT's, and LED's, but, for devices using donor diffusion the research is minimal. The diffusion of donor impurities such as S, Se, Te, and Sn is slower than Zn, making higher temperatures necessary. The result is a greater probability of dissociation of the surface of the semiconductor and compound formation.

The significant differences among diffusion, ion implantation, and epitaxy are manifested in the doping profile results. Epitaxy is essentially a uniform doping process; junctions are necessarily abrupt, whereas the other processes give a profile described by a particular function. The procedural stipulation obviously depends on the required terminal characteristics of the device and on the limitations imposed by processing.

key paper

Electrical Properties Of N-Type Layers In GaAs Prepared By Solid-To-Solid Sulphur Diffusion

S. Asai, S. Okazaki, and H. Kodera
(Central Research Laboratory, Hitachi Ltd., Tokyo, Japan)
Institute of Physics and Physical Society Symposium on GaAs (1972)
Pages 130-139

Abstract. The electrical properties of n-type layers and p–n junctions in GaAs prepared by a newly proposed method of solid-to-solid sulphur diffusion are reported. The surface electron concentration was $1\text{–}3 \times 10^{18}$ cm^{-3} depending on diffusion temperature in agreement with the authors' previous results based on junction depth measurements. The diffusion consists of two components being slower at concentrations higher than 5×10^{17} cm^{-3}, and the profile cannot be described by a simple complementary error function. The electron mobility in the diffused layer was reasonably high. The p–n junction diodes were linearly graded and showed injection luminescence near the band edge, although the current was dominated by generation–recombination mechanisms. The diffusion of S was carried out into high purity epitaxial material with no appreciable deterioration. These results indicate the applicability of the proposed method to the fabrication of GaAs devices.

1. Introduction

Gallium arsenide has been used as the material for a variety of electron devices such as Gunn oscillators, infrared diodes, injection lasers, mm-wave diodes and transistors of several types. With the increasing interest in GaAs devices, considerable development has been made in the technology of processing the material. Impurity diffusion is one of the most important techniques in semiconductor technology. Diffusion of acceptor Zn into GaAs has been extensively investigated and has often been used for the preparation of p^+–n junctions in varactors, IMPATT diodes etc. However, there have been few known examples of GaAs devices fabricated by means of donor diffusion. This is because several problems are involved in donor diffusion. Major difficulties in diffusing group VI impurities into GaAs have been vapour etching and compound formation on the surface (Goldstein 1961, Yeh 1964, Frieser 1965, Kendall 1968, Young 1969). These difficulties are emphasized by the slower diffusion of donor impurities such as S, Se, Te and Sn as compared to Zn, because it is necessary to use high temperatures which are likely to cause dissociation of the constituents of the compound semiconductor.

Previously, we proposed a new method of sulphur diffusion into GaAs (Asai and Kodera 1972). It utilizes solid-to-solid redistribution and diffusion of species between GaAs and Ga_2S_3 below the eutectic temperature (940 °C) of the quasi-binary GaAs–Ga_2S_3 system. This method is more reliable and reproducible than that of conventional diffusion from the vapour phase because it is free from the effects due to vapour etching and compound formation.

The present work was directed towards application of the above method to device fabrication. For this purpose, electrical properties of the S-diffused GaAs layers have been investigated. The electron concentration and mobility in the n-type diffused layers have been determined using both p-type and semi-insulating substrates. Electrical properties and injection luminescence of diffused n^+–p junction diodes have also been studied. Furthermore, donor diffusion has been carried out into high purity epitaxial GaAs wafers.

2. Experimental

2.1. Diffusion procedure

The possibility of the proposed method of solid-to-solid S diffusion was inferred from a consideration on the Ga–As–S ternary phase diagram, shown in figure 1(*a*), which had been estimated by Young (1969). The phase diagram for the quasibinary GaAs–Ga_2S_3 system as derived from figure 1(*a*) is shown in figure 1(*b*). It is seen

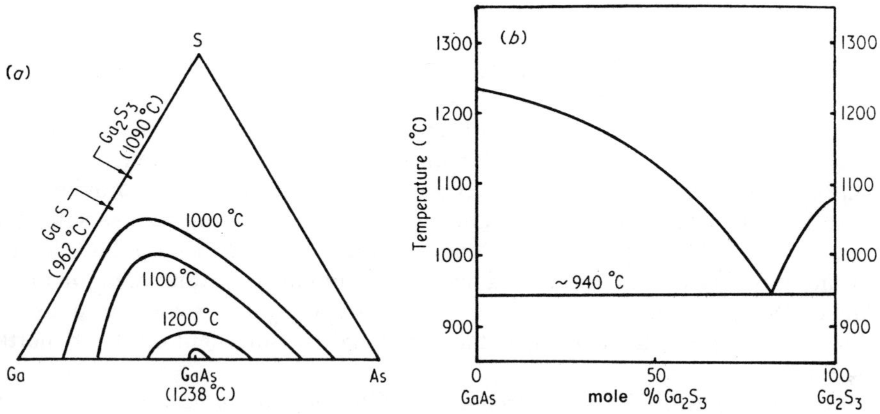

Figure 1. (*a*) Ga–As–S ternary phase diagram (Young 1969) and (*b*) GaAs–Ga_2S_3 quasibinary phase diagram.

that no liquid phases exist below the eutectic temperature at about 940 °C. It should therefore be possible to diffuse S into GaAs from the solid source by placing a GaAs wafer in contact with Ga_2S_3 at high (but below 940 °C) temperature. That this is true was shown in the previous paper (Asai and Kodera 1972), and the diffusion coefficient D and surface concentration n_s of active donors S were determined from the junction depth measurement on wafers with different acceptor concentrations to be

$$D = 240 \exp\left(-\frac{3{\cdot}4\ \text{eV}}{kT}\right)\ \text{cm}^2\,\text{s}^{-1} \tag{1}$$

and

$$n_s = 1{\cdot}7 \times 10^{20} \exp\left(-\frac{0{\cdot}5\ \text{eV}}{kT}\right)\ \text{cm}^{-3} \tag{2}$$

respectively.

Several kinds of GaAs substrate were used in this experiment. The semi-insulating substrate was doped with Cr, of resistivity higher than $10^8\ \Omega\,\text{cm}$. The Zn-doped wafers had carrier concentrations about $1{\cdot}2 \times 10^{17}\ \text{cm}^{-3}$. These were mechanically polished with 0·3 μm alumina powder and etched in $4H_2SO_4:1H_2O_2:1H_2O$ solution. Epitaxial wafers on heavily doped n-type material as well as on high resistivity substrates were used. Their carrier concentrations were around 2×10^{15} cm^{-3}. All these wafers were oriented in the [100] direction. Thin films of Ga_2S_3, about 3000 Å thick, were evaporated onto the wafers from the powdered source in an alumina-coated tungsten basket. The samples were coated with SiO_2 thin film to avoid decomposition of the source and deterioration of the crystal. The diffusion was carried out in an evacuated quartz ampoule at a temperature in the range 800–900 °C. After the heat treatment, the SiO_2 film was removed using buffered HF solution. The solid S source, Ga_2S_3 films, can be removed by boiling the wafer in concentrated H_3PO_4. The surface of the wafer was in general smooth and shiny for diffusion temperatures below 850 °C. However, it sometimes became dull when higher temperatures were used.

2.2. Measurements

Clover-leaf-shaped specimens were cut out from the diffused wafers using an ultrasonic tool for the Hall effect measurement (van der Pauw 1958). Ohmic contacts were provided by alloying with small In balls (for semi-insulating substrate) or by evaporating Au/Ge/Ni alloy at an elevated temperature (for a p-type substrate). The measurement was repeated after successively removing thin (0·1–1 μm) layers from the surface. This was done by etching the specimen in $NaOH+H_2O_2+H_2O$ solution with the contact areas coated with wax. The electron density and mobility profile in diffused layers were obtained in this way (Johansson and Mayer 1970). Carrier concentration in the n-type layer was measured also by combination of step etching and capacitance–voltage measurement on Au–Schottky barrier diodes.

The p–n junction diodes were prepared as follows. The ohmic back contact to the p-type substrate was evaporated Au:3% Zn. The contact to the n-type side was Au/Ge/Ni alloy evaporated through a metal mask having circular holes with a diameter of 150 μm. The diodes were isolated by etching in $4H_2SO_4:1H_2O_2:1H_2O$ solution into mesa structures 400 μm square and by scribing. They were then mounted on TO-18 headers.

3. Experimental results

3.1. Electron concentration and mobility

The variation of electron concentration and mobility in the S-diffused (850 °C, 72 hr) region in a p-type crystal is shown in figure 2. These quantities, being obtained from the layer removal Hall method, represent the average within each layer. Therefore, they are shown in figure 2 with short horizontal bars. The surface concentration of carrier or uncompensated active S donors is of the order of low 10^{18} cm^{-3}. The junction depth obtained by angle-lapping and staining (indicated with an arrow in figure 2) is in reasonable agreement with the result of Hall effect measurement. The mobility has an extremum at a depth of 1–2 μm from the surface and drops sharply towards the junction and decreases slightly towards the surface.

Similar results obtained for high resistivity substrate are shown in figure 3. The diffusion time and temperature are the same as for the p-type substrate shown in figure 2. The electron concentration determined from the *C–V* measurement is also shown (with an asterisk) and in good agreement with the values from the Hall measurement. The behaviour of carrier concentration and mobility is qualitatively the same as in p-type crystals. The electron density in semi-insulating substrate is

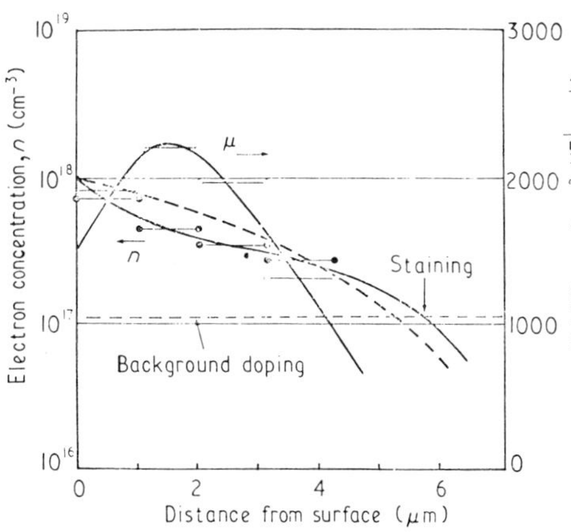

Figure 2. Electron concentration n (●) and mobility μ (○) in a S-diffused n-type layer in p-type GaAs (diffusion conditions: 850 °C, 72 hr). The heavy broken line is a complementary error function (see § 4.2).

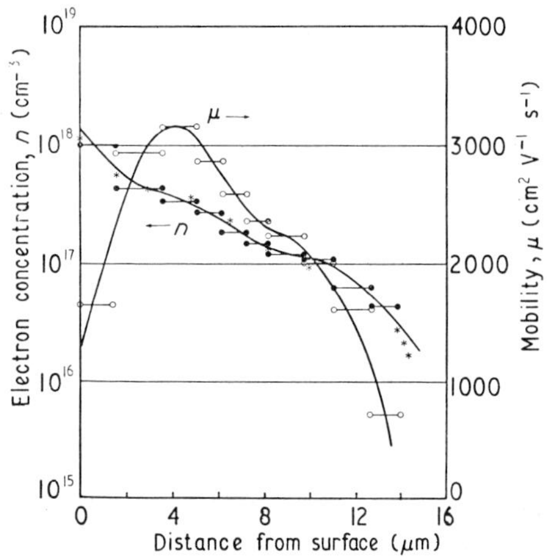

Figure 3. Electron concentration n (●) and mobility μ (○) in a S-diffused n-type layer in semi-insulating GaAs (diffusion conditions: 850 °C, 72 hr). The asterisks indicate electron concentration from C–V measurement.

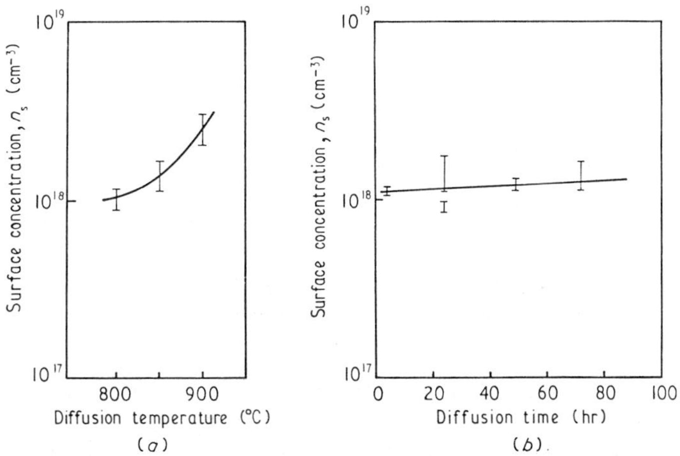

Figure 4. Surface electron concentration n_s as a function of diffusion temperature (a) and time (b).

in general slightly higher than in p-type substrate. This is natural because compensation by background doping is absent in the former case.

From figures 2 and 3 it seems that the distribution of sulphur donors S^+ cannot be described by a simple complementary error function. Apparently, the diffusion coefficient at concentrations above about 5×10^{17} S^+/cm^3 is smaller than at lower concentrations. The surface concentration varies as a function of diffusion temperature and time as shown in figure 4. It increases from 1×10^{18} to 3×10^{18} cm^{-3} as the temperature is varied from 800 to 900 °C, but is not strongly dependent on diffusion time. Figure 5 shows how the maximum electron mobility in the diffused layer varies with diffusion conditions. It is seen that the mobility is higher when the

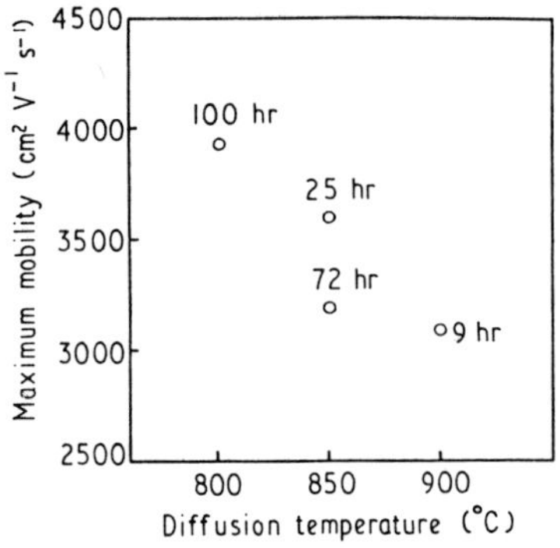

Figure 5. Maximum electron mobility in the diffused layer as a function of diffusion temperature.

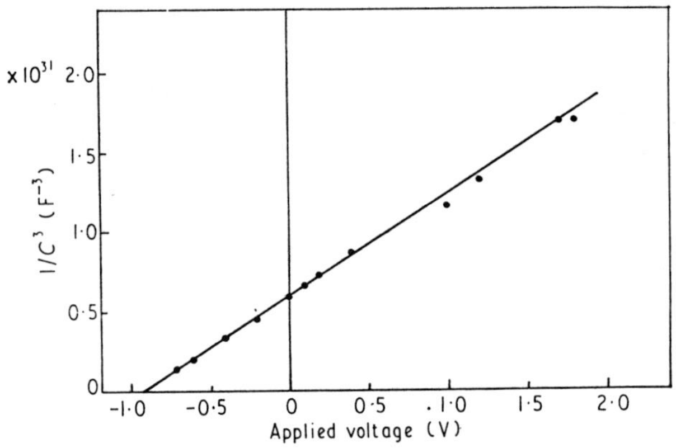

Figure 6. Capacitance–voltage characteristics of a S-diffused GaAs p–n junction diode (diffusion conditions: 800 °C, 49 hr).

diffusion is carried out at a lower temperature in a short time. The highest mobility of 3930 $cm^2 V^{-1} s^{-1}$ was obtained for diffusion at 800 °C at a concentration of $2{\cdot}2 \times 10^{17}$ cm^{-3}. This is quite comparable to the bulk mobility.

3.2. p–n junction diodes

The capacitance–voltage characteristics of a p–n junction diode fabricated by diffusing S at 800 °C for 49 hr are shown in figure 6. It can be seen that C^{-3} varies linearly with the bias voltage V, that is,

$$C^{-3} = \frac{12}{qa\epsilon_S^2}(V + V_{bi}). \qquad (3)$$

This means that the junction is linearly graded, and the concentration gradient a

is determined as

$$a = 2{\cdot}9 \times 10^{20}\ \mathrm{cm}^{-4}. \tag{4}$$

The built-in voltage

$$V_{\mathrm{bi}} = 0{\cdot}92\ \mathrm{V} \tag{5}$$

is somewhat lower but in reasonable agreement with the theoretical value of 1·06 V (Sze 1969) for the value of a given above.

The forward current–voltage characteristic of the diode is found to be given by

$$I = 4{\cdot}75 \times 10^{-8} \exp\left(\frac{qV}{1{\cdot}93kT}\right) \quad \mathrm{A\,cm}^{-1} \tag{6}$$

indicating that generation–recombination is the dominant current-carrying process through the space charge region. The reverse breakdown voltage of the diode was typically 40 V.

4. Discussion

4.1. Electrical properties of the diffused layer

The Hall measurement was carried out down to liquid nitrogen temperature. The carrier concentration was essentially independent of temperature of measurement. The mobility in each layer had a peak at a temperature between 150 and 250 °K.

When ionized impurity scattering is dominant, it is possible to estimate the total number of ionized impurities using the Brooks–Herring expression for the mobility (Brooks 1951). The acceptor density N_A and compensation ratio N_A/N_D are determined in this way and plotted in figure 7. Acceptor concentration, which is

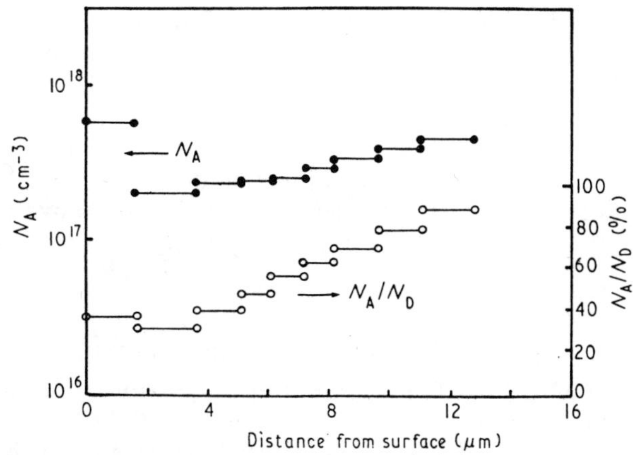

Figure 7. Acceptor concentration N_A (●) and compensation ratio N_A/N_D (○) in the diffused layer (same sample as in figure 3).

high near the surface, decreases sharply and then increases gradually with the depth. The decrease in mobility towards the inside (figures 2 and 3) is due to this increase in N_A. The variation of N_A across the diffused layer might be the result of possible redistribution of acceptor impurities (Cr, Zn, etc).

Naturally, donors are fully compensated with acceptors well inside the semi-insulating substrate. Similar behaviour of N_A and N_A/N_D was found in p-type crystals, in which N_A/N_D approached 1 near the chemically delineated junction depth.

4.2. Diffusion of S in GaAs

The mechanism of diffusion of S in GaAs has already been discussed by several authors (Kendall 1968, Goldstein 1962, Young 1969, Young and Pearson 1970). We describe here briefly a model relevant to our solid-to-solid diffusion. When the source compound Ga_2S_3 is dissolved in GaAs, it forms a neutral complex V_{Ga}–S_3 (Young 1969). The front of this solid solution will migrate rather slowly. Meanwhile, the complex is decomposed through the reaction

$$V_{Ga}\text{–}S_3 \rightleftharpoons 3S^+_{As} + 3e^- + V_{Ga} \tag{7}$$

into isolated donors, which can diffuse much faster than the complex. Within the solid solution, the above reaction controls the formation of donor (through, for example, outdiffusion of V_{Ga}), while diffusion of ordinary isolated donor S^+_{As} dominates deep inside the crystal. This model can explain two different diffusion constants at concentrations higher and lower than 5×10^{17} cm^{-3} as seen in figures 2 and 3.

A complementary error function distribution calculated using D and n_s given in equations (1) and (2) is shown in figure 2 for the same diffusion conditions. Reasonable agreement is seen with the profile obtained directly from the electrical measurement. The surface concentrations especially agree excellently. Since in the previous experiment we used p-type crystals with carrier concentrations of $5{\cdot}7 \times 10^{17}$ cm^{-3} and $1{\cdot}2 \times 10^{17}$ cm^{-3}, the junction depth measurement must have given a certain average between the diffusion coefficients at high and low concentrations. However, it became clear in this experiment that the actual distribution of active donors S^+_{As} cannot be described by a simple complementary error function. There are two experimental results that are consistent with the observed profile which is steep near the surface and gradual inside. Firstly, the concentration gradient a at the p–n junction given in equation (4) was rather anomalously low considering the surface concentration and junction depth. Secondly, the breakdown voltage, 40 V, was lower than the theoretical value for the above magnitude of a, 70 V (Sze 1969).

4.3. Applicability to device fabrication

The main purpose of this work was to survey the applicability of the proposed method of S diffusion to device fabrication. A direct possible application of diffusion into GaAs would be to fabricate injection luminescent diodes. Especially, if we could have n^+–p junctions, efficiencies of infrared emitters might be improved because of more efficient electron injection than in usual p^+–n diodes. The luminescence spectra of a diode (diffused at 850 °C for 72 hr) at a current density of $7{\cdot}9$ A cm^{-2} are shown in figure 8. The edge emission is observed although the external efficiency is not very high. An improvement in the efficiency should be possible by more careful design of diodes and removal of deep level impurities. The presence of these impurities is clear from the fact that the generation–recombination current is dominant in the junction.

Another important application of diffused layers has been to provide low resistance contacts to various devices. We diffused sulphur into epitaxial GaAs wafers with low carrier concentrations. Figure 9 shows the electron concentration profile of an epitaxial crystal grown from the vapour phase on a Cr-doped high resistance sub-

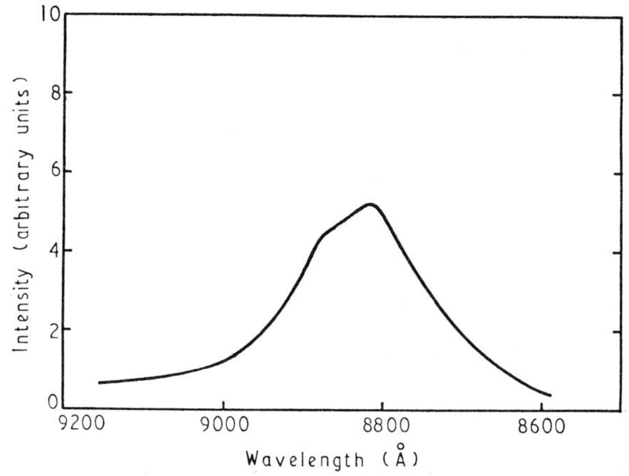

Figure 8. Injection luminescence spectra of a S-diffused GaAs diode (diffusion conditions: 840 °C, 72 hr).

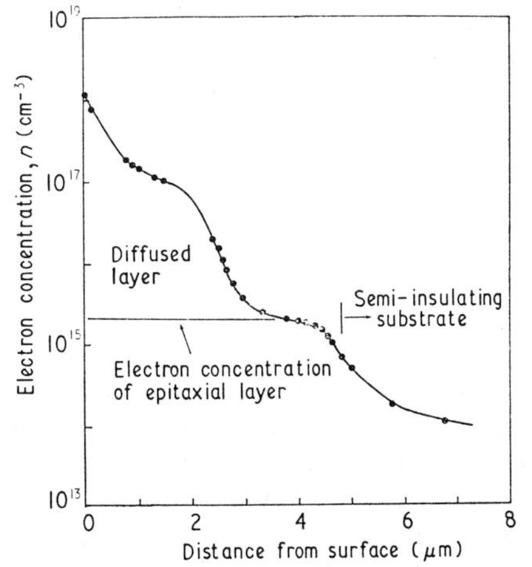

Figure 9. Electron concentration profile in S-diffused epitaxial GaAs grown on a semi-insulating substrate.

strate after diffusion of S at 850 °C for 4 hr. The initial carrier concentration in the epitaxial layer was $1{\cdot}5 \times 10^{15}$ cm^{-3}, which is preserved after the diffusion process. There seem to be no signs of deterioration in the epitaxial material. Moreover, Gunn oscillators with epitaxial active regions on n^+ substrates and diffused n^+ electrodes have been made successfully. These results are encouraging in the use of this technique to provide n^+ electrode areas to GaAs devices which require low resistance contacts. However, the maximum donor concentration is limited to low 10^{18} cm^{-3}, which is much lower than in Si.

The technique of selective diffusion is essential to planar technology. A preliminary study of selective diffusion of S into GaAs has been made. Figure 10 shows the junction depths in p-type crystals with $p = 1{\cdot}2 \times 10^{17}$ cm^{-3} as functions of the thickness of a sputtered SiO_2 diffusion mask under various diffusion conditions.

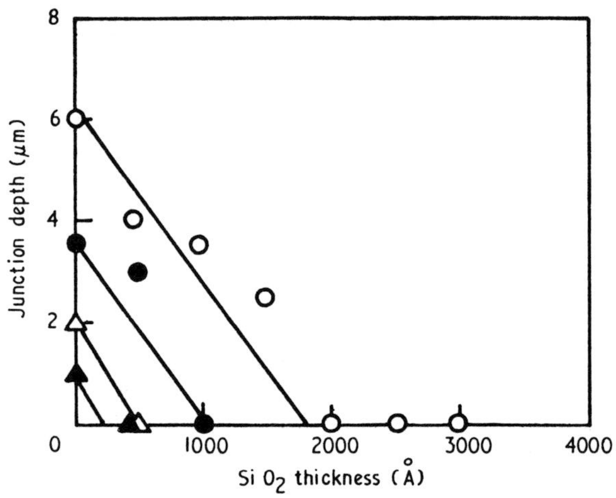

Figure 10. Junction depths in p-type GaAs through a sputtered SiO_2 diffusion mask as functions of mask thickness. ○ 850 °C, 72 hr; ● 850 °C, 24 hr; △ 800 °C, 72 hr; ▲ 800 °C, 25 hr.

It is seen, for example, that S diffusion at 850 °C for 24 hr would be masked with a 1500 Å-thick sputtered SiO_2 film.

5. Conclusions

We have studied the feasibility of applying a newly proposed method of S diffusion to the fabrication of GaAs devices. The method utilizes the redistribution of S between deposited Ga_2S_3 film and the substrate wafer. The electrical properties of the diffused n-type layers were comparable to those of bulk crystals. The diffusion process was shown to be compatible with a high purity epitaxial technique. The p–n junctions were linearly graded and showed the band-edge injection luminescence. From these experimental results, we think this technique might be useful in GaAs planar technology.

It would be worth while noting here that the proposed diffusion method is applicable in principle to any other combination of a semiconductor and a source compound if the lowest liquidus line of the system lies at a sufficiently high temperature.

Acknowledgments

We thank T Tsukada and T Toyabe for helpful discussions and A Seto for technical assistance.

References

Asai S and Kodera H 1972 *J. Japan. Soc. Appl. Phys.* **41** suppl p231
Brooks H 1951 *Phys. Rev.* **83** 879
Frieser R G 1965 *J. Electrochem. Soc.* **112** 697
Goldstein B 1961 *Phys. Rev.* **121** 1305
—— 1962 *Compound Semiconductors* vol 1 ed R K Willardson and H L Goering (New York: Rheinhold) p345
Johansson N G E and Mayer J W 1970 *Solid St. Electron.* **13** 317
Kendall D L 1968 *Semiconductors and Semimetals* vol 4 ed R K Willardson and A C Beer (New York: Academic Press) p198
van der Pauw L J 1958 *Philips Res. Rep.* **13** 1
Sze S M 1969 *Physics of Semiconductor Devices* (New York: Wiley–Interscience)
Yeh T H 1964 *J. Electrochem. Soc.* **111** 253
Young A B Y 1969 *Stanford Electronics Laboratory, Stanford University Technical Report* No 5116-1
Young A B Y and Pearson G L 1970 *J. Phys. Chem. Solids* **31** 517

1.4 ion implantation

Ion implantation has many potential advantages in the formation of junction structures:

1./The processing temperature can be considerably lower than that required for diffusion.
2./There is almost no limit to the dopant source due to chemical or physical limitations which exist with other junction-formation methods.
3./The final distribution of impurities can be controlled.

It seems apparent that, with the difficulties of doping compound semiconductors by diffusion or epitaxy illustrated thus far, ion implantation should be the optimum method. Much work has been done with compounds, but results have not been too encouraging. There are no commercial devices that have been made from GaAs or InP using impurity implantation.

The most interesting application has been the use of proton bombardment converting GaAs into a semi-insulating form using 400 KeV protons. This application was used to form a guard ring around conventionally fabricated GaAs IMPATT diodes[16]. It is ironic that the most useful application is dependent on the damage introduced by the implantation process. However, Zn, Cd, S, Se, and Te have all been implanted successfully into GaAs substrates held at 200°C, with a subsequent anneal between 400 and 800°C.

Factors determining the range of the implant are the energy, the atomic number of the implant species, and the host lattice. The atoms penetrate the lattice and, in ideal circumstances, have a Gaussian distribution with a maximum below the semiconductor surface. The range also depends markedly on both the degree of orientation of the host crystal and whether channelling of the incident beam is used. The maximum range is proportional to the (beam energy)$^{1/2}$. Because the atomic number of GaAs is approximately twice that of silicon, beam energies must be high to achieve significant penetration in the compound semiconductors. Typically, 150 keV beam gives about 0.5 μm penetration channelled on the $< 110 >$ axis. Lattice disorder created must be a function of beam energy; thus, the damage created by the implantation is considerably greater than that for silicon.

An extension of the implantation process is proton enhanced diffusion (PED). Controlled bombardment with protons can modify or progress impurities already diffused into the lattice. Application of this technique to compound semiconductors has not yet been reported.

It is not the intention of this book to imply that ion implantation is not a viable method for doping compound semiconductors; but there are unique difficulties in the processing of these materials. These problems undoubtedly will be overcome and implantation will find applications in device manufacture, making optimal use of the advantages offered by the procedure. It is inevitable that implantation will be used to form High-Low and Low-High-Low IMPATT structures discussed previously.

1.5 surface passivation

Oxide films are required on compound semiconductors as diffusion and metallisation masks, in common with Si devices. But unlike Si, the compounds do not easily self-oxidise with films that are compatible with subsequent processing. Oxides must be deposited on to the wafer surface indirectly. Sputtering and pyrolytic deposition of SiO_2 are the most significant methods.

Sputtering has been used successfully to deposit dielectric layers on to GaAs. Careful attention must be taken in substrate cleaning, control of temperature during the sputtering process, and in the subsequent heat treatment. Problems occur due to "crazing" of the deposited layers caused by differences in thermal expansion; also, there always appear to be marked discrepancies between layers grown at different laboratories, primarily in their chemical properties.

Pyrolytic deposition can be obtained over large temperature range. The most common technique uses the decomposition of tetraethylorthosilicate under vacuum at temperatures around 700°C. This temperature is usually higher than is desirable, particularly for other substrate materials. An extension of the same procedure which incorporates extra oxygen reduces the growth temperature to 400°C or less. The apparatus is seen in Figure 1-8. Argon, bubbled through the TEOS, is subsequently mixed with oxygen before entering the reaction tube. The exhaust leaves through another bubbler (containing silicone oil) to atmospheric pressure.

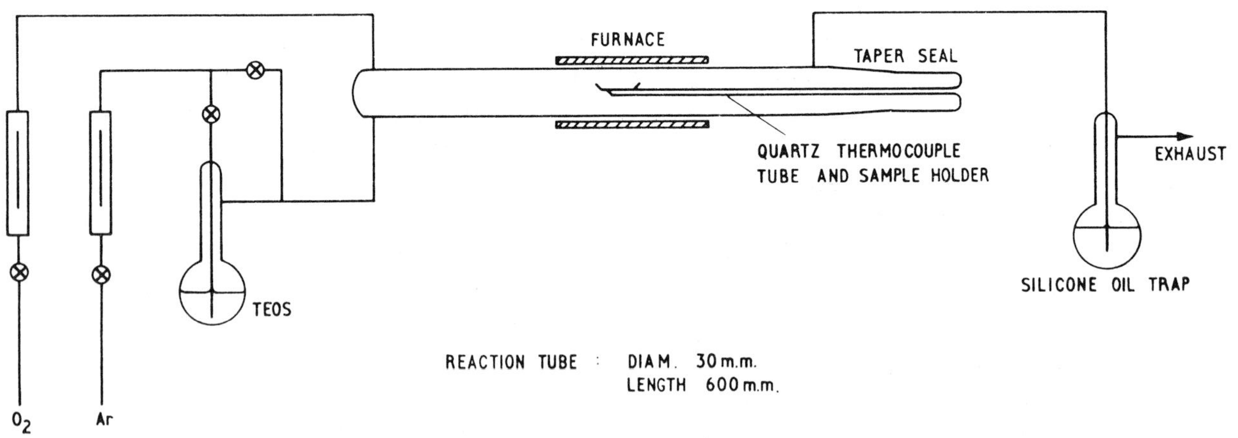

Figure 1-8/Silicon Dioxide Deposition Equipment

The deposition temperature and the flow rate and composition of the TEOS/O_2/Ar mixture determine the growth rate. In the RRE laboratory a temperature of 400°C with Argon/oxygen ratio of 2:1 by volume is used and the TEOS is at room temperature; the results are a minimum measured thickness of 300Å and a growth rate of 2000Å/hr[20]. The films produced are excellent and appear to be stable throughout device processing.

Another pyrolytic method of note is that using gaseous mixtures of silane and oxygen that are stabilised with an inert gas. Again, relatively low deposition temperatures are necessary. The process is typical of the other pyrolytic techniques; however, there is the added hazard of handling silane.

1.6 summary

In summation, this section has evaluated how the various materials growth and doping procedures contribute to the technology of devices made from compound semiconductors. The significance of bulk and epitaxial growth of material is obvious both in present and future devices. The technology is well developed and it is unlikely that any major changes of these basic methods will be introduced. Recent technologies, such as ion implantation, promise much for the future, especially where greater control is required in regions near the surface of the epitaxial wafer. Yet to be discussed is the extent to which ion implantation will become a feature of the large scale production of devices.

The diffusion of impurities into the compound semiconductors requires more experimentation in order to establish the mechanisms and controls required for sophisticated devices. In many areas such diffusion is being overtaken by epitaxy. The materials growth of the compound semiconductors is more clearly understood and controlled than that of silicon. It is apparent, though, that the processing for the compounds is considerably more difficult than for silicon.

chapter two

materials characterisation and assessment

The number of available assessment techniques is vast and continually increasing. Virtually each new device type generates a new assessment technique. The development of a device involves a continuous feedback from the device engineer to the materials scientist, with referral to problems associated with the performance of the completed device. The relevant parameter is isolated by appropriate measurement and the materials technology is modified to allow the next stage in the development of the device. Thus, materials assessment progresses from highly sophisticated and specialised techniques to simple, routine, on-line characterisation for production.

Generally, measurements first can be separated into chemical, physical, and electrical analyses, then into three subsequent stages relating to the research, development, and production stages of the programme. Chemical analysis is essential for the initial purification of all starting materials and the bulk crystals prepared from them. A special process is necessary for thin layers, because of the minimal material available for analysis. These techniques are primarily extensions of the bulk method; however, there exist specific procedures for investigating surface layers.

Physical analysis provides information on the crystal structure and crystallographic properties of the semiconductor, along with its mechanical and dimensional characteristics.

Electrical analysis determines parameters which directly influence the performance of devices made from the material. The methods used are dependent on the restrictions imposed by device structures.

There is not a clearly defined sequence of investigation, as all of the techniques are inter-related. Each method is examined within the framework of the device objective in order to select the optimum assessment procedure.

The major criticism of the semiconductor business is that much effort is expended measuring and recording useless or redundant information. It cannot be disputed that measurement consumes time and costs money; thus, the value of measurement in terms of the useful information obtained should be frequently reassessed.

2.1 chemical analysis

Table 2-1 compares the techniques for chemical analysis.

Instrument	Bulk/Surface	Details	Sensitivity	Measurement Frequency	Cost
Mass Spectroscopy	Bulk	a) Chemical analysis of starting materials b) Analysis of bulk crystals	0.02-200 ppm atomic, high vacuum low temperature Destructive	a) Could be 100% b) Research	High
Ion Probe Analysis	Thin film	Chemical analysis of grown layers	1 μm x-y and 50-100Å in z direction, sensitivity much as for mass spec.	Research	High
Photoluminescence	Surface could be extended by successive etch	Identification of impurities, activation energy and activity	Needs high purity background for good resolution Not quantitative, non-destructive	Random sampling or Research	Medium
Cathodoluminescence	Thin film	As for photoluminescence, different excitation source	Up to 6 μm depth. Information also available on lifetimes, diffusion length and surface recombination	Random sampling or Research	Medium/High
Auger Spectroscopy	Surface	2 keV electron	Qualitative analysis, most sensitive for low mass elements Non destructive	Research	High
Rutherford Scattering	Surface	High energy scattering up to 4 MeV	Can detect monolayers of impurity with mass greater than host, stoichiometry, non destructive	Research	High
Electron Microprobe	Thin film	X ray spectrometer fitted to conventional SEM	10-1000 ppm. Best sensitivity with high mass elements. Non destructive	Random sampling	Medium
Radio Tracers	Thin film	Analysis carried out as film is stripped off	Satisfactory for diffusion profiles	Research	Medium

Table 2-1/Comparison of Chemical Analysis Techniques

2.1.1 bulk materials

Mass spectroscopy is the most common method for analysis of bulk material. The basic procedure is well known — the instrument utilizes a spark source with the spectra displayed on a photographic plate. The absolute limiting sensitivity of the instrument is about one part in 10^9. Elements such as Si, O_2, N_2, and C can cause severe problems. Such difficulties provide severe limitations as they are often due to just the impurities of interest. As with many other analytical techniques, diligent interpretation of results is necessary to avoid discrepancies between different laboratories. A thorough investigation of impurities in bulk InP revealed impurities of approximately the 10^{15} level. These results were obtained by ensuring that the spark chamber was cooled to liquid helium temperatures; the residual C and O levels were then found to be in the range of 0.1-2 parts per million. The method of mass spectroscopy has produced purer starting materials and purer bulk crystals; however, it is capable only of identifying impurities quantitatively — their activity must be assessed in different ways. The application of mass spectroscopy to thin layers is discussed later.

Emission spectroscopy is one of the oldest means for trace analysis. Like mass spectroscopy, it can use spark excitation. The emission spectrograph employs a grating to separate the emitted light into a usable spectrum. The sensitivity of emission spectroscopy is generally inferior to mass spectroscopy.

2.1.2 thin film

The assessment of these grown layers presents a different set of problems, primarily because the devices available use epitaxially grown material. A good review of many of the following techniques is given in Reference 28.

Ion Probe Analysis[28]

Ion probe analysis follows from the adaptation of the mass spectrometer to the study of thin films. Ion probing is accomplished by bombarding the sample with a beam of 1-20 keV ions, causing the upper atomic layers to be sputtered or stripped off. Most of the material leaves as neutral atoms, but a small fraction is ejected as positive or negative ions (which are extracted into a mass spectrometer for mass/charge separation). A full mass spectrum can be taken, or an individual secondary ion mass can be monitored, as the primary ions erode the sample. The spectrometer is a very useful extension of mass spectrometry as it now allows for a three-dimensional analysis, essential for deposited or diffused layers. The resolution claimed (dependent on the instrument) is 1 μm in x-y directions, the ability to examine fractional surface monolayers, and an in-depth resolution of 50-100Å with the acquisition of secondary ion images. The sensitivity is between 10^{-15} to 10^{-19} gram, depending on the element analysed. The variation in sensitivity over four orders of magnitude presents one of the problems associated with the instrument.

Commercial instruments take a number of forms, but all are expensive (approximately £100,000 or $200,000). As with mass spectrometry, the analysis is destructive.

The application of this form of analysis to GaAs is described in the key paper by Kim et al.

Photoluminescence

Photoluminescence has only recently been applied to the study of GaAs[22] and InP[23]. The technique is straightforward. The sample under investigation is maintained at a temperature from 1.8-77°K, and is illuminated by a suitable source, such as a laser. The luminescence spectrum is observed either with photomultiplier methods or directly on a photographic plate. The key paper by White describes the photoluminescent analysis of GaAs. Procedural problems occur when the resulting spectra are analysed. A good survey of the various recombination mechanisms is given by Dean[31].

Qualitative assessment is established, by deliberate incorporation of known impurities and by comparison with mass spectroscopy results[20]. The additional information extrapolated from photoluminescence is the activity of the impurity. The assessment is essentially of the semiconductor surface — to obtain information as a function of depth into the layer requires successive stripping. The successive etching of InP has not, as yet, given any conclusive results; it may well be that the exposed free surface effectively masks changes that might be present in the bulk. Luminescence measurements have been significant in the development of luminescent devices; however there has not yet been any comparable correlation between luminescent behaviour of the material and performance of microwave devices.

Cathodoluminescence

Cathodoluminescence is similar to photoluminescence but excitation is achieved by electron beams in the energy range 1-60 keV. Qualitative assessment is relatively easy but quantitative determinations are not. The advantage of cathodoluminescence over photoluminescence arises from the continuous range of excitation energies available. The penetration of the electron beam is proportional to $E^{1.7}$ which gives working depths from 10^{-2} to 6 μm without any need for successive removal of material. Because this method is non-destructive, regions well removed from any disordered surface can be studied. In addition to impurity analysis, minority carrier lifetimes, diffusion lengths, and surface recombination can also be studied. Also, emission wavelengths from ultra-violet to 5 μm can be measured.

X-Ray Techniques

Impurity analysis is made possible when X-rays are used in conjunction with the electron microscope. X-ray spectrometers are fitted to the basic microscope, providing the dual facility of direct observation of the area under investigation. The resolution can be as fine as 1 μm. The limitation, though, of the electron microprobe is the lack of good crystal analyzers to give optimum sensitivity over the whole mass range. Sensitivity is from about 10 ppm down to 1000 ppm for the lightest elements.

Auger Spectroscopy

The inner orbits of some of a material's atoms become ionized when it is X-irradiated, or bombarded with electrons of sufficient energy. The primary electron transfers some of its energy to the atom, and then departs, being observed externally as an inelastically reflected electron. The ionized atoms return to their normal state by absorption of electron from upper levels; the energy released can be accompanied by external fluorescence radiation (if the energy is transferred to a photon) or by ejection of an electron (if the interaction is with an electron). The ejected electron is an Auger electron. The Auger electron is characteristic of the particular atom species. The energies can be predicted within certain limits imposed by other mechanisms. The method is highly sensitive for elements of low atomic number; positive identification of heavy elements is possible, assuming only one or two are present for analysis in the sample.

The source energy is about 2 keV so that Auger spectroscopy is essentially an approach for surface analysis, with current work extending to fraction of a monolayer. The ultimate sensitivity though, of the device has not yet been established. Quantitative analysis is difficult, and the instrument requires very good vacuum systems. It is interesting that the combination of Auger and electron microprobe spans the entire mass range.

Rutherford Scattering and Channelling

Rutherford scattering of energetic ions from crystal surfaces has been widely used for surface studies, thin film analysis, and for determining the concentration profile of impurities. The procedure has matured with that of ion implantation; it is particularly valuable in the assessment of ion implanted layers. Only recently has the Rutherford scattering been applied to compound semiconductors as an analytical tool for the study of stoichiometry, impurity content, and crystal perfection[24]. A beam of up to 4 MeV alpha particles is guided onto the surface of a single crystal. The yield of particles backscattered into a solid state detector is measured — first, with the incident beam aligned parallel to one of the open axes (channelled); then, in a random direction. (See Figure 2-1.) Particles scattered from the crystal surface region in the Rutherford scattering have an energy

$$E_1(O) = k^2 E_o$$

where E_o is the incident ion energy and k^2 the kinematical factor, defined as:

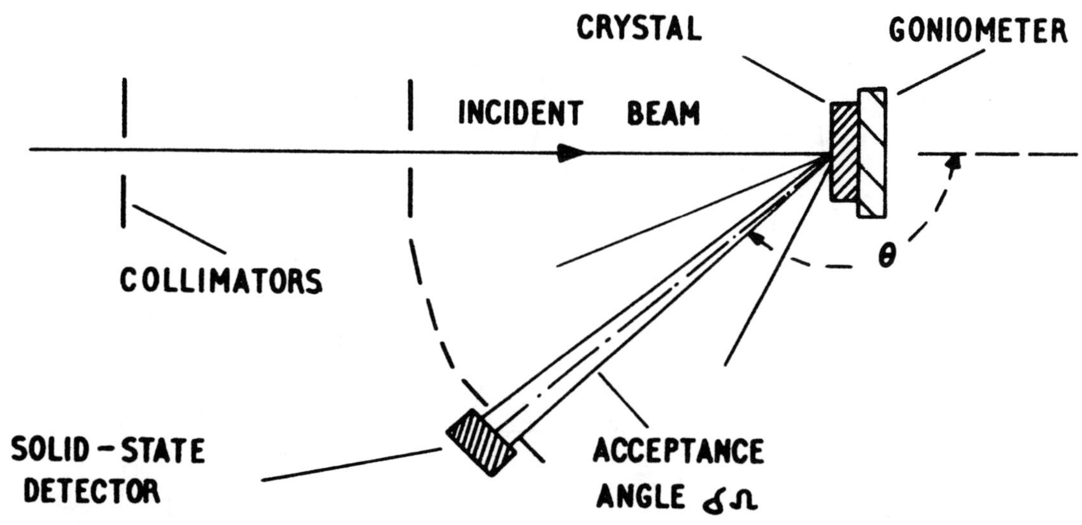

Figure 2-1/Rutherford Scattering Experimental Arrangement

$$k^2 = \frac{M_1 \cos\theta + (M_2^2 - M_1^2 \sin^2\theta)^{1/2}}{M_1 + M_2}$$

where M_1 is the mass of the incident ion, M_2 is the mass of the scattering atoms, and θ is the scattering angle.

Each surface impurity has its own characteristic value of k^2 defined uniquely by the mass M_2, yielding a unique value of E. For normal incidence, the energy scale of the host spectrum may be converted into a depth scale:

$$E_2(z) \sim k^2 E_o - Z[S(E_o)\, k^2 + S(E_1)/\cos\theta']$$

$S(E_o)$ and $S(E_1)$ are the stopping powers of the ion at these energies. The energy scale thus can be converted into both an impurity mass or a depth scale. A typical spectrum of GaAs is seen in Figure 2-2. To calculate the impurity concentration, the normal method is to compare the impurity yield to that of the non-aligned host spectrum. Hence, the surface impurity concentration N_x is given by:

$$N_x = N_a \frac{A_x}{Y_a} \left(\frac{Z_a}{Z_x}\right)^2$$

where A_x is the area under the impurity peak, Y_a is the non-aligned yield, and N_a is the number of host atoms per monolayer per cm.

In the case of a binary compound, it is possible to calculate the ratio of the two constituents using the same method:

$$\frac{N_a}{N_b} = \frac{A_a}{A_b} \left(\frac{Z_b}{Z_a}\right)^2$$

The restrictions on the Rutherford scattering technique are:

1./The differences in the atomic masses of the two constituents must be large enough to enable the two surface peaks to be resolved.

2./The atomic numbers must not differ by too great a value; if they do, the yield from the lighter species can get lost in the background from the heavier sub-lattice.

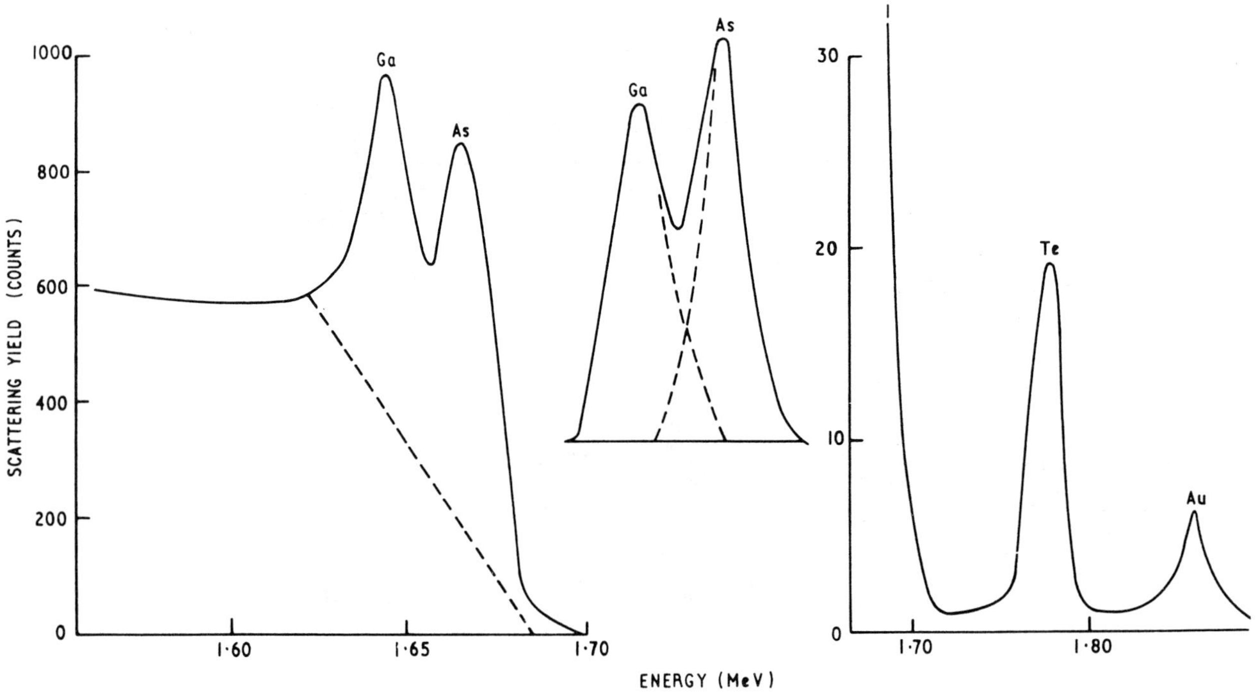

Figure 2-2/Rutherford Scattering Spectrum for GaAs.

3./The sensitivity decreases rapidly as the atomic number of the impurity atom decreases, causing an abrupt reduction of atoms lighter than the host atoms.

For GaAs this fact means that an impurity with a mass greater than 75 can be detected, such as Te, Br, Cd, Ge, Au, In, Pb, Ag, and Sn, Al, C, O, S, and Zn cannot be perceived. InP, with its higher mass, permits the detection of even fewer impurities.

Rutherford scattering also is used to determine the degree of surface disorder. The key paper by Morgan and Wood discusses the application of Rutherford backscattering.

Radio-Tracers

The most convenient method for analysis of diffusion profiles in compound semiconductors is through the use of radio-tracers. The diffusion source is a radioactive isotope of the required impurity. The subsequent analysis of the diffusion profile is executed by successive removal of the layers, with the specific activity of each measured.

key paper

Analysis Of The 'I-Layer' Region In Epitaxial GaAs By Direct Image Mass And Spark Mass Spectrographic Methods

H.B. Kim, D.L. Barnett, and G.G. Sweeney
(Westinghouse Research Laboratories, Pittsburgh, Pennsylvania, US)

Institute of Physics and Physical Society Symposium on GaAs (1972)
Pages 88-97

Abstract. The high resistance region in the net carrier concentration profile at the epitaxial–substrate interface was investigated by direct image mass (ion probe) and spark-source mass spectrographic analysis. Two types of 'i-layers' were identified; one associated with accumulated oxide, and the other without oxide accumulation and caused by stoichiometric imbalance during transient growth conditions.

1. Introduction

The high resistance region in the net carrier concentration profile at the epitaxial–substrate interface is referred to henceforth as the 'i-layer'. The i-layer can be a slight depression of 30–40% or a dip of a decade or more in the net carrier concentration profile with a width from submicrometres to several micrometres. Since the early observations of the i-layer phenomenon (Monsanto Report 1967, Wolfe *et al* 1968, Hasegawa and Saito 1968), studies on the subject have been stimulated by upsurging interest in microwave solid state device development.

Several methods have been employed to investigate the i-layer and to correlate it to a chemical identity forming the i-layer complex. Differential capacitance against voltage (Hildebrand and Gold 1960, Thomas *et al* 1962, Meyer and Guldbrandsen 1963, Amron 1964) or directly plotting Copeland methods (Copeland 1969) are frequently used to obtain the net carrier concentration profile with high resolution (Debye length $\simeq 0{\cdot}015$ μm at 10^{17} cm^{-3} to $0{\cdot}5$ μm at 10^{14} cm^{-3}). Radio tracer (Blakeslee and Lewis 1969), photoluminescence (Nakashima and Hirao 1970, Iwasaki and Sugibuchi 1971), photo-ionization (Hollan *et al* 1971) on a reverse-biased Schottky barrier diode, and secondary ion mass spectrometry (also known as direct imaging mass analysis, DIMA) (DiLorenzo *et al* 1971, DiLorenzo 1971) have been used to identify the chemical complex of the i-layer. Of these, the ion probe method complements best the Copeland measurements with high resolution (~60 Å).

Explanations of the chemical identity of the i-layer complex have been numerous. Wolfe's proposed Si out-diffusion or autodoping (Wolfe *et al* 1968) of the epitaxial layer from the Si-doped n^+ substrate was countered by Hasegawa and Saito (1968) who observed the i-layer as a defect complex formed during the initial transient growth condition. Some control over the i-layer phenomenon was attained by the introduction of excess arsenic (Saito and Hasegawa 1971) or by vapour etching (Hirao and Nakashima 1970) during the initial stage of growth. Defect complexes

(Iwasaki and Sugibuchi 1971, Hollan *et al* 1971) of Si–As vacancy and Ga vacancy-donor and copper contamination (Nakashima and Hirao 1970) were found to cause the i-layer. By the ion probe method, DiLorenzo *et al* (1971) and DiLorenzo (1971) observed a high concentration ($\sim 10^{21}$ cm^{-3}) of Si at the i-layer, and Na, K and Ca were also found localized at the i-layer.

A Si concentration of 10^{21} cm^{-3} cited by DiLorenzo exceeds the solubility limit of Si in GaAs and should segregate out to form a second phase or precipitates, since this concentration is nearly 5% of either the Ga or As lattice atoms. In this work, therefore, we examined the anomalously high Si concentration at the i-layer by employing the ion probe technique and a dendrite-probed spark-source mass spectrographic analysis (Sweeney and Berkey 1971) technique and identified two types of the i-layer complexes.

2. Experimental method

2.1. Epitaxial growth procedure

The epitaxial GaAs materials analysed here were grown by using the Ga–$AsCl_3$–H_2 vapour transport system (Knight *et al* 1965). Figure 1 shows a schematic diagram

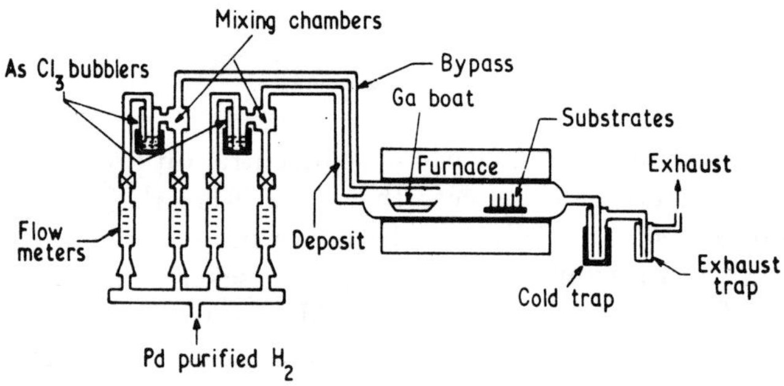

Figure 1. The Ga–$AsCl_3$–H_2 epitaxial growth system.

of the epitaxial GaAs growth system. High purity epitaxial GaAs growth procedures similar to recent advances in the epitaxial growth technology have been employed (Effer 1965, Wolfe *et al* 1968, Maruyama *et al* 1969, Barry 1970, Cairns and Fairman 1968, 1970).

The substrate wafers were Si-doped (n = 1–3 × 10^{18} cm^{-3}), of (001) orientation, lapped and chemical-mechanically polished using as the etchant a solution of $16H_2SO_4 : 1H_2O_2$ followed by rinsing in double quartz distilled water. The substrate wafers were *in situ* etched at 750 °C prior to deposition by HCl produced using the bypass bubbler system ($AsCl_3$ with H_2). A Hall mobility of 7500 $cm^2 V^{-1} s^{-1}$ at room temperature with 10^{15} cm^{-3} net carrier concentration was achieved routinely. Thickness of the epitaxial layer was controlled from 0·5 μm and thicker. The growth rate was controlled at 0·03 to 0·1 μm min^{-1} on (001) GaAs substrates. The net carrier concentration profiles were measured by the direct-plotting Copeland (CIP) method on reverse-biased Al or Au Schottky barrier diodes.

2.2. Mass analysis

2.2.1. Secondary ion mass spectrometry (*or* DIMA). The instrument employed was the Cameca SMI 300, a special mass spectrometer with a sputtering source and unique analytical capability. Three types of information can be obtained; *the mass spectrum* of all elements from hydrogen to uranium can be obtained with a resolution of the order of 1000; *the ion image* of each element giving a distribution of corresponding ions with a resolution limit of about 1 μm; and *the concentration profile* of any constituent with a resolution of the order of 10 atomic layers (~60 Å for GaAs) having an erosion rate governed by the primary ion beam. Of the three, the mass spectrum and the concentration profile of selected chemical constituents were extensively used.

Ar^+ and O_2^+ were used as the primary ion beams, which eroded the sample surface with about 300 μm in crater diameter at sputtering rates of about 1 μm min^{-1}. The aperture of the ion probe was about 150 μm in diameter. The depth of the crater was checked by interference microscope.

2.2.2. Spark-source technique (SSMS). The instrument employed was the AEI-702 spark-source mass spectrometer. Unlike the ion probe, SSMS erodes the solid sample surface by an RF spark using a counter electrode and has only elemental analytic capability. In order to circumvent the contamination encountered by a metal electrode probe, a dendrite of GaAs (Sweeney and Berkey 1971) was used as the probing electrode. The high purity dendrites were grown in the epitaxial reactor by self-nucleation on the quartz substrate holder and the quartz reactor tube wall. Further details and capability of the SSMS techniques may be found elsewhere (Ahearn 1972).

3. Results and discussions

The Si-concentration profiles obtained by ion probe methods were compared with the net carrier concentration profiles obtained by the Copeland method. Figure 2 is a composite of these profiles measured on sample A. An i-layer width of about 0·5 μm was observed in the Si-concentration profile compared with about 1 μm in the Copeland net carrier concentration profile. Figure 3 shows the same concentration profile on sample A at higher sensitivity. Note that the Si-concentration

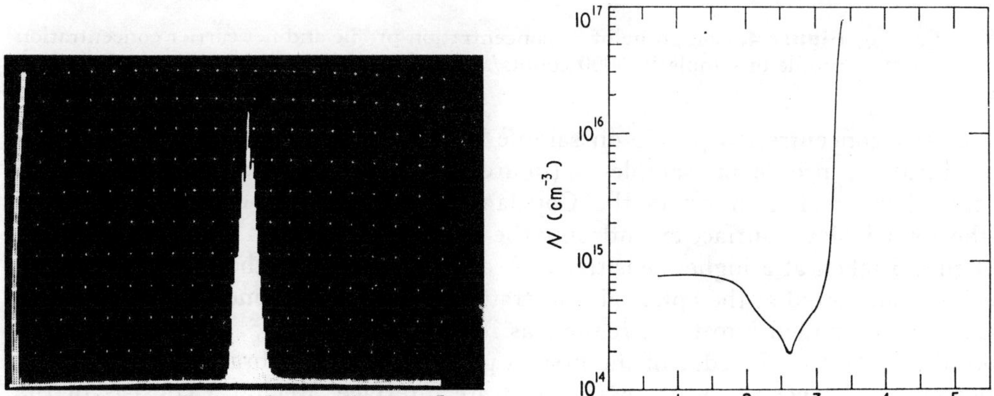

Figure 2. Composite of Si-concentration profile and net carrier concentration profile of sample A. 2500 counts/s FS, O_2^+ ion beam.

level is lower in the epitaxially grown layer than in the substrate, in which the net carrier concentration level was about 2×10^{18} cm^{-3}. Also note that there are two Si-concentration peaks, one at the surface of the epitaxial layer with about 0·1 μm width and the other corresponding to the i-layer region near the epitaxial–substrate interface. Figure 4 shows a composite of the Si-concentration profile and Copeland

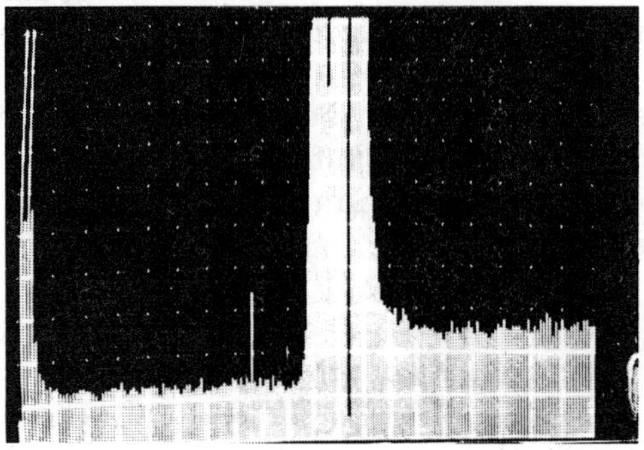

Epitaxy ← | → Substrate

Figure 3. Si-concentration profile of sample A. 1000 counts/s FS, O_2^+ ion beam.

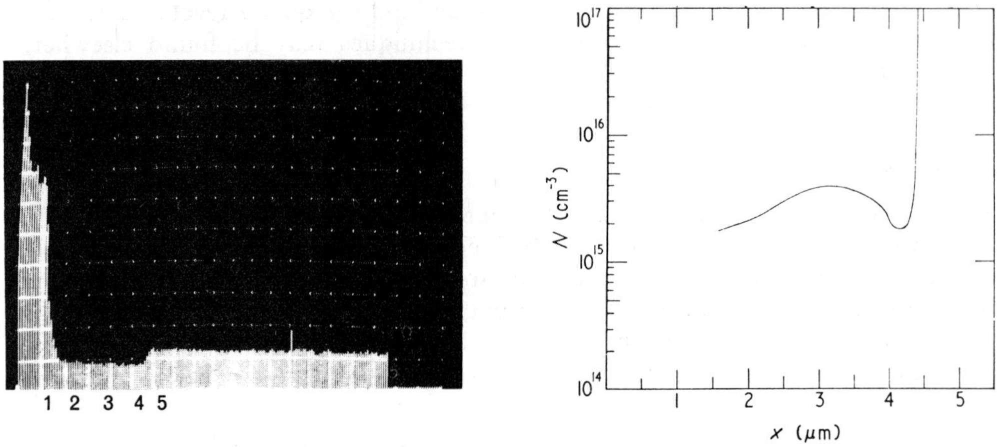

Figure 4. Composite of Si-concentration profile and net carrier concentration profile of sample B. 5000 counts/s FS, O_2^+ ion beam.

net carrier concentration profile on sample B. Unlike sample A (figure 2) the Si-concentration, profile on sample B (figure 4) shows no Si-concentration peak at the i-layer region shown in the Copeland profile. The Si-concentration peak at the epitaxial layer surface extends into the layer about 1 μm on sample B as shown in figure 5 taken at a higher sensitivity. A distinct change in the Si-concentration level is again noted at the epitaxial–substrate interface. The same Si-concentration profile is subdivided into three regions as indicated in figure 5. Region 1 extends from the surface to the edge of the first Si peak. Region 2 embraces the remaining inner epitaxial layer to the epitaxial–substrate interface. Region 3 consists of the substrate. Figure 6 shows the mass spectrum taken at the depth corresponding to the three regions designated above on sample B in figure 5. The first region shows the necessary surface contamination oxides and hydrocarbon. The second shows the pure spectrum of the epitaxial layer, while the third shows a slight increase in mass 28 of Si at region 3 representing the higher Si concentration level in the substrate GaAs.

All of the Si-concentration profiles and the mass spectra presented in figures 2–6 were taken by sputtering the samples using O_2^+ as the primary ion beam. The most obvious difference in Si-concentration profiles between samples A and B is that sample A displays a high Si-concentration peak corresponding to the i-layer region but sample B has no Si-concentration peak near the i-layer region. Sample A was re-examined using Ar^+ instead of O_2^+ as the sputtering ion beam. The Ar^+ sputtering ion beam enabled us to look at oxygen and oxide species such as SiO. Figure 7(*a*) shows the oxygen concentration profile, and figure 7(*b*) shows the SiO concentration profile on sample A. Both profiles show identical peaks at the locations

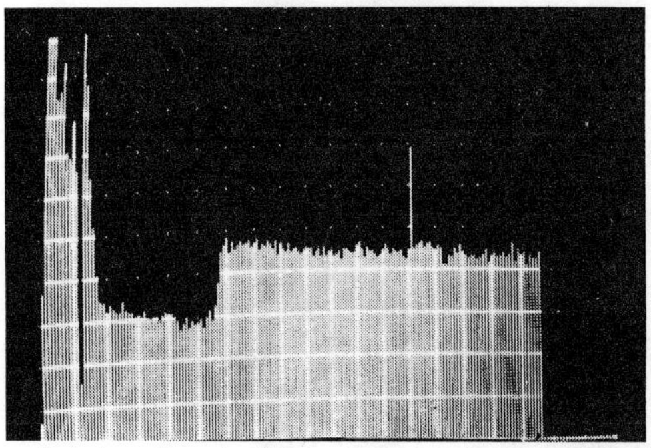

Figure 5. Si-concentration profile of sample B. 1250 counts/s FS, O_2^+ ion beam.

corresponding to that shown for the Si-concentration profile in figure 2. The position of the interface was determined by measuring the depth of the sputtered crater in each analysis and establishing a uniform sputtering rate. The presence of O_2 (or SiO) peaks observed at the surface and the i-layer region of sample A causes an enhancement phenomenon which is known as a chemical effect. This is a result of the difference in the ionization probability of Si and its oxide species. The anomalously high Si concentration ($\sim 10^{21}$ cm^{-3}) at the i-layer region reported by DiLorenzo could have been the same type of Si-concentration peak, which was actually an oxide enhancement phenomenon as observed here on sample A. The same effect can be seen at the surface of sample B.

The Si–O complex behaves as a deep acceptor (0·2 eV and 0·4 eV below the conduction band) in the upper half of the forbidden gap according to the proposed model of Weiner and Jordan (1972). The electrically active SiO acceptor complex should change n-type epitaxial material into a more compensated high resistivity material as observed in the net carrier concentration profiles at the i-layer region in figure 2 on sample A and at the epitaxial surface, region 1, in figures 4 and 5 on sample B.

Table 1 lists the results of dendrite-probed spark-source mass spectrographic analysis (SSMS). Some hydrocarbons were found on all sample surfaces. Carbon and oxygen were present in the spectra but no significant quantitative values were worthy, since high values were obtained mostly at the surface. Impurities other than Si, found at levels below 5×10^{16} cm^{-3}, were F, P, Cl, K, Ca and S. The presence of hydrocarbons, F, Cl and S were due to surface contaminants left after

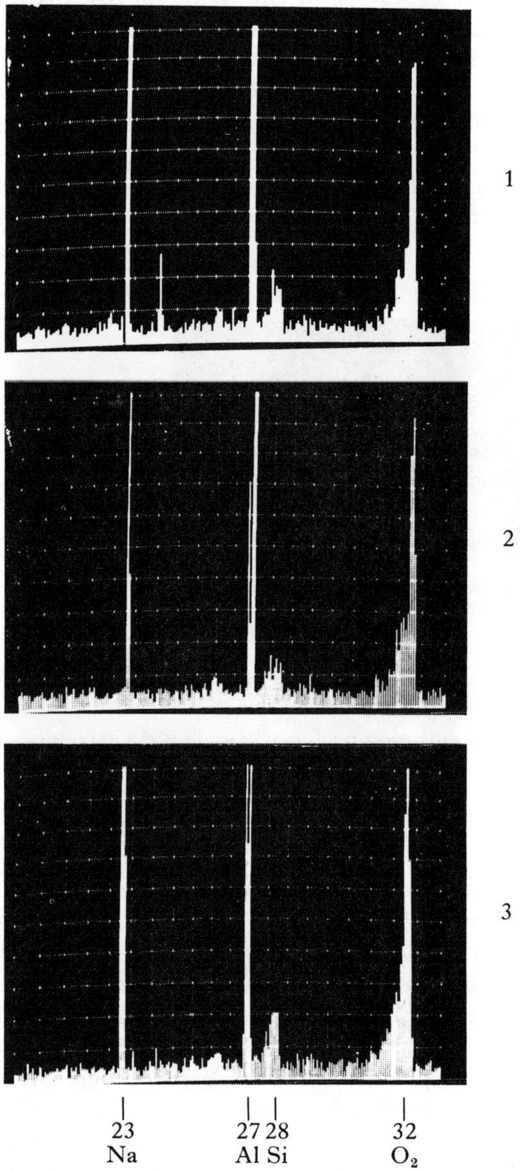

Figure 6. Ion probe mass spectrum of (1) the surface, (2) epitaxial and (3) substrate regions of sample B as shown in figure 5.

Table 1. Dendrite probed SSMS

	F†	Si	P	Cl†	K	Ca	S†	Hydrocarbon	Oxygen
Sample A	3·6E16	8·9E14	3·6E16	4·4E16	1·3E16	4·4E15	1·3E16	yes	yes
Sample B	3·6E16	8·9E14	4·4E14	6·7E16	1·3E16	4·4E15	1·8E16	yes	yes

† Surface contaminants due to cleaning. (Notation: 3·6E16 $\equiv 3{\cdot}6 \times 10^{16}$ cm^{-3}.)

cleaning. Si-concentration levels of $8{\cdot}9 \times 10^{14}$ cm^{-3} were observed in both samples A and B. Wolfe *et al* (1970) has reported impurities measured in high purity epitaxial GaAs layer by SSMS. The Si-concentration levels ranged from 1 to 4×10^{15} cm^{-3} in their samples, which agrees well with our SSMS results. The low concentration level ($< 1 \times 10^{15}$ cm^{-3}) of Si observed in the epitaxial layer by the SSMS method supports the 'oxide enhancement' interpretation of the high Si peak observed at the i-layer region.

Attempts have been made to avoid the occurrence of the i-layer by SiO. Figure 8(*a*) shows the reflection electron diffraction (RED) pattern of the substrate surface often observed on the polished GaAs. Some amorphous oxide was found to exist on the sample surface by the RED pattern. Careful 'final cleaning' in

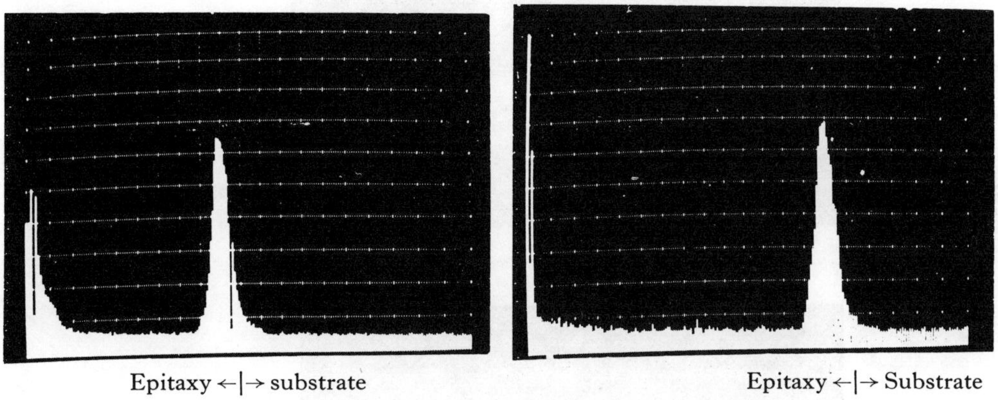

(*a*) (*b*)

Figure 7. With Ar^+ sputtering ion beam, (*a*) oxygen concentration profile, 1000 counts/s FS, (*b*) SiO (mass 44) concentration profile, 200 counts/s FS.

$5H_2SO_4 : 1H_2O_2 : 1H_2O$ etchant for 2 min followed by rinsing in double quartz distilled water just prior to the insertion and *in situ* etching before deposition helped to eliminate the oxide-type i-layer. Figure 8(*b*) shows the RED pattern on the 'final cleaned' GaAs substrate surface, and indicates the removal of the amorphous layer by the 'final cleaning' procedure.

The i-layer of nonoxide type as in sample B was more difficult to avoid. Since the i-layer normally occurs near the epitaxial–substrate interface, it is reasonable to assume that the initial transient growth conditions and the condition of the exposed growth surface of the substrate governed the i-layer phenomenon. The substrate surface preparation was shown to be important in eliminating the oxide-type i-layer. Harris *et al* (1969) reported that a high resistivity layer forms on the surface of the GaAs substrate during anneal, owing to loss of As. The As vacancies become acceptors or form acceptor complexes with impurity atoms. With amphoteric impurities (ie Ge, Si or Sn filling the As vacancies), the complex acts as an acceptor, which changes the n-type material into high resistivity. The frequent occurrence of the i-layer in epitaxial layers grown on Si- or Sn-doped, but not on Te- or S-doped substrates suggests the acceptor complex is an As vacancy–Si or –Sn complex.

To prevent loss of As, the substrate should be heated in as short a time as possible, and the high resistance layer should be removed by *in situ* etching prior to growth. Alternatively an excess As pressure can be provided to prevent As loss during the heating period to reach the growth temperature. Saito and Hasegawa (1971) suppressed the i-layer by introducing excess arsenic vapour pressure during the initial growth condition. The case of As vacancy formation in the substrate surface layers is merely the first part of a Ga/As stoichiometric imbalance within the GaAs crystal. The Ga/As stoichiometric balance is most difficult to provide during

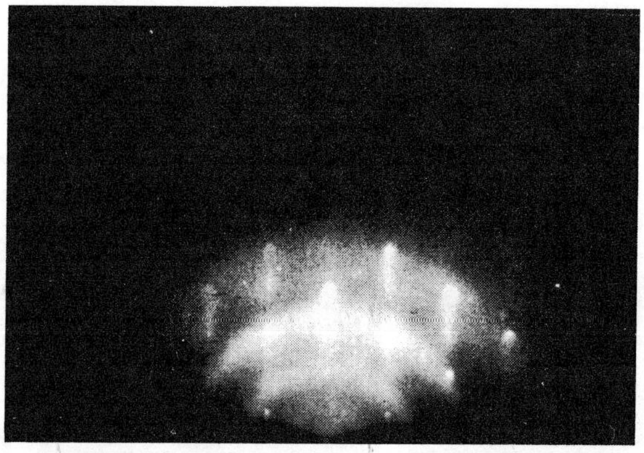

(*a*)

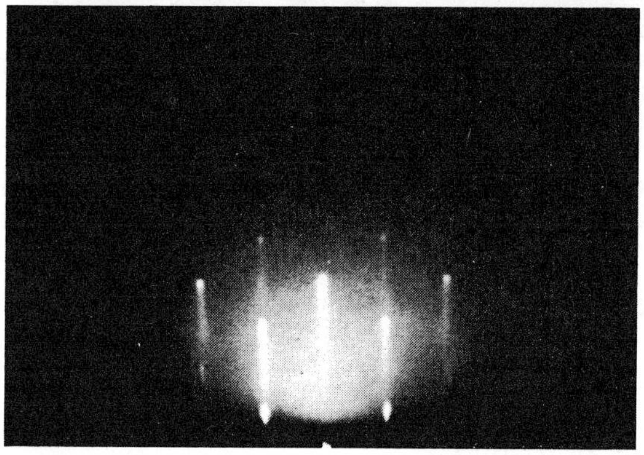

(*b*)

Figure 8. Reflection electron diffraction (RED) patterns of substrate surface showing (*a*) amorphous oxide layer prior to 'final cleaning', (*b*) final cleaned surface with no oxide layer.

the transient heating period to reach growth temperature. As-depletion of the saturated Ga adds to the already complex problem of maintaining the changing Ga to As component ratio when the etching condition is abruptly changed to the growth condition. Even if the vapour phase can be maintained to give the desired Ga/As stoichiometric ratio, the GaAs growth condition may be kinetically limited. Applying an atomistic model to the morphology of the epitaxial GaAs layer, Kim (1972) has assessed the normal growth conditions to be a Ga species-rich–stoichiometric imbalance. This crystal growth condition tends to enhance the As vacancy level, which in turn causes the As vacancy–acceptor or Si–As (or Sn–As) vacancy acceptor complex.

Total elimination of the i-layer is, indeed, a complex problem. It involves the final cleaning of the substrate surface to remove the amorphous oxide layer, ***in situ*** etching to remove the As-depleted surface of the substrate, and maintaining a one-to-one Ga to As stoichiometric ratio throughout the entire growth period. All of these conditions must be optimized for each individual reactor system and for each desired growth rate and growth morphology.

4. Summary

The i-layer and the high resistance region in the net carrier concentration profile were correlated with the peaks in the oxygen and SiO-concentration profile, where the SiO acceptor complex compensates n-type GaAs material. The anomalously high Si concentration ($\sim 10^{21}\ \mathrm{cm}^{-3}$) at the i-layer region reported by DiLorenzo (1971) could have been due to an oxide enhancement phenomenon. The low concentration level of Si ($< 10^{15}\ \mathrm{cm}^{-3}$) observed in the dendrite-probed SSMS method agreed with the results of the ion probe analysis.

Two types of i-layer were identified, one associated with the SiO complex and the other of nonoxide type, which was mainly caused by Ga/As stoichiometric imbalance which may occur during the transient initial growth stage. Control of 'oxide'-type i-layers was obtained. Some solutions to the complex problems of 'nonoxide'-type i-layer control were advanced.

Acknowledgment

The assistance of A F Lovas, R E Blose and D P Nebel in epitaxial growth is gratefully acknowledged. We wish to thank A J Noreika for RED measurements, and Bernard Autier (of Cameca Instruments, Inc) for ion probe measurements.

References

Ahearn A J (ed) 1972 *Trace Analysis by Mass Spectrometry* (New York: Academic Press)
Amron I 1964 *Electrochem. Tech.* **12** 327
Barry B E 1970 *Proc. 3rd Int. Symp. on GaAs: Inst. Phys. Conf. Ser. No 9* p172
Blakeslee A E and Lewis J E 1969 *Electrochem. Soc. May Meeting, New York* (Princeton: Electrochemical Society) abstract 121
Cairns B and Fairman R 1968 *J. Electrochem. Soc.* **115** 327C
—— 1970 *J. Electrochem. Soc.* **117** 197C
Copeland J A 1969 *IEEE Trans. Electron. Dev.* **ED-16** 445
DiLorenzo J V 1971 *J. Electrochem. Soc.* **118** 1645
DiLorenzo J V, Marcus R B and Lewis R 1971 *J. Appl. Phys.* **42** 729
Effer D 1965 *J. Electrochem. Soc.* **112** 1020
Harris J S, Nannichi Y, Pearson G L and Day G F 1969 *J. Appl. Phys.* **40** 4575
Hasegawa F and Saito T 1968 *Jap. J. Appl. Phys.* **7** 1342
Hildebrand J and Gold R D 1960 *RCA Rev.* **21** 245
Hirao M and Nakashima H 1970 *Proc. 2nd Conf. on Solid State Dev., Tokyo* (Tokyo: Japanese Society of Applied Physics) p46
Hollan L, Hallais J and Schiller C 1971 *J. Crystal Growth* **9** 165
Iwasaki H and Sugibuchi K 1971 *Appl. Phys. Lett.* **18** 420
Kim H B 1972 *J. Crystal Growth* to be published
Knight J R, Effer D and Evans P R 1965 *Solid St. Electron.* **8** 178
Maruyama M, Kikuchi S and Mizuno O 1969 *J. Electrochem. Soc.* **116** 413
Meyer N I and Guldbrandsen T 1963 *Proc. IEEE* **51** 1631

Monsanto Report 1967 *Air Force Avionics Lab., Wright–Patterson Air Force Base Interim Eng. Prog. Report* Contract F-33615-C-1432
Nakashima H and Hirao M 1970 *Jap. J. Appl. Phys.* **9** 1495
Saito T and Hasegawa F 1971 *Jap. J. Appl. Phys.* **10** 197
Sweeney G G and Berkey E 1971 *Presented at 10th National SAS Meeting, St Louis, Mo* (Society of Applied Spectroscopy) paper 102
Thomas C O, Kahng D and Manz R C 1962 *J. Electrochem. Soc.* **109** 1055
Weiner M E and Jordan A S 1972 *J. Appl. Phys.* **43** 1767
Wolfe C M, Foyt A G and Lindley W T 1968 *Electrochem. Tech.* **6** 208
Wolfe C M, Stillman G E and Owens E B 1970 *J. Electrochem. Soc.* **117** 129

The Photoluminescence Spectrum Of Bound Excitons In Indium Phosphide And Gallium Arsenide

A.M. White, P.J. Dean, L.L. Taylor, R.C. Clarke, D.J. Ashen, and J.B. Mullin
(Royal Radar Establishment, Great Malvern, Worcs., UK)

Journal of Physics C: Solid State Physics, Volume 5 (1972)
Pages 1727-1737

key paper

Abstract. Photoluminescence spectra of refined epitaxial indium phosphide and gallium arsenide show clearly a close correspondence of bound exciton transitions. Classified in both materials are emissions due to excitons bound to neutral acceptors and donors, free exciton peaks and various types of phonon coupling. Excitons bound to neutral acceptors in both materials give rise to a sharp doublet emission which readily reveals the presence of strain in the sample. The magnitude of the strain is estimated to be $\sim 10^{-3}$ in certain regions of typical epitaxial layers. Reflectivity experiments provide new estimates of the exciton gaps—1·4182 eV in indium phosphide and 1·5150 eV in gallium arsenide.

1. Introduction

Following the recently renewed interest in indium phosphide resulting from its applications in the microwave (Colliver *et al* 1972) and electroluminescence (Blom and Woodall 1970) fields, the quality of available material, particularly that grown by vapour phase epitaxy, has increased markedly. While several reports of photoluminescence studies of indium phosphide exist (Turner *et al* 1964, Gobel *et al* 1971 and references therein, Leite 1967, Roder *et al* 1970 and references therein, Heim 1969, Dean *et al* 1971), we believe that recent advances merit a closer examination of some aspects of the photoluminescence emission. The purpose of this paper is to report on those features of the spectrum which owe their origin to the occurrence of bound excitons and to point out the close correspondence with the spectra of the related III–V compound gallium arsenide. The luminescence spectra of these excitons, which contain very sharp no-phonon lines in sufficiently dilutely doped samples ($N_a + N_d < 10^{16}$ cm^{-3}) can be used to fingerprint the donors and, particularly, the acceptors to which they bind.

2. Technique

We have observed photoluminescence spectra from vapour phase grown indium phosphide with N_d–N_a between 1×10^{14} cm^{-3} and 9×10^{15} cm^{-3} and with mobilities between 2000 cm^2 V^{-1} s^{-1} and 5000 cm^2 V^{-1} s^{-1} at room temperature. The samples were freely suspended (still on substrates) in liquid helium at 1·8 K with provision for application of magnetic fields of up to 36 kG. Excitation was with 488 μm light from an argon ion laser focused to a spot of diameter ~250 μm. Spectra were recorded photographically on Kodak IN plates with a Baird Atomic model GWR1 spectrograph. Although the plate sensitivity deteriorates rapidly for photon energies below 1·4 eV they allow rapid recording and facilitate intercomparison of typical spectra down to about 1·37 eV. This range of sensitivity is sufficient to allow full study of the dominant near-gap luminescence of both indium phosphide and gallium arsenide at low temperatures. We emphasize our results on indium phosphide in this paper, drawing upon comparison with gallium arsenide particularly when our own data on the latter material seem either inconsistent with results published by other authors, or shed additional light on them.

3. The photoluminescence spectra

The main features of the typically observed photoluminescence spectrum of indium phosphide are shown in figure 1*a*. We show for comparison the equivalent regions of the photoluminescence spectrum of gallium arsenide in figure 1*b*. In the discussion which

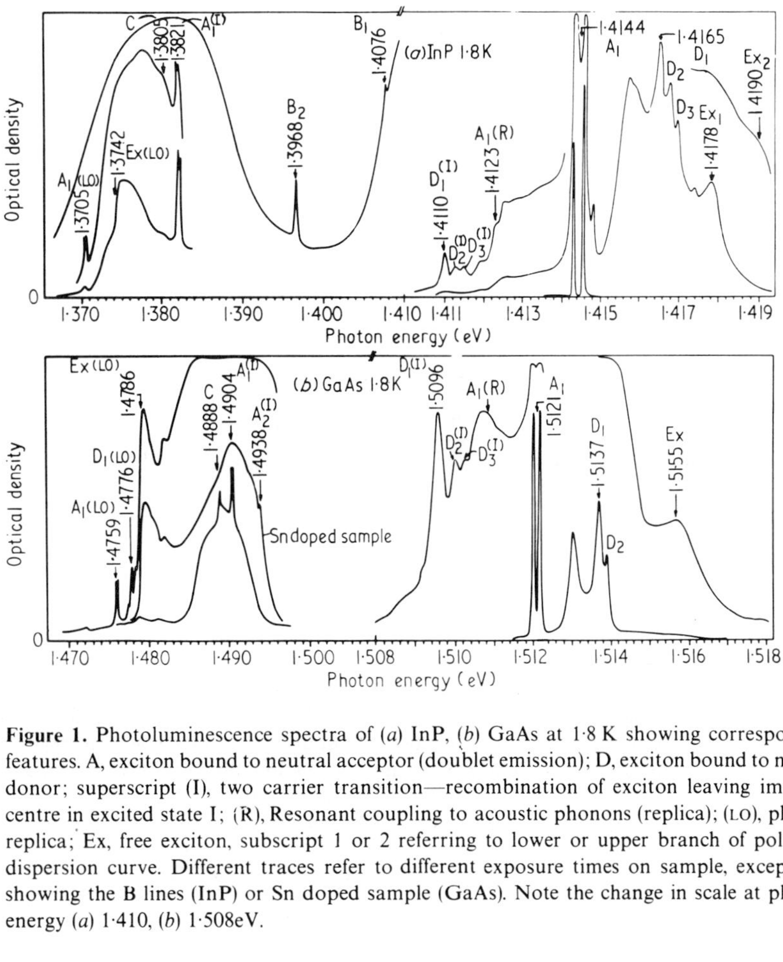

Figure 1. Photoluminescence spectra of (*a*) InP, (*b*) GaAs at 1·8 K showing corresponding features. A, exciton bound to neutral acceptor (doublet emission); D, exciton bound to neutral donor; superscript (I), two carrier transition—recombination of exciton leaving impurity centre in excited state I; (R), Resonant coupling to acoustic phonons (replica); (LO), phonon replica; Ex, free exciton, subscript 1 or 2 referring to lower or upper branch of polariton dispersion curve. Different traces refer to different exposure times on sample, except that showing the B lines (InP) or Sn doped sample (GaAs). Note the change in scale at phonon energy (*a*) 1·410, (*b*) 1·508eV.

follows our interest will generally be directed at the indium phosphide spectrum, adding where appropriate corresponding energies or remarks for gallium arsenide in parentheses.

3.1. The strong doublet emission A_1

In all our samples the doublet centred at 1·4144 eV with spacing ~0·24 meV is the strongest emission. Low temperature photoluminescence spectra of vapour grown gallium arsenide show a strikingly similar feature in the near gap emission at 1·5121 eV with a spacing of ~0·19 meV and it appears very likely that both have a similar generic origin. Sharp lines near 1·5121 eV in gallium arsenide have been attributed to the recombination of excitons bound to neutral acceptors. A doublet has been reported by Shah *et al* (1969) and also by Schairer (1971) and Dingle (R Dingle, private communication; see also Bergh and Dean 1972 figure 45), the latter also reporting a much weaker and broader line 0·5 meV higher in energy. Gilleo *et al* (1968)† observed a clear triplet, and Bimberg *et al* (1971) observed a doublet and a triplet from different samples. Rossi *et al* (1970) show a spectrum with four lines in this region. The total binding energy (see Appendix) of the electron and hole

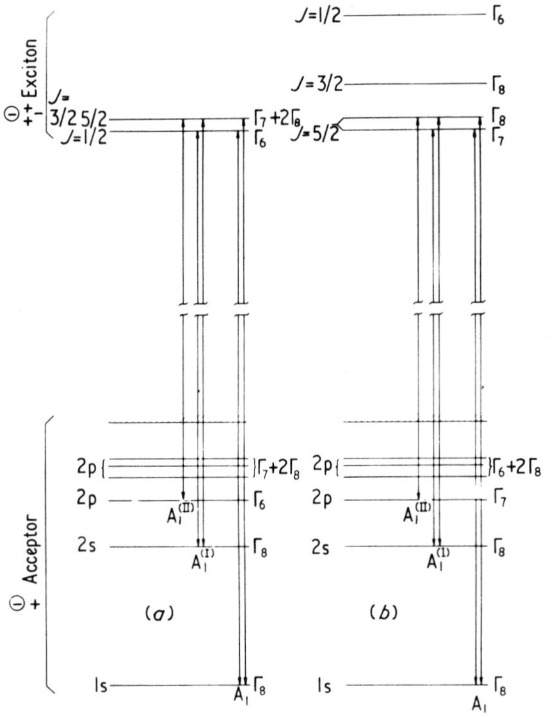

Figure 2. Alternative possible energy level structures for exciton bound to a neutral acceptor, showing corresponding energy levels for the terminal (acceptor) level. In (*a*) the doublet splitting is due to the hole–hole interaction, splitting due to electron–hole interaction or crystal field being unresolved. In (*b*) the doublet splitting is due to the crystal field interaction, other levels of the exciton being unobserved. In both cases the corresponding acceptor levels are shown, arranged to show how a single line two-hole replica could occur. Forbidden electric dipole transitions are $\Gamma_6 \rightarrow \Gamma_6$ and $\Gamma_7 \rightarrow \Gamma_7$.

† After etching the material yielded doublet emission, the change being presumed to be due to relief of strain (Hill 1971). This is also our experience in many cases.

in indium phosphide is 8·6 meV (6·7 meV). This corresponds to an exciton localization energy E_{BX} of 3·8 meV (2·9 meV), pointing to the binding of an exciton to a neutral acceptor rather than donor site†. That the components of the doublet originate from the same site is proved by magnetic field experiments (White *et al* 1972b, to be published) which show strong interaction between the Zeeman subcomponents from each line. This situation is not possible in the case of an exciton bound to a neutral donor, since each species of donor should provide just one independent emission line assuming the the electron is in the orbital ground state of the donor. Complete analysis of the Zeeman splittings has not yet proved possible, principally due to the closeness of the pair in zero field, an especially difficult problem for gallium arsenide.

The nature of the emission in gallium arsenide, in triplet or doublet form has been

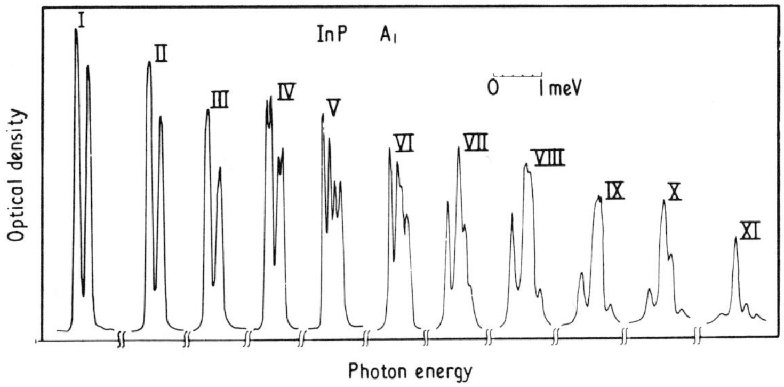

Figure 3. Variation of the appearance of the A_1 doublet emission in InP due to strain. Distance apart on the sample surface between successive spectra ~0·2 mm. The magnitude of the strain corresponding to spectrum V ~ 4×10^{-4}.

universally attributed to the effects of *j–j* coupling between the two holes and one electron attendant at the acceptor site. Two identical holes with $j = \frac{3}{2}$ can couple to form $J = 0$ or 2. The further (small) interaction with the electron yields $J = \frac{1}{2}, \frac{3}{2}$ and $\frac{5}{2}$, the latter two levels being closely spaced and allegedly unresolved when the doublet emission is observed. However we have not seen any evidence which excludes the possibility that the components on the doublet may be due, say, to the Γ_7 and Γ_8 levels which derive from the $J = \frac{5}{2}$ level of the exciton in the crystal field of T_d symmetry (see figure 2). It would also be a little surprising that the high energy component ($J = \frac{1}{2}$) with only twofold degeneracy should appear as intense as the low energy component ($J = \frac{3}{2}, \frac{5}{2}$) which has a total degeneracy of ten, although no definite assertion can be made on this point until explicit calculations are carried out using reliable wavefunctions. The terminal state of the transition is the ground state of the acceptor, which in T_d symmetry is a single Γ_8 level, supporting the assertion that the doublet splitting is in the exciton structure. This is further confirmed by the observation of thermalization of the doublet

† This is so because the observed value of E_{BX}/E_d ~ 0·50 (~0·48) (assuming we are dealing with donors) significantly exceeds the value, ~0·33, predicted for excitons bound to neutral donors when $m_e/m_h = 0$. For nonzero m_e/m_h still lower values of E_{BX}/E_d are predicted. Taking m_e/m_h ~ 0·16 we expect $0{\cdot}1 > E_{BX}/E_a > 0{\cdot}055$ in fair agreement with the observed ratio ~0·09 (~0·10). The predicted energy ratio is very sensitive to the mass ratio in this region. (See Hopfield 1964.)

components, when it is observed that for pumping powers below 5 mW the intensity of low energy component increases significantly, becoming ~2 times stronger than the upper component. In all, four exciton levels might be anticipated (figure 2), possibly more in the event of any built-in axial or lower symmetry strain. We now present evidence which shows clearly that triplet and quadruplet forms in indium phosphide spectra, like

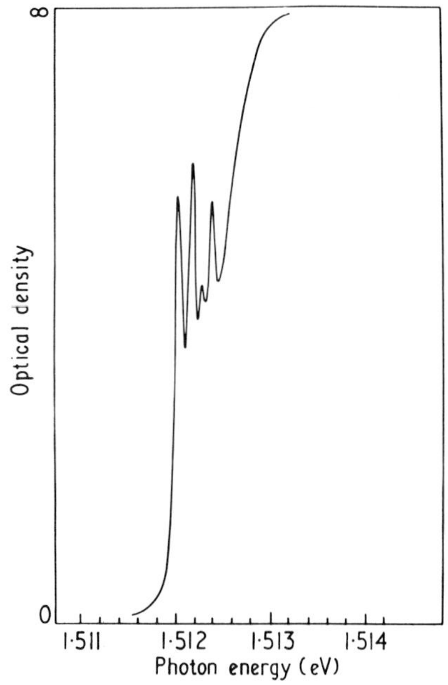

Figure 4. A_1 emission in Sn doped, liquid phase grown GaAs. The strongest pair of lines correspond exactly to A_1 emission in vapour grown samples and consitutes the only A_1 emission in *vapour* grown Sn doped GaAs.

those observed in gallium arsenide by Gilleo *et al* (1968), Bimberg *et al* (1971) and possibly Rossi *et al* (1970), can be generated from a doublet spectrum by the presence of internal strain. Figure 3 shows the variation of the form of the A_1 emission at intervals as the focused laser spot exciting the emission is traversed across the surface of a layer. This type of variation is frequently observed both in epitaxial indium phosphide and epitaxial gallium arsenide. It is normally smooth and continuous although not always so. The sharpness of the splitting found typically indicates that the internal strain is essentially uniform over the ~250 μm diameter region directly excited by the laser beam. Slight shifts in the centres of gravity of the strain split doublets also occur. The detailed form of the perturbed spectrum is not always exactly the same and clearly depends on the precise nature of the strain inadvertently introduced into the material, which in some cases may be spatially inhomogeneous in direction as well as in magnitude. We estimate the magnitude of the strain corresponding to type V spectra (figure 3) to be $\sim 4 \times 10^{-4}$ taking average deformation potentials for the ground state splitting of shallow acceptors as in GaAs (Bhargava and Nathan 1967).

The doublet form of the emission occurs more frequently than any other, is often independent of position over a wide region of the free crystal surface and has provided the narrowest emission lines. In our vapour grown gallium arsenide sharp doublet emission is almost always observed, although both triplet and quartet forms have been recognized. On the other hand our best quality liquid grown epitaxial layers of gallium arsenide (figure 4) all give type V (figure 3) emission which could be explained by a uniform and repeatable strain which does not affect to any observable extent the sharp emission at 1·5064 eV due to excitons tightly bound to deep neutral tin acceptors (White *et al* 1972d, Bimberg *et al* 1970). This latter line and its replicas are not shown in figure 1. Such disparate strain splittings for shallow and deep acceptors could be understood by a strong dynamic Jahn–Teller coupling to asymmetric vibrational modes for the tin acceptor (Morgan 1970). However, for other reasons discussed below, we believe that the quartet emission in our tin doped samples (figure 4) is comprised of two independent doublets corresponding to two different acceptors.

Allowing for these effects of strain we uncover a striking feature of the spectra of our undoped samples of indium phosphide, as well as our undoped vapour epitaxial gallium arsenide, namely that there is evidence for only *one* dominant shallow acceptor in refined undoped forms of both materials. On a straightforward comparison in terms of the intensities of observed bound exciton lines, which should be valid for acceptors with ionization energies lying within the reasonably narrow range claimed for many of the group II and group IV substituents in GaAs (Williams and Bebb 1972), we conclude that the concentration of any subsidiary shallow acceptors is at most $\sim 1\%$ of that of the principal acceptor. The weak line at 1·4148 eV is ascribed to a second acceptor but it is not clear whether it is a singlet or an overlapping doublet. A corresponding line is sometimes to be seen in vapour grown gallium arsenide at 1·5125 eV. Although we observed bound exciton luminescence only from neutral acceptors, this is not a limitation in diagnosis since an appreciable fraction of the ionized acceptors present in n type material become neutralized under the intense photo-excitation we employ. As far as can be judged from published data the same acceptors appear in the spectra of indium phosphide and especially gallium arsenide reported by other workers, with subsidiary acceptor concentrations usually at least an order of magnitude lower. There is appreciable general evidence that silicon is the impurity which controls the electrical properties of refined gallium arsenide (Di Lorenzo and Machala 1971), though preliminary evidence strongly implicates the zinc acceptor in our undoped vapour–epitaxial gallium arsenide (Dean *et al*, to be published). The likely dominant shallow acceptor in indium phosphide is somewhat less certain, although present evidence again favours zinc. We shall see below that the situation is different for donor related luminescence. Typical spectra from both gallium arsenide and indium phosphide show evidence of two dominant donors, often apparently present with comparable concentrations. Obvious possible candidates, on general grounds, are silicon and sulphur but no clear proof of this can be given yet for either semiconductor.

3.2. Replicas of the A_1 emission

Clearly allied to the A_1 doublet are the lines at 1·3705 eV (1·4759 eV) marked A_1 (LO) in figure 1 and at 1·3821 eV (1·4904 eV) marked $A_1^{(I)}$. The former is identified as the LO^{Γ} phonon replica of the A_1 doublet, the phonon energy obtained from the energy displacement being 43·9 meV (36·2 meV) close to the reported (Mooradian and Wright 1966) value of 43·8 meV (36·7 meV). Both the phonon replica and the $A_1^{(I)}$ emission are often resolved into doublet structure and reproduce faithfully the appearance of the $A_1^{(I)}$ emission under varying strain conditions. This provides strong evidence that the A_1

emission is also intimately related to the main doublet A_1. The energy displacement is 32·3 meV (21·7 meV), roughly three quarters of the hydrogenic acceptor binding energy. It seems very likely that this emission in indium phosphide, as in gallium arsenide (Schairer 1971) is due to a 'two hole' transition in which the exciton bound to the neutral acceptor recombines to leave the hole in an excited state of the acceptor, hence the superscript (I) in our notation. The similarity of the strain variation would indicate that either the whole of the strain splitting arises in the energy level structure of the exciton, or that the excited state of the acceptor splits exactly like the ground state. In the former event, since up to five lines are seen under strain conditions, corresponding to at least five levels, each of which must carry a Kramers degeneracy of at least two, the total degeneracy of the A_1 doublet must be at least tenfold, implying that the doublet emission comprises at least the $J = \frac{3}{2}$ and $\frac{5}{2}$ levels (or their crystal field derivatives $\Gamma_7 + 2\Gamma_8$), and that both exciton levels are each at least fourfold degenerate. However we believe that both the ground and first excited states of the acceptor are Γ_8 levels (figure 2) which split similarly under strain, so that we do not obtain any conclusive evidence regarding the exciton level structure. This contrasts with the situation in gallium phosphide and silicon (Manchon and Dean 1970 and references therein) where the lowest lying acceptor excited state is shown to be Γ_7. Transitions to low-lying Γ_7 acceptor levels should appear in the 'two-hole' spectra if such levels existed in gallium arsenide or indium phosphide.

The sharp single line at 1·4888 eV in gallium arsenide frequently occurs at the same time as $A_1^{(I)}$ (figure 1*b*) but has been observed in spectra where $A_1^{(I)}$ is absent yet with A_1 still sharp. The absence of $A_1^{(I)}$ implies an impurity concentration sufficiently high to degrade the purity of the wider $n = 2$ orbit appropriate to the $A_1^{(I)}$ spectrum (the maximum impurity concentration allowed before perturbation of the nth Bohr orbit becomes serious varies as n^{-6}) and further implies that the single sharp line must represent a relatively closely confined system. It is possibly due to an exciton bound to a relatively deep centre, in many ways similar to the exciton bound to the tin acceptor in gallium arsenide. Without clear evidence we prefer merely to label the line as C in figure 1*b*. The corresponding feature in indium phosphide is weak and broad at 1·3805 eV; apparently conditions are rarely favourable for its distinct observation. Returning to gallium arsenide, Schairer has reported the doublet which we observe at 1·4938 eV at the same time as the $A_1^{(I)}$ emission, and accordingly identifies it as being due to a *shallower* excited state of the A_1 exciton. In contrast we never observe it and the $A_1^{(I)}$ emission in the same sample, and indeed find it only in our tin doped liquid epitaxy grown gallium arsenide samples which show more additional structure at A_1 (figure 4). We accordingly label this line as $A_2^{(I)}$. It is unexpected that $A_2^{(I)}$ would broaden more quickly than $A_1^{(I)}$ under the influence of inter-impurity or other interactions if both are associated with a single acceptor (1 = 2) since the relevant acceptor state (then responsible for $A_1^{(I)}$) would be appreciably *more* localized, while the initial (exciton) state would be the same. Such differences in excited states as $A_2^{(I)}$–$A_1^{(I)}$ in similar circumstances have been found in indium phosphide (White *et al* 1972c, to be published). The fact that the only two hole replica observed in our tin doped gallium arsenide samples is a doublet leads us to believe that the quartet observed near the A position (figure 4) is composed of two independent doublets, the lower energy one A_2 at the same energy as A_1 in other samples, and in fact providing $A_2^{(I)}$. The higher energy doublet is slightly weaker and narrower and is presumably associated with a yet unobserved two hole replica. Weak single lines at 1·3968 eV and 1·4076 eV infrequently observed in indium phosphide which gives sharp A_1 but not $A_1^{(I)}$ emission, could possibly be assigned to the same type of origin, that is the lower energy line might be the two hole replica of the other (zero phonon) line associated with a different centre. In view of the meagre data on these lines and the small energy separation between them, 10·8 meV, we have merely labelled these lines B_1 and B_2 in figure 1*a*.

A striking feature is the great breadth of the two hole replicas compared with the fundamental doublet A_1. This must reflect the greater radius of the hole state in excited orbits of the acceptor, and is the exact opposite of the trend expected if these relatively lowlying luminescence lines were no-phonon lines of independent bound exciton luminescence. For still more weakly bound acceptor terminal states the broadening must be even greater and may account for some of the faint structure perceptible in the broad band at 1·38 eV (1·49 eV).

In both indium phosphide and gallium arsenide a relatively broad line occurs within these broad bands which shifts rapidly to higher energy in increasing magnetic fields. Rossi *et al* (1970a) have shown that such emission bands in gallium arsenide can be due to donor–acceptor pair transitions or 'free-to-bound' transitions involving the capture of a free electron by a neutral acceptor. The near linear variation in a magnetic field favours the latter identification. In zero field the line is too diffuse to observe but is close to the position of line C. Recombinations at distant donor–acceptor pairs also producing broad luminescence bands (see also Heim 1969, Dingle 1969) apparently contribute only a small proportion of the total luminescence from many of our undoped crystals under the intense photoexcitation at 1·8 K.

The weak doublet at 1·4123 eV (not clearly resolved in gallium arsenide at 1·5107 eV) labelled A_1(R), also reproduced faithfully the strain variation of the main doublet. Its displacement is 2·1 meV (1·4 meV), much too small for its origin to be due to a characteristic phonon of the lattice, since $\hbar\omega_{TA} > 7$ meV at the density of states maxima. We believe it very likely that the doublet is due to a 'resonance' in the coupling to an acoustic phonon arising when the lattice vibration wavelength is of the same order in size as the exciton complex. General coupling to acoustic phonons is then responsible for the broad plateau extending to the A_1(R) doublet.

3.3. Donor related peaks

Of a totally different nature is the emission appearing near 1·4165 eV in indium phosphide (1·5137 eV in gallium arsenide). Although this commonly appears as a doublet, its relation in intensity to the acceptor doublet is arbitrary. Indeed on occasion it appears as a singlet with at least one weaker line lying a little to higher energy. It is in the energy range expected for excitons bound to neutral donors†. It is assumed at this stage that each component is due to a different donor as mentioned above, and the lines are labelled accordingly D_1, D_2 etc in figure 1. We discount the possibility that the higher energy components may represent excited states of the exciton bound to the donor because of the apparent randomness in intensity ratios together with the failure to observe any thermalization effects and also the observation that in a magnetic field each line appears to split similarly (White *et al* 1972a, to be published), with no interactions, indicating a similar, but independent, origin for each one. If this interpretation is correct we must expect the binding energies of the individual donors to differ by ~1 meV†, a surprisingly large amount for these narrow, direct gap semiconductors. The conclusion that the lines are related to the presence of donors is supported by the observation of a replica spectrum at energies lower by 5·62 meV ($D_1^{(1)}$, $D_2^{(1)}$). As in gallium arsenide (where the corresponding shift is 4·42 meV (Rossi *et al* 1970b)) these lines are attributed to 'two electron' transitions, that is,

† For donors, again taking $m_e/m_h \sim 0{\cdot}16$ we expect $E_{BX}/E_d \sim 0{\cdot}2$ compared with the observed ratio ~0·22 (~0·22) for D_1. See footnote to page 1730.

to the decay of excitons bound to neutral donors which leave the donor in its first orbital excited state. The displacement in strain-free indium phosphide is close to 5·74 meV as reported by Chamberlain *et al* (1971) from far infrared photoconductivity experiments. In increasing magnetic fields the two-electron lines are observed to split rapidly (White *et al* 1972a, to be published). At 25 kG we observe the shift of the lowest energy component to be 2·92 ± 0·1 meV compared with the value of 2·94 meV reported by Chamberlain *et al* (1971). This component shows some structure which appears to reproduce the splitting of the parent line, in contrast to the situation in gallium arsenide where structure is not seen. Broader, weaker lines moving to even lower energies are associated with higher excited states of the donor centre in the final state of the luminescence transition.

The broad emission peak at 1·415 eV (1·5130 eV) never exhibits any fine structure. For this reason we do not propose to discuss it further save to note that it is believed to be associated with the recombination of free holes with neutral donors. No well defined bound exciton state has been seen as an intermediate step in such recombinations, although excitons may bind weakly to *ionized* donors in these semiconductors, where $m_e/m_h \ll 1$.

3.4. *Free exciton emission*

The peak labelled Ex_1 at 1·4178 eV is identified as the free exciton peak possibly to correspond with the peak at 1·5155 eV in gallium arsenide, commonly identified as the free exciton, but recently more specifically identified as emission from the upper polariton branch of the exciton dispersion curve (Sell *et al* 1971). We believe that the poorly resolved feature labelled Ex_2 at 1·4190 eV in figure 1*a* should strictly correspond to the main peak in gallium arsenide (see Appendix). Ex_1 should then correspond to the lower polariton branch emission at 1·5149 eV in gallium arsenide (Sell *et al* 1971), apparently unobserved in our samples due to excessive donor emission.

The relatively sharp cut-off seen at 1·3742 eV (1·4896 eV) is identified with phonon assisted recombination of 'cold' free excitons. To be consistent, one should measure the LO phonon energy from the low energy cut-off of the free exciton peak, but owing to the interference from the donor spectrum and to the complications of the exciton spectrum (see Appendix) this feature is not clearly defined. However the free exciton gap can be estimated by reflectivity measurements which then yield a LO phonon shift of 44·0 meV (36·4 meV) (Appendix) which can be compared with the shift A_1–A_1(LO) of 43·9 meV (36·2 meV). At least some of the emission intensity in the 1·38 eV (1·49 eV) band must be due to phonon assisted recombination of warmer excitons which utilize phonons other than at the zone centre in order to conserve both energy and momentum. The free exciton cut-off at Ex(LO) sharpens considerably in a magnetic field of over 10 kG. It is likely that the weaker feature showing a cut-off near 1·4810 eV in the GaAs spectrum represents Ex(TO) since $\hbar\omega(\mathrm{LO}) - \hbar\omega(\mathrm{TO}) \sim 2$ meV (Mooradian and Wright 1966).

4. Conclusions

Photoluminescence spectra have been presented which show clearly the one to one correspondence between indium phosphide and gallium arsenide as regards both free and shallow bound exciton transitions. The analysis has been aided by the fact that excitons bound to neutral acceptors dominate the spectrum in both compounds, and the effects of inadvertently introduced strain help to identify related lines. On the other hand, built-in strain can give a very misleading spectrum and one must be careful to distinguish between strained and unstrained regions of the specimen either from observation of the spatial variation or by magnetic experiments which can for example distinguish between a quartet formed by strain and a quartet formed by two independent doublets. It is indeed fortunate that in indium phosphide at least, only one acceptor appears to be introduced to any great extent during undoped growth. Its effect in the spectrum serves as a landmark which helps to avoid errors in identification which can arise because of shifts in energy sometimes caused by uniform hydrostatic stress. Identification of this inadvertently present acceptor should be possible in the near future using the photoluminescence techniques we have described, and will give valuable assistance in the chemical control of indium phosphide (and gallium arsenide) at low impurity levels, $<10^{15}$ cm^{-3}, which are extremely difficult to assess by other means.

Acknowledgments

The authors thank Miss B Day, Mr I Hinchliffe and Mr P Porteous for their technical assistance. They also thank Dr P D Greene of the Standard Telecommunications Laboratories Ltd, Harlow, for supplying samples of tin doped, liquid phase epitaxially grown gallium arsenide. This paper is communicated with the permission of the Director, Royal Radar Establishment, Malvern, Worcestershire.

Appendix. The energy gaps of indium phosphide and gallium arsenide

The free exciton energy gap of gallium arsenide is readily observed by means of reflectivity measurements. Figure A.1*b* shows measurements taken from one of our samples. The high energy peak is identified as being the free exciton gap 1·5150 eV in close agreement with 1·5151 eV obtained by Sell *et al* (1971). They also obtained structure in the luminescence spectrum, according to which the peak labelled Ex in figure 1 is identified as the upper polariton branch of the exciton. We do not observe the lower branch. The phonon assisted exciton cut-off Ex(LO) is shifted 36·4 meV from the free exciton gap. If we take the free exciton binding energy to be 3·8 meV (Bimberg and Schairer 1972 and references therein) we find the band gap of gallium arsenide to be 1·5188 eV at 1·8 K.

The position in indium phosphide is a little more confused in that the reflectivity spectrum in our samples does not show the distinctive high energy peak (figure A.1*a*) and also the principal high energy luminescence peak lies *below* the main reflectivity dip, although a weak, rather poorly resolved luminescence peak is sometimes observed at a high energy (~1·4190 eV) (figure 1*a*). At quite low magnetic fields (~2 kG) this feature sharpens considerably revealing a low energy cut-off characteristic of free exciton emission. For consistency this emission should be identified with free exciton decay, corresponding to that seen in gallium arsenide, and the free exciton gap should lie near the high energy shoulder of the sharp dip in the reflectivity spectrum, near 1·4182 eV (figure A.1*a*). This result differs significantly from the free exciton gap E_{gx} obtained from the absorption of very thin crystals by Turner *et al* (1964) presumably because of the effects of strain on the absorption spectra of very thin crystals. The phonon assisted free exciton cut-off, more diffuse in indium phosphide, is shifted by 44·0 meV. We expect the free exciton binding energy to be in the region of 4·8 meV, exceeding that found for gallium arsenide in the ratio of the donor binding energies in the two materials. This is

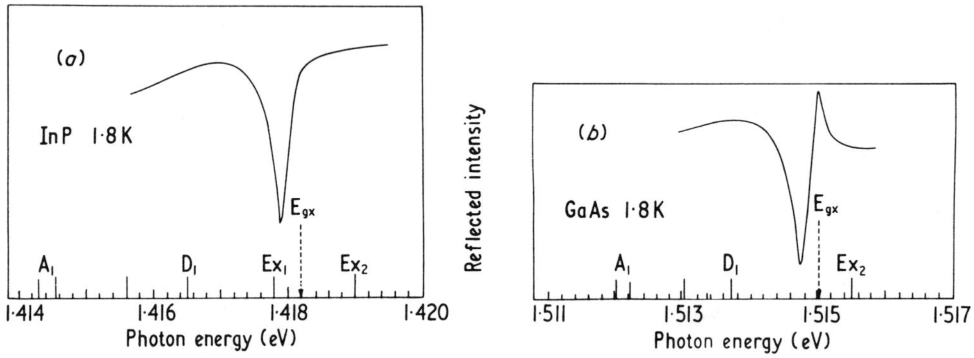

Figure A.1. Reflectivity spectra of InP and GaAs at 1·8 K. Positions of the main luminescence features are marked on the energy axis. Weak photoluminescence of the D_1 and A_1 bound excitons, also observed in these spectra, have been disregarded.

significantly greater than the earlier reported value of 4·0 meV (Turner *et al* 1966). The larger binding energy and large E_{gx} would point to an energy gap for indium phosphide of 1·4230 eV at 1·8 K.

References

Bergh A A and Dean P J 1972 *Proc. IEEE* **60** 156–224

Bhargava R N and Nathan M I 1967 *Phys. Rev.* **161** 695–8

Bimberg D and Schairer W 1972 *Phys. Rev. Lett.* **28** 442–5

Bimberg D, Schairer W, Sondergeld M and Yep T O 1970 *J. Luminesc.* **3** 175–84

Bimberg D, Sondergeld M and Grobe E 1971 *Phys. Rev.* B **4** 3451–5

Blom G M and Woodall J M 1970 *Appl. Phys. Lett.* **17** 373–6

Chamberlain J M, Ergun H B, Gehring K A and Stradling R A 1971 *Solid St. Commun.* **9** 1563–6

Colliver D J, Gray K W and Joyce B D 1972 *Electron. Lett.* **8** 11–2

Dean P J, White A M, Williams E W and Astles M G 1971 *Solid St. Commun.* **9** 1555–8

DiLorenzo J V and Machala A E 1971 *J. Electrochem. Soc.* **118** 1516–7

Dingle R 1969 *Phys. Rev.* **184** 788–96

Gilleo M A, Bailey P T and Hill D E 1968 *Phys. Rev.* **174** 898–905

Gobel E, Queisser H J and Pilkuhn M H 1971 *Solid St. Commun.* **9** 429–31

Heim U 1969 *Solid St. Commun.* **7** 445–7

Hilsum C and Rees H D 1970 *Electron. Lett.* **6** 277–8

Hill D E 1970 *Phys. Rev.* B **4** 1863–4

Hopfield J J 1964 *Proc. Int. Conf. Phys. Semicond., Paris 1964* (Paris: Dunod) 725–35

Leite R C C 1967 *Phys. Rev.* **157** 672–6

Manchon D D Jr and Dean P J 1970 *Proc. Int. Conf. Phys. Semicond., Cambridge 1970* (Oak Ridge, Tenn: USAEC) pp 760–6

Mooradian A and Wright G B 1966 *Solid St. Commun.* **4** 431–4

Morgan T N 1970 *Phys. Rev. Lett.* **24** 887–90

—— 1968 *J. Luminesc.* **1** 420–34

Roder O, Heim U and Pilkuhn M H 1970 *J. Phys. Chem. Soc.* **31** 2625–34

Rossi J A, Wolfe C M and Dimmock J O 1970a *Phys. Rev. Lett.* **25** 1614–7

Rossi J A, Wolfe C M, Stillman G E and Dimmock J O 1970b *Solid St. Commun.* **8** 2021–4

Schairer W and Yep T O 1971 *Solid St. Commun.* **9** 421–4

Sell D D, Dingle R, Stokowski S E and DiLorenzo J V 1971 *Phys. Rev. Lett.* **27** 1644–7

Shah J, Leite R C C and Nahory R E 1969 *Phys. Rev.* **184** 811–20

Turner W J, Reese W E and Pettit G D 1964 *Phys. Rev.* **136** 1467–70

White A M, Dean P J, Day B, Clarke R C and Taylor L L 1972a to be published

White A M, Dean P J, Day B, Taylor L L and Clarke R C 1972b to be published

White A M, Dean P J, Fairhurst K M, Bardsley W and Day B 1972c to be published

White A M, Hinchliffe I, Dean P J and Greene P D, 1972d *Solid St. Commun.* **10** 497–500

Williams E W and Bebb H B 1972 *Semiconductors and Semimetals* ed R K Willardson and A C Beer (London: Academic Press) **8** 321–92

key paper

Surface Studies By Means Of α Particles Of High Energy (2-4 MeV)

D.V. Morgan and D.R. Wood (Communicated by D. Tabor)
(Department of Electrical and Electronic Engineering, University of Leeds, Leeds, UK)

Proceedings of the Royal Society, Volume A335 (1973)
Pages 509-523

In this paper we describe how the channelling phenomenon may be used for surface studies. The origin of the 'surface peak' of an alined energy spectrum is discussed, and the procedure for estimating the total concentration of unscreened surface atoms contributing to this peak is described. The interpretation of this area and the impurity peaks in terms of quantitative measures of surface stoichiometry, surface 'disordered-layer' thicknesses and absolute impurity concentration, are described.

In order to illustrate the applications and limitations of the technique, we describe the application of the technique to three compound semiconductors, InP, GaAs and GaP.

1. Introduction

During the last 10 years we have seen the advent of a wide variety of new techniques developed for the study of solid surfaces (see, for example, Chang 1971; Reuter 1971; Socha 1971); methods such as scanning electron microscopy (s.e.m.), electron diffraction (l.e.e.d. and r.h.e.e.d.), Auger spectroscopy and ion microanalysis, to name but a few. Each of these techniques has its own special attributes enabling one or more surface property to be measured. S.e.m. is a direct imaging technique which shows surface topography, whereas l.e.e.d. and r.h.e.e.d. are diffraction techniques which show the arrangement of surface atoms. Auger spectroscopy, on the other hand, detects qualitatively the presence of foreign atoms at the surface, while ion microanalysis yields similar information, but with the added facility of being able to erode away the surface and hence record the distribution in depth of the various constituent atoms. Hence we see that to enable a complete study of a surface to be conducted it is advantageous to utilize a number of different techniques, enabling the limitations of any one technique to be overcome by another. In this paper we describe how MeV α particles may be used in conjunction with the Rutherford scattering-channelling phenomenon for surface studies, and discuss the application and limitations of the technique.

2. Rutherford scattering and channelling techniques

Rutherford scattering of energetic ions from solid surfaces has been widely used as an analytical tool for lattice site location studies (Mayer, Eriksson & Davies 1970), surface studies (Morgan & Bøgh 1972; Ball, Buck, Macnair & Wheatley 1972; Rubin 1959), thin film analysis (Mitchell, Kamoshida & Mayer 1971*a, b, c*; Morgan & Gittins 1972), and for determining the concentration profile of impurities in solids (Zeigler & Baglin 1971). We now consider how, in conjunction with the

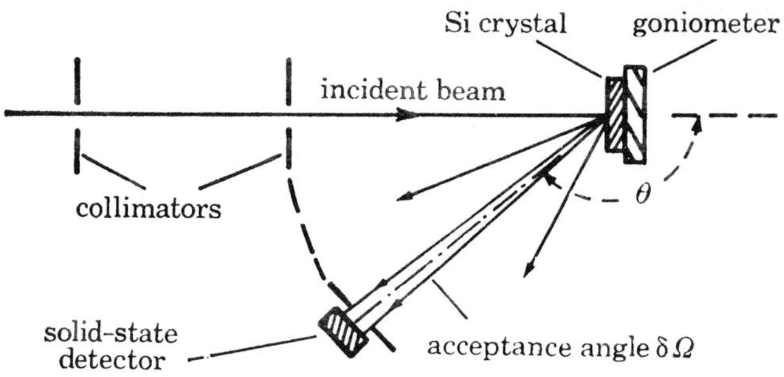

FIGURE 1. A schematic illustration of the scattering geometry.

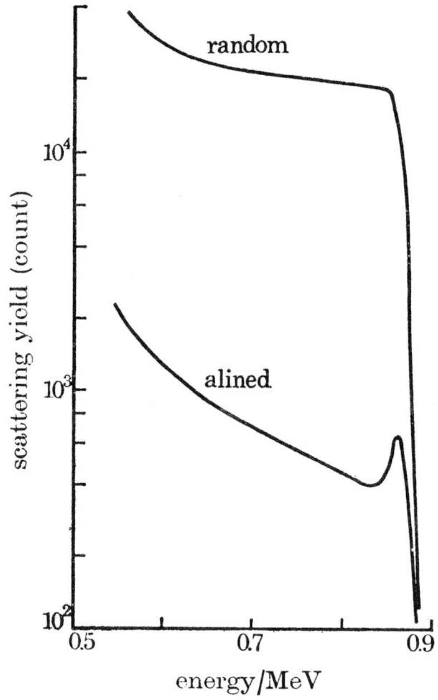

FIGURE 2. A random and alined spectrum from a silicon crystal polished in a mixture of $3HF : 5HNO_3 : 3CH_3COOH$. The incident beam was 1.5 MeV α particles and the scattering angle θ, 150°.

channelling phenomenon, Rutherford scattering can be applied to the study of the surfaces of binary solids, and show how the technique may be used as a quantitative analytical tool for the study of the stoichiometry†, impurity content and perfection of the crystal surface. The technique can be used to study both clean surfaces and surfaces prepared in the atmosphere ('real' surfaces). The work described in this paper falls into the latter category.

The principle of the technique is illustrated in figures 1 and 2. Figure 1 illustrates the experimental arrangement, and figure 2 the Rutherford scattering yield of

† The term 'stoichiometry' is used here in a loose sense since it refers to the ratio of the two constituent atoms, a and b, in the surface region of the solid.

1.5 MeV α particles scattered from a silicon surface. The two spectra, figure 2, correspond to the incident beam alined to a crystallographic axis (channelled) and with the beam incident in a random direction. In the former case, a drastic reduction in the scattering yield occurs behind the surface layer due to the screening of the atoms – a characteristic of the channelling phenomenon. The ratio of the alined yield to random yield just behind the surface peak (χ_{min}) gives a measure of the perfection of the crystal surface over a distance comparable with the depth resolution of the detector system (i.e. $\simeq$ first 5–30 nm).† The theoretical values of χ_{min} obtained from the continuum model (Lindhard 1965; Picraux 1969) are an underestimate, principally because they ignore the influence of surface contaminant films (i.e. oxides and carbon deposits). The oxide films on freshly prepared surfaces are typically *ca.* 1.5 to 3 nm. Added to this is a hydrocarbon deposit which results from the adsorbed oil molecules from the diffusion pump 'cracking' on the surface under the influence of the probing beam. This is a troublesome problem, which can be reduced by surrounding the target with a cold shield to trap the oil deposits. With no shielding, a high quality ⟨110⟩ GaAs crystal, for example, will have at best a minimum yield *ca.* 0.04 to 0.05. The same crystal with a liquid nitrogen temperature cold shield will have $\chi_{min} \simeq 0.035$ to 0.04, while, with a cryoshield cooled to 25 K, a value of less than 0.02 is typical (Bøttiger, Davies, Whitton & Lori 1972). The difference between the last value and the theoretical value of ~ 0.01 is readily accounted for by the surface oxide. We thus see that the minimum yield is a sensitive function of the environment of the crystal before and during analysis and hence should only be used as a relative estimate of the crystal perfection.

Consider now the Rutherford scattering from the crystal surface. Particles scattered from the surface region will have an energy

$$E_1(0) = k^2 E_0, \tag{1}$$

where E_0 is the incident ion energy and k^2 the kinematic factor, defined as

$$k^2 = \left[\frac{M_1 \cos\theta + (M_2^2 - M_1^2 \sin^2\theta)^{\frac{1}{2}}}{M_1 + M_2}\right]^2, \tag{2}$$

where M_1 is the mass of the incident ion and M_2 the scattering atoms and θ the scattering angle in the laboratory system. We may draw two conclusions from these equations. First, each surface impurity has its own characteristic value of k^2 uniquely defined by the mass M_2 and hence will yield a unique value of E_1. Secondly, for normal incidence, the energy scale of the host spectrum may be converted into a depth scale, using the relation (Bøgh 1973)

$$E_2(z) \simeq k^2 E_0 - z[S(E_0)\, k^2 + S(E_1)/\cos\theta'], \tag{3}$$

where $S(E_0)$ and $S(E_1)$ are the stopping powers of the ion at the energies E_0 and E_1 respectively and $\theta' = (180 - \theta)$. Hence the energy scale in figure 2 can be converted into both an impurity mass scale and a depth scale.

† 1 nm = 10 Å.

The scattering yield, Y_x, from a particular atomic species, x, with a density N_x, is given by†

$$Y_x = [(Z_1 e^2/4E_0)\,\mathrm{cosec}^2\,\tfrac{1}{2}\theta_c]^2\, N_x Z_x^2\, \delta\Omega, \tag{4}$$

where Z_1 and Z_x are the atomic numbers of the incident ions and the scattering atoms respectively; $\delta\Omega$ the solid angle of the detector, and θ_c the scattering angle in the centre of mass coordinates. The identity of surface impurities may be determined by using equations (1) and (2), and then equation (4) may be used to determine the concentrations of the various impurities present at the solid surface. The necessity of calculating absolute yields may be avoided by comparing the impurity yield with that from the non-alined host spectrum. Hence the surface impurity concentration N_x is given by

$$N_x = N_a \frac{A_x}{Y_a}\left(\frac{Z_a}{Z_x}\right)^2, \tag{5}$$

where A_x is the area under the impurity peak, Y_a the non-alined yield and N_a is the number of host atoms per monolayer per cm^2. There is some uncertainty in interpreting the impurity peak in terms of a surface concentration. For thin films ($\lesssim 25$ nm) the shape of the impurity peak is determined by the detector resolution and hence yields no information about the thickness of the region containing the surface impurities. Thus, it is possible that the impurities are distributed throughout the surface oxide film. We measure the concentration of impurity atoms per unit area of the surface. It is simply a matter of convenience that we choose to normalize our results by comparing this concentration with the number of atoms per unit area, in one monolayer of host crystal and hence quote N_x in terms of a fraction of a monolayer.

In the case of a binary compound, equation (4) may be used to calculate the ratio of the two constituents, a and b, at the surface, thus

$$\frac{N_a}{N_b} = \frac{A_a}{A_b}\left(\frac{Z_b}{Z_a}\right)^2, \tag{6}$$

where A_a and A_b are the respective areas under the surface peaks, calculated after a suitable background correction. (This is discussed in detail below.)

It will be seen from the above discussion that in the analysis of the energy spectra, we are confronted with the problem of estimating the area under the appropriate peak of the host yield and the impurity yield. For heavy surface impurities, the problem is trivial, unless two or more peaks overlap. However, from equation (4) we see that the Rutherford scattering cross-section decreases as the square of the atomic number of the impurity atoms, hence the sensitivity of the technique decreases rapidly for light impurities. This problem is accentuated by the presence of the host background spectrum at low energies, which increases rapidly with

† This equation ignores the recoil of the target atom. When the mass of the incident ions $M_1 \simeq M_2$ (the mass of the target atoms) a correction is needed to equation (4). When α particles are used for the probing beam then this correction is less than 2% for target atoms heavier than oxygen.

decreasing energy (figure 1*b*). Surface impurities lighter than the host atoms will have their peaks superimposed upon this background level, which introduces a further uncertainty in the measured area under the peak. Clearly, the combination of the reduced cross-section and increasing background makes the technique insensitive for measuring small concentrations of light impurities such as carbon or oxygen on crystals such as Ge or GaAs. In special cases, alternative techniques may be used to overcome this difficulty. G. Dearnaley (private communication) for example, he has used the broad resonance in Rutherford scattering of protons on oxygen (Laubenstein, Laubenstein, Koester & Mobley 1951), centred around 4.0 MeV, to selectively obtain more than an order of magnitude increase in the yield from oxygen and hence to study oxide films on metal surfaces. Even with a substrate as light as silicon, without this enhanced yield the oxygen peak is barely visible above the background. Another technique based on the excitation of nuclear reactions can also be used to look for specific impurities. Cachard, Roger, Pivot & Dupuy (1971), for example, have used the reactions (^{16}O (d, p) ^{17}O and ^{28}Si (d, p) ^{29}Si) to study the stoichiometry of silicon oxide films, whereas J. Whitton & I.V. Mitchell (private communication) have used the oxygen reaction to look at oxide films formed on GaAs crystals. Although closely related to Rutherford scattering, nuclear reaction techniques will not be discussed further in this part of the paper.

In principle, the relation (3) can be used directly to study the solid surface. In practice, however, to obtain good enough depth resolution to enable the surface scattering to be isolated from the bulk and studied, demands an energy resolution from the detector beyond that normally obtainable in practical systems. For example, with a silicon surface barrier detector, a system resolution of 15 to 20 keV (for MeV α particles) is typical and this yields a depth resolution of around 50 atomic layers. Using a magnetic spectrograph as a particle detector improves the depth resolution to around 5 to 10 atomic layers. (This is discussed in detail by Bøgh (1968).) There is an alternative way of isolating the surface scattering which does not rely on improving the detector resolution; that is to use the so-called 'surface peak' in the channelled spectrum (figure 2). Now the surface peak is a direct result of the Rutherford scattering of the ion beam from the *unscreened* host atoms in the surface region of the solid. The concentration N of these unscreened atoms contributing to the surface peak may be divided into two groups: (i) the host atoms which are a constituent part of the surface oxide film; and (ii) host atoms in the first few monolayers of the perfect crystal. Clearly, in the case of an oxide free surface prepared under ultra-high vacuum conditions the only contribution will be from atoms in group (ii) and the interpretation would be unambiguous. However, in order to be able to interpret meaningfully the spectra of chemically prepared surfaces, it is important to be able to estimate the relative contribution from each group. Bøgh (1973 and unpublished work) has recently considered the problem and calculated the screening of the host atoms as a function of depth in the crystal. The surface atoms (first monolayer figure 3*a*) are unscreened and hence the yield from these atoms will be the same as the non-alined yield. In a crystal, devoid of thermal vibrations, all subsequent atoms are screened from the channelled ions. In a real crystal, because of thermal vibrations, the channelling effect takes some distance to build up and hence the second, third, ..., *n*th atoms in a string will not be fully screened, and consequently will make a small but finite contribution to the surface

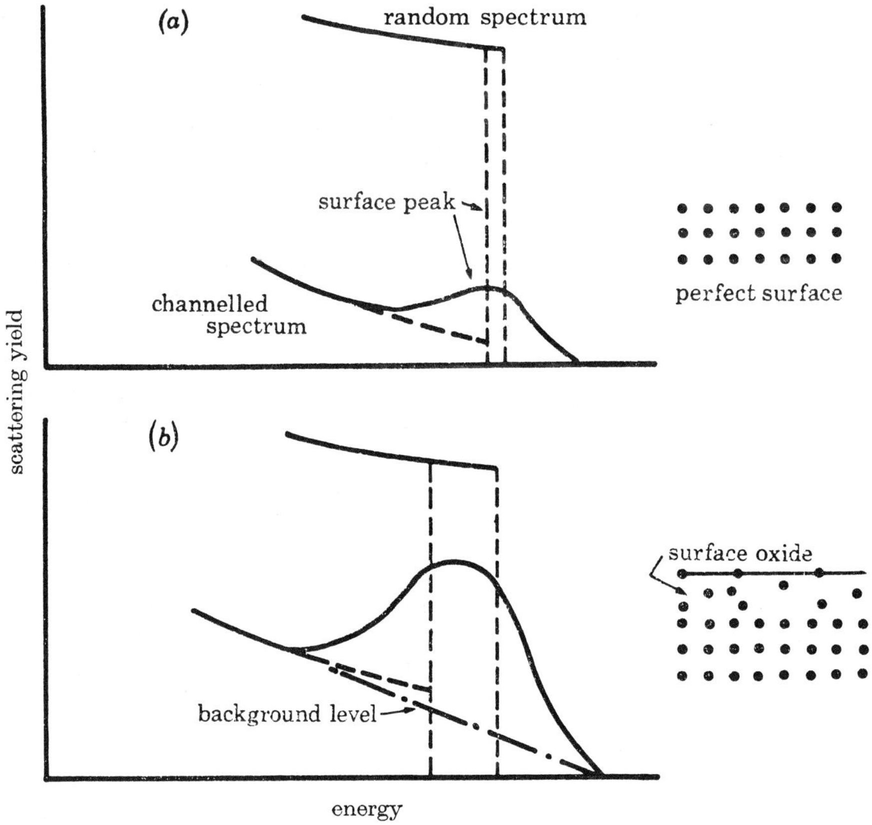

FIGURE 3. A schematic illustration of the scattering yield (logarithmic scale) from (*a*) a clean surface and (*b*) a surface covered with a thin surface oxide. The broken line corresponds to the alined spectrum that would be obtained if the particle detector had infinitely good resolution; its height would be equal to the random yield and its thickness would correspond to the thickness of unscreened surface atoms. The full line in the channelled spectrum shows the same peak distorted by the finite detector resolution. The surface oxide increases both the area under the surface peak and the minimum yield just behind the surface peak.

peak. The total *effective number* of atoms contributing to the surface peak is therefore a function of the crystal and its temperature (thermal vibrational amplitude). Bøgh (1973) has calculated a theoretical effective thickness of unscreened atoms for both silicon and gold to be *ca.* 1.3 monolayers, while the experimental values are *ca.* 1.8 and 4 monolayers respectively. The differences between these values can be accounted for by the uncertainty in the value of vibrational amplitude to be used for the surface atoms and the possible introduction of defects by the surface preparation. The large increase for the gold value arises because the surface examined was the anomalous (100) plane. In general, the yield from the second and succeeding atoms in a string is very small and similar (in theory) for two dissimilar materials. Now, since the materials discussed in this paper, GaAs, GaP, InP, have the same crystal structure as Si and the problem of an anomalous surface structure does not arise, we propose that the value for silicon, *ca.* 1.8 monolayers is a reasonable estimate of the effective thickness for these materials.

In elemental semiconductors, such as silicon, it is convenient to normalize the concentration of unscreened atoms N, by comparing it with the number of atoms per unit area in one monolayer of host crystal. Hence, for fixed geometry, the area under the surface peak A may be used directly to estimate a 'thickness', t, of the disordered surface layer from the equation

$$t \simeq \frac{A}{Y_a[k^2S(E_0)+S(E_1)/\cos\theta']}. \tag{7}$$

With chemically prepared surfaces the values of t obtained are of the order of 10 monolayers. If we now assume that the contribution of the underlying perfect crystal is *ca.* 2 monolayers then 80% of the atoms contributing to A reside in the oxide film. Now t must be interpreted as the thickness of crystal subtrate we would obtain if these 'host' atoms were condensed into an epitaxial film on the surface. If we subtract the contribution of the crystal proper (*ca.* 2 monolayers) then the true thickness of oxide film can in principle be determined, provided the concentration of surface oxygen atoms and the density of the oxide films (which depends on the stoichiometry) are known. In binary compounds this procedure is only acceptable if $N_a = N_b$. Clearly, if for example the concentrations of gallium and arsenide in the surface peaks are not equal, then it is not meaningful to talk of the thickness of an effective epitaxial layer. In these circumstances the individual concentrations, N_a and N_b, must be given directly in atoms per unit area.

The problem of determining the area under the surface peak A, has difficulties due to the uncertainty in the position of the background level. The finite resolution of the detector has two effects on the spectrum: (i) it spreads the surface peak over the background from the host (this background yield χ_{min} is itself a strong function of the thickness of the surface film), (ii) it also rounds off the leading edge of the background spectrum. The net result is illustrated in figure 4. The dot–dash line shows the true position of the background level, and the broken line a straight line approximation to this curve. The procedure for estimating a straight line approximation is subjective and is responsible for the relatively large error ascribed to the resulting ratio N_a/N_b. In principle, a better estimate of the true curve is obtained by generating the background line shape from the random spectrum. In this instance, the random spectrum is scaled by an appropriate factor, such that, at the position of minimum yield, the yields from the alined and the random spectra are equal. Now, the edge of the random spectrum has the same shape as the background curve because both are step functions modified by the detector resolution. Thus, the leading edge of the scaled random spectrum gives directly the approximate background curve illustrated in the energy spectra of figures 5, 7 and 8. However, a dechannelling component present on the alined spectrum causes a larger increase in yield with depth than on the random spectrum. Thus, a problem can arise when trying to determine the point on the alined spectrum at which to normalize the random spectrum. The latter approximation has been used entirely in the work reported here. The errors involved with these approximations are difficult to

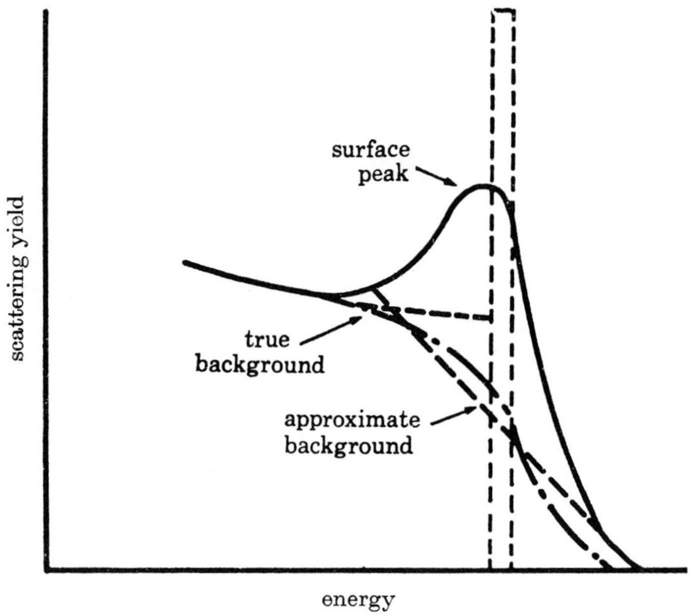

FIGURE 4. A schematic illustration of the surface peak together with the true background and the straight line approximation.

quantify since the true position of the background cannot be determined. These procedures differ from that reported earlier by Morgan & Bøgh (1972) who, in effect, ignored the change in background level due to the finite detector resolution. Their analysis gave $N_{As}/N_{Ga} = 2.0 \pm 0.2$ compared to the value of 1.7 ± 0.2 obtained using the straight line approximation and 2.3 ± 0.2 using the line shape of the random spectrum. (The GaAs ⟨110⟩ crystal has been etched in IHF:$3HNO_3$:$2H_2O$ and from the energy spectrum (see Morgan & Bøgh 1972, figure 3), the concentrations are $N_{As} = 1.14 \times 10^{16}\,cm^{-2}$ and $N_{Ga} = 4.92 \times 10^{15}\,cm^{-2}$). In the case of surface stoichiometry measurements, this uncertainty in background level introduces the largest error into the measurement. In GaAs for example, even in the most favourable case where the Ga and As surface peaks are clearly resolved, the uncertainty in the Ga to As ratio is at best *ca.* 10 % and in general it is nearer to 15 %.

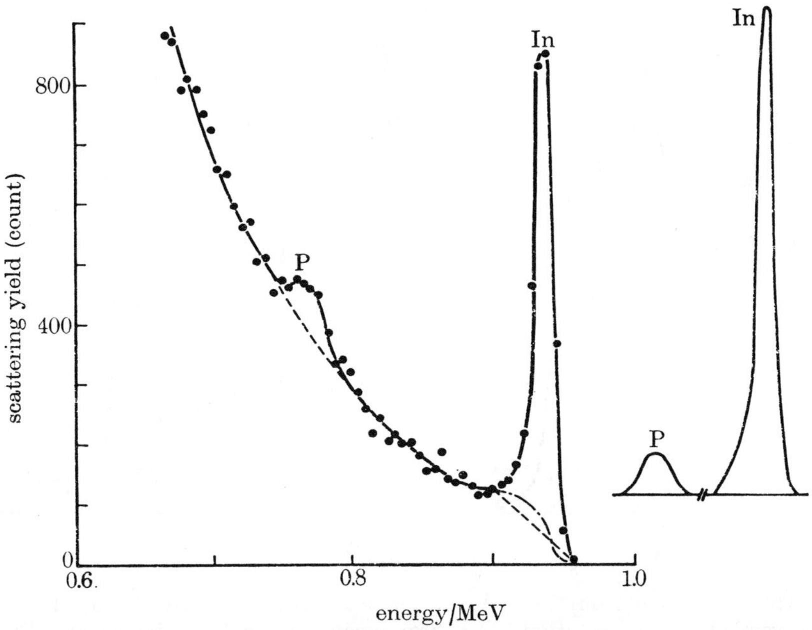

FIGURE 5. The energy spectrum for 1 MeV α particles scattered through 90° from an 'as grown' epitaxial InP crystal. The incident beam and the detector were both alined to ⟨100⟩ axes (i.e. double alinement). The inset shows the In and P peaks after the background correction.

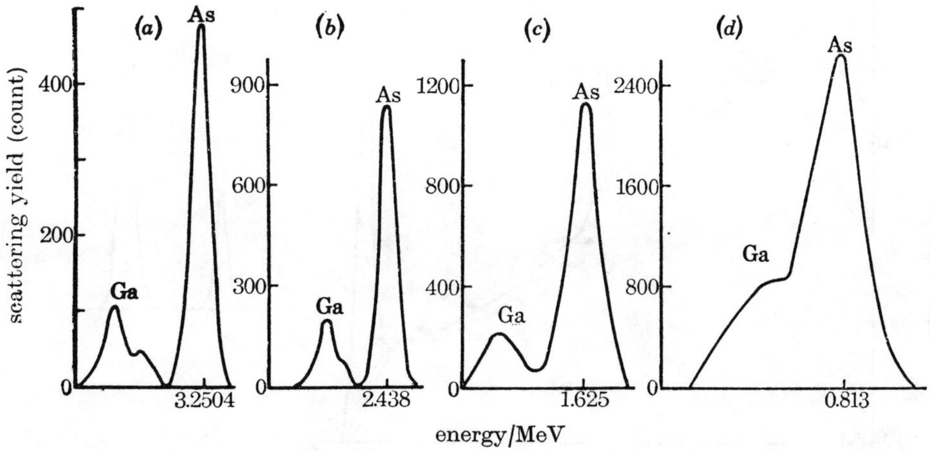

FIGURE 6 (*a–d*). A series of four spectra for helium ions scattered through 160° for a GaAs crystal alined to ⟨110⟩. Each spectrum shows the surface peak after removal of the background and corresponds to incident ion energies of 4, 3, 2 and 1 MeV respectively. The Ga and As peaks can be resolved when the incident ion energy reaches approximately 2 MeV.

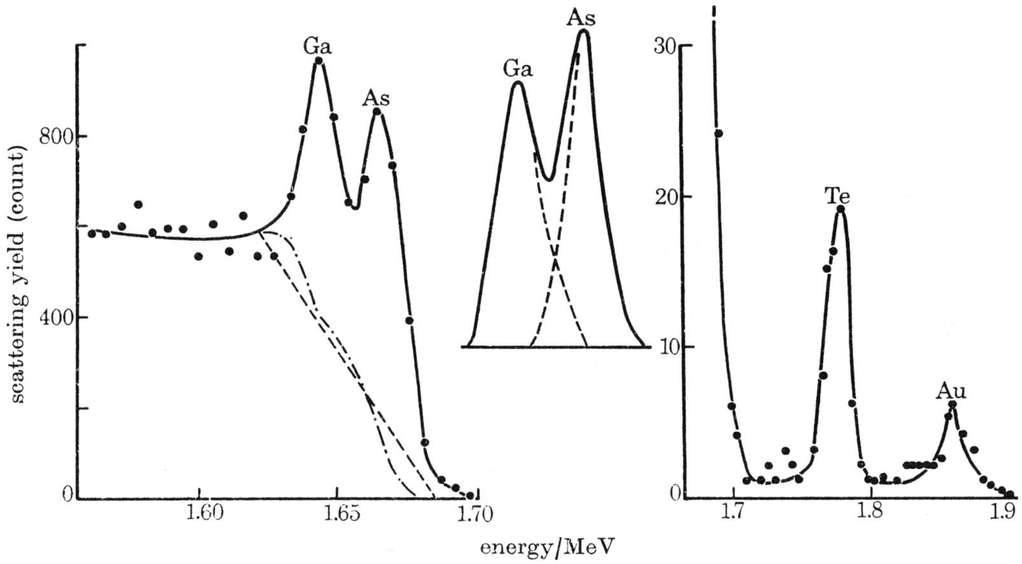

FIGURE 7. The energy spectrum for 2 MeV α particles scattered through 135° from a ⟨110⟩ GaAs crystal etched in a solution of 4% bromine in methanol. The inset shows the surface peaks after the background correction and the broken line shows the separate Ga and As peaks. From the areas of the peaks, $N_{As}/N_{Ga} = 1.2 \pm 0.1$ with a concentration $N_{As} = 2.9 \times 10^{15}\ cm^{-2}$.

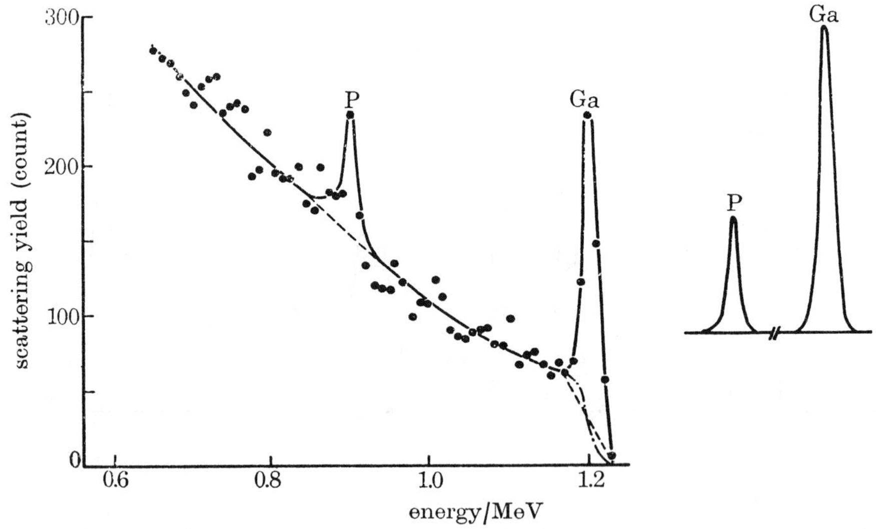

FIGURE 8. The energy spectrum for 1.5 MeV α particles scattered through 135° from a ⟨110⟩ GaP crystal vibratory polished and chemically etched in a solution of $2HNO_3:1HCl$. The inset shows the Ga and P peaks after a background correction.

3. Application to real surfaces

In order to illustrate some of the practical problems and limitations of the technique we now consider its application to three compound semi-conductors GaAs, GaP and InP. These materials were chosen, partly because of their practical interest to the authors and partly because they illustrate the advantages and limitations of the technique. Indium phosphide is a material where the *atomic numbers* of the two sublattices (49 and 15 respectively) differ appreciably, resulting in a large disparity in the scattering yields from the two constituents (see equation (4)). Gallium arsenide, on the other hand, is a case where the constituents have comparable *atomic masses* (^{69}Ga, ^{71}Ga and ^{75}As) making it difficult to resolve the three surface peaks. Finally, we consider gallium phosphide, a material which represents the optimum case where neither of these two problems occur.

Consider first the case of indium phosphide. Using equation (4), the ratio of the indium and phosphorous yields is

$$Y_{In}/Y_{P} = (Z_{In}/Z_{P})^2 = 10.67. \tag{8}$$

Consequently, the phosphorous peak, which is superimposed upon the indium spectrum, is barely visible above the background level. One way of improving the situation is to reduce this background level. This can be achieved by using the double alinement procedure, i.e. the experimental arrangement is modified such that both the incident beam and a detector whose acceptance angle is less than the critical angle are both alined to a crystal axis. Figure 5 illustrates such a spectrum for an indium phosphide crystal, together with the indium and phosphorous peaks after removing the background levels. The crystal used for this analysis was an epitaxial layer with the surface in the 'as grown' condition. The minimum yield χ_{min} for single alinement ($\langle 110 \rangle$) was 0.034 and for double alinement ($\langle 100 \rangle - \langle 100 \rangle$) was 0.008, illustrating a high degree of perfection in the crystal structure. Using the area under the surface peak, we estimate $N_{In}/N_{P} = 0.73 \pm 0.1$ with

$$N_{In} = 2.86 \times 10^{15}\,\text{cm}^{-2}.$$

We also see from figure 5 that, despite the use of double alinement, the phosphorous peak is only just visible above the background level. The major source of error associated with the measurement of stoichiometry lies in the relatively poor statistics of the phosphorous peak relative to the background level. For the case shown in figure 5 we estimate this to be *ca.* 15 %. A further difficulty in this case is the complication of using the double alinement procedure. This technique is considerably more difficult and time-consuming than single alinement; however, this objection may be partly overcome by the use of the unidirectional geometry described by Appleton & Feldman (1970).

The second example we consider is gallium arsenide, a material whose constituent atoms lie close together in the periodic table. Clearly, in this case the yield problem experienced in indium phosphide does not exist. Instead, we are confronted with the difficulty of resolving the surface peaks. Using equation (2) the difference in the value of k^2 for gallium and arsenic is $\Delta k^2 \simeq 0.0124$ (for $\theta = 160°$ and $^4He^+$); hence in order to resolve the surface peaks the energy separation between the two peaks must be greater than the detector resolution ΔE_{res}, i.e.

$$\Delta E = 0.0124 E_0 > \Delta E_{res}. \tag{9}$$

The necessary resolution may be achieved by one of two procedures: (i) increasing the detector resolution, or (ii) maximizing ΔE. In practice, the limit to the detector resolution is set by the available equipment. Using silicon surface barrier detectors a system resolution of around 15 to 20 keV for MeV α particles is typical, although, in special cases, resolution as good as 9 keV has been reported. Alternatively, by using a magnetic detector, the energy resolution can be reduced to around 1 keV. A problem in this latter case is that detectors of this type subtend very small solid angles and hence reduce the scattering yield, thus in order to obtain good statistics in the results, the integrated beam needed causes excessive radiation damage of the crystal. E. Bøgh (unpublished work) has recently overcome this problem by installing a small magnetic analyser very close to the crystal surface. Unfortunately, with this equipment the maximum α particle energy that may be analysed is 0.5 MeV, which itself introduces a new problem: at this low energy the multiple scattering of the back-scattered beam by the surface film smears out the gallium and arsenic peaks. With clean surfaces where the effective film thickness is two to three monolayers the multiple scattering is negligible and the two surface peaks can be resolved. Chemically prepared surfaces, on the other hand, have an effective film thickness $\gtrsim$ 10 monolayers and hence the gallium and arsenic peaks cannot be resolved.

The second possibility is to increase ΔE. This can be achieved by either increasing Δk^2 or alternatively increasing the beam energy E_0. Considering the first possibility; from equation (2) we see that for a fixed value of M_2 (scattering atom), k^2 increases with both θ and M_1. Clearly, the maximum value of θ is 180°. On the other hand M_1 may be increased by using a nitrogen or carbon beam in place of a helium beam. Unfortunately, when using surface barrier detectors, the increase in k^2 is compensated by a decrease in the detector resolution. Added to this is the problem of the deterioration of the crystal surface caused by the probing beam. In practice, the use of a helium beam is found to be a convenient compromise. The second possibility of resolving the surface peaks arises from the dependence of ΔE on E_0, hence for good resolution the maximum beam energy is desirable. For most accelerators at present being used for solid state studies, the upper limit is around 2 MeV with singly charged ions. However, even when this limit does not exist, care must be exercised when working with α particles at energies above this, since deviations from coulombic scattering occur due to the onset of nuclear terms in the scattering cross-section (Zeigler & Baglin 1971; Morgan & Gittins 1972).

In figure 6 (*a–d*) we show a series of spectra obtained from a helium beam scattered through 160° from GaAs ⟨110⟩ crystal after polishing in $1HF:3HNO_3:2H_2O$. The spectra correspond to incident beam energies of 4, 3, 2 and 1 MeV respectively (with a surface barrier detector system resolution of *ca.* 18 keV). We see that two peaks begin to appear between 1.0 and 2.0 MeV. However, in order to obtain a reasonable measurement of the Ga and As areas it is not sufficient to simply resolve two peaks, but also, the scattering from the gallium and arsenic must be separated from the resulting spectrum. Gallium has two stable isotopes, 69 and 71, with relative abundances of 60 and 40 % respectively. Clearly, analysis is much simpler and the accuracy much improved when two separate peaks (from Ga and As) are visible, and the difficulty of subtracting the contribution of ^{71}Ga from the ^{75}As peak is removed. In these circumstances it is much easier to obtain a true measure of the arsenic peak. This point may be illustrated by comparing figures 6 and 7. A detailed study of GaAs surfaces prepared by various chemical polishes will be described in a later paper.

The third example to be considered is gallium phosphide, which represents the optimum case. The ratio of the Rutherford scattering yields from the two sublattices is $Y_{Ga}/Y_P = 4.27$, hence both peaks are able to stand clear of their respective backgrounds. At the same time, the mass disparity is sufficiently large (i.e. $\Delta k^2 = 0.1961$) to enable the two surface peaks to be easily resolved with relatively low energies. For a system resolution of *ca.* 20 keV the two peaks can be resolved with helium ions with an incident energy greater than 0.1 MeV.

The spectrum obtained from a GaP⟨110⟩ crystal vibratory polished and chemically etched for 2 min in a solution of $2HNO_3$:1HCl is illustrated in figure 8, together with the gallium and phosphorous peaks after a background correction (inset). The helium beam energy was 1.5 MeV. The ratio $N_{Ga}/N_P = 0.78 \pm 0.1$ with concentrations $N_{Ga} = 5.34 \times 10^{15}\ cm^{-2}$ and $N_P = 6.86 \times 10^{15}\ cm^{-2}$.

Thus far we have confined our discussion to the analysis of the channelled surface peaks. By comparison, the measurement of surface impurities is simpler. When the impurities are heavier than the host crystal, the peak will appear to the right of the leading edge of the spectrum, as is illustrated for GaAs⟨110⟩ crystal polished in a mixture of 4 % Br_2 in methanol in figure 7. There are two peaks visible, one for tellurium (*ca.* 0.005 monolayer) which arises from the accumulation at the surface of the Te dopant during the chemical polishing. The second peak corresponds to *ca.* 0.0005 monolayer of gold, a common feature of crystals which have been washed in organic solvents (in this case Analar grade methanol). The small background count arises from pulse pile-up in the pulse height analyser and decreases rapidly as the energy increases beyond the leading edge of the host spectrum. The case of impurities which are lighter than the host atoms is more difficult, since their peaks will be superimposed upon the relatively large background spectrum. This is illustrated for the case of a ⟨111⟩ bulk grown indium phosphide crystal polished in a mixture of 10 % Br_2 in methanol, in figure 9. The indium surface peak is visible, together with a bromine peak (*ca.* 0.11 monolayer) and a peak corresponding to a value of $k^2 = 0.859$ (i.e. an impurity with a mass around M_0 (*ca.* 0.087 monolayer)).

The sensitivity of the technique depends strongly on the impurity mass. Clearly, as a result of the host background spectrum, there is an abrupt division between those atoms which are lighter and those heavier than the host. Furthermore, we must remember that the cross-section increases with the square of the atomic number of the impurity, hence the technique is more sensitive to gold than to tellurium (figure 7). For the same reason, the surface bromine on indium phosphide stands clearly above the background level, whereas the oxygen in the surface oxide cannot be detected. A further difficulty arises when two peaks are close together making it difficult to resolve the two components. One such example is bromine on the surface of gallium arsenide. When the energy of the analysing α particle beam is less than 3 MeV the relatively small bromine peak falls on the trailing edge of the surface arsenic peak and cannot be observed. When the beam energy is increased to 4 MeV this difficulty is removed.

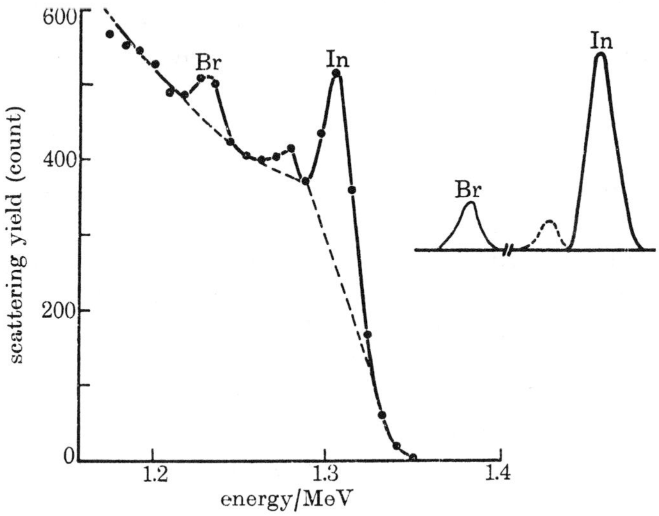

FIGURE 9. The energy spectrum for 1.5 MeV α particles scattered through 150° by a ⟨111⟩ InP crystal which had been polished in a solution of 10 % bromine in methanol. Three peaks are visible near the edge of the spectrum, the In surface peak together with peaks corresponding to bromine and to molybdenum, two surface impurities.

4. Conclusions

In this paper we have described how Rutherford scattering in conjunction with the channelling phenomenon may be used to study the surfaces of binary compounds. From this discussion we conclude that the technique may be used to measure the relative concentrations of the two constituent atoms at the surface of the solid. The restrictions on the technique are: (i) the difference in the atomic masses of the two constituents must be large enough to enable the two surface peaks to be resolved (GaAs represents this limit); (ii) the atomic numbers must not differ by too great a value otherwise the yield from the lighter species will be lost in the background from the heavier sublattice (InP represents this limit). With regard to impurity analysis, the technique yields direct the absolute concentration averaged over an area of the surface defined by the probing beam ($\lesssim 1.0\,mm^2$). The sensitivity decreases rapidly as the atomic number of the impurity atom decreases, with an abrupt reduction in sensitivity for atoms lighter than the host atoms. The technique provides us with a valuable tool to be used in conjunction with other surface analysis techniques for the study of solid surfaces.

We wish to thank our colleagues, Drs E. Bøgh (Aarhus), G. Dearnaley (A.E.R.E. Harwell), J. Whitton (Chalk River) and D. Colliver (R.R.E. Malvern) for many valuable discussions; the S.R.C. for providing financial support for accelerator time and N.A.T.O. and C.V.D. for the award of research grants. Finally, one of us (D. R. W.) would like to thank the S.R.C. for the award of a studentship.

References

Appleton, B. R. & Feldman, L. C. 1970 *Atomic collisions phenomena in solids* (eds. D. W. Palmer, M. W. Thompson and P. D. Townsend). Amsterdam: North Holland Publishing Co.
Ball, D. J., Buck, T. M., Macnair, D. & Wheatley, G. H. 1972 *Surf. Sci.* **30**, 69.
Bøgh, E. 1968 *Can. J. Phys.* **46**, 653.
Bøgh, E. 1973 *Channelling: its theory, observation and applications* (ed. D. V. Morgan). New York: J. Wiley.
Bøttiger, J., Davies, J. A., Whitton, J. L. & Lori, J. 1972 *Progress Report A.E.C.L.* no. 4351. Chalk River.
Cachard, A., Roger, J. A., Pivot, J. & Dupuy, C. H. S. 1971 *Phys. Stat. Sol.* (*a*) **5**, 673.
Chang, C. C. 1971 *Surf. Sci.* **25**, 53.
Laubenstein, R. A., Laubenstein, M. J. W., Koester, L. J. & Mobley, R. C. 1951 *J. appl. Phys.* **84**, 12.
Lindhard, J. 1965 *Mat. fys. Meddr.* **34**, no. 14.
Mayer, J. W., Eriksson, L. & Davies, J. A. 1970 *Ion implantation in semiconductors.* New York: Academic Press.
Mitchell, I. V., Kamoshida, M. & Mayer, J. W. 1971*a* *Appl. Phys. Lett.* **18**, 292.
Mitchell, I. V., Kamoshida, M. & Mayer, J. W. 1971*b* *J. appl. Phys.* **42**, 4378.
Mitchell, I. V., Kamoshida, M. & Mayer, J. W. 1971*c* *J. appl. Phys.* **43**, 1717.
Morgan, D. V. & Bøgh, E. 1972 *Surf. Sci.* **32**, 278.
Morgan, D. V. & Gittins, R. P. 1972 *Phys. Stat. Sol.* (*a*) **13**, 517.
Picraux, S. T. 1969 Ph.D. Thesis, California Institute of Technology.
Reuter, W. 1971 *Surf. Sci.* **25**, 80.
Rubin, S. 1959 Ion scattering methods. In *Treatise on analytical chemistry.* New York: Interscience.
Socha, A. J. 1971 *Surf. Sci.* **25**, 147.
Zeigler, J. F. & Baglin, J. E. E. 1971 *J. appl. Phys.* **42**, 2031.

(*Communicated by D. Tabor, F.R.S. – Received* 6 *April* 1973)

2.2 physical analysis

The techniques of physical analysis are seen in Table 2-2. The information obtained from physical analysis of a sample includes orientation, crystal perfection, defect and dislocation densities, extent of cutting or polishing damage, layer thickness measurement, and studies of surface contamination. In most cases there is no clear distinction between bulk and surface procedures. Where a distinction exists, note is made in the text.

Etching and Microscopy

The most important aspect of physical analysis is the direct observation of the material, using conventional or electron microscopy. Defect structure can often be enhanced by suitable etching procedures. The overwhelming advantage of the method is that it is efficient and does not require expensive equipment.

InP bulk material has been examined in this way[2] to study etch pits, twins, dislocations, and striations using H_2O: HNO_3: HCl etch in the proportions 6:6:1 by volume. Little research has been found on epitaxial InP, mainly because of difficulty in finding suitable dislocation etches.

For GaAs, the work is well documented[25, 26]. One of the most interesting dislocation etches is that developed by Abrahams and Buiocchi (AB):

2 ml. H_2O; 8 mg. $AgNO_3$:1 g. CrO_3; 1 ml. HF

This etch is used fresh, and it brings out the projections on the surface plane of the original dislocation. It also shows up unusual features at epitaxial layer interfaces and propagation of defects from substrates. Some of these effects are

Instrument	Bulk/Surface	Details	Sensitivity	Measurement Frequency	Cost
Etching and Microscopy	Bulk or thin film	Etchants depend on material. Optical and electron microscopy	Resolution limited by microscopy, destructive	Routine	Cheap
X-Ray Topography	Penetration limited to ~ 50 μm	Transmission or reflection diffraction	Orientation, and crystallographic investigation. Measurement of strain	Research	High
Electron Diffraction LEED, RHEED	First few nonlayers		Very sensitive to adsorbed impurities, qualitative non-destructive	Research	Medium/High
Rutherford Scattering	50-300Å below surface	See Table 3 for details of chemical analysis	Gives a measure of the surface disorder but not absolute	Research	High
Thickness Measurement	layered structures	a) Cleave and Stain etch dependent on material	Resolution ± 0.5 μ no limit on max thickness	Routine	Cheap
		b) IR interference	0.5-15 μ for most layers does not work well with high purity on SI substrates	Routine	Cheap
		c) X ray interference	Thickness up to 0.1 μm	Semi-routine	Medium
Ellipsometry	Thin dielectric layers	Subject to errors unless set up carefully	5-5000 Å, good resolution. Difficult computation	Research-semi routine	Medium
Interferometry	Thin dielectric Layers	Easier to set up than ellipsometer	100 Å upwards. Not so accurate as ellipsometer (~ 1%)	Routine	Cheap

Table 2-2/Comparison of Physical Analysis Techniques

not fully understood, particularly the etching at the interface. Scanning electron microscopy can be used advantageously in these studies.

X-Ray Techniques

X-ray diffraction patterns are informative in various ways. The Laue method is generally used for orientation of a single crystal, where the crystal is bathed in a beam of white X-radiation. The necessary variable has to be provided by the range of wavelengths in the beam in order to ensure that the Bragg law is satisfied for all planes in the crystal. A back reflection photograph generally is taken and the pattern of spots produced lie on hyperbolae. All the spots are reflections from planes of a single zone. This Laue pattern indicates the symmetry of the crystal. The bulk crystal is cut approximately to a given orientation, then mounted on a goniometer mount and aligned on the X-ray equipment. After orientation, the complete assembly can be mounted on a cutting machine and cut to the given orientation.

Transmission X-ray topography is another powerful method; not only for the observation of dislocations, but also for the accurate measurement of lattice parameters and the distribution of strain. For GaAs and InP, samples must be thinned to around 70 μm to allow transmission of the X-radiation. Reflection topography overcomes this thickness limitation, enabling measurements to be executed on shallow surface layers. The penetration is a function of the reflecting planes, which determine the inclination to the surface and the Bragg angle. For GaAs, the penetration is between 0.5 and 22 μm, and for InP 0.3 to 11 μm using Cu Kα radiation[27].

Electron Diffraction

The effects of electron diffraction are used primarily for the study of the first few atomic layers of the crystal surface. Either low or high energy electrons can be used; we first consider the low energy case — LEED.

The electron beam is backscattered from the crystal surface to a suitable screen or photographic plate; the incident energy of the beam can be as low as 100 volts. The advantages of the LEED technique are easy to appreciate. The structural changes undergone by surfaces (initially shown to be clean) with the adsorption of gases may be viewed simply as adsorption processes. These procedures are non-destructive, but can only provide a qualitative physical picture of the surface. A major problem is the lack of a sound theoretical foundation for scattering effects, thereby making analysis difficult. High-energy diffraction — RHEED — utilizes a beam of about 30 keV which strikes the specimen at as near grazing an incidence as possible. The diffraction patterns originate from atomic layers near the surface; their interpretation in terms of surface structure is similar to that for LEED, although the precise depth of penetration is not well defined. RHEED is used in molecular beam epitaxy to ensure that the GaAs substrate surface is clean before the molecular beam is allowed to impinge on the surface. As with X-rays, electron beams can be used in transmission microscopy and diffraction.

Rutherford Scattering

Additional information about Rutherford scattering is available from the comparison of random and channelled spectra. Taking the ratio of the aligned yield to the random yield immediately behind the surface peak provides a measure of the perfection of the crystal surface over a distance comparable with the depth resolution of the detector system (50-300Å). For a fixed geometry, the area under the surface peak (A) is used to estimate the thickness (t) of the disordered surface layer from the equation:

$$t \cong \frac{A}{Y_a\ [k^2 S(E) + S(E_1)/\cos\theta']}$$

The terms are as previously defined. These measurements apply for both GaAs and InP, and have also been extended to measurements taken on a

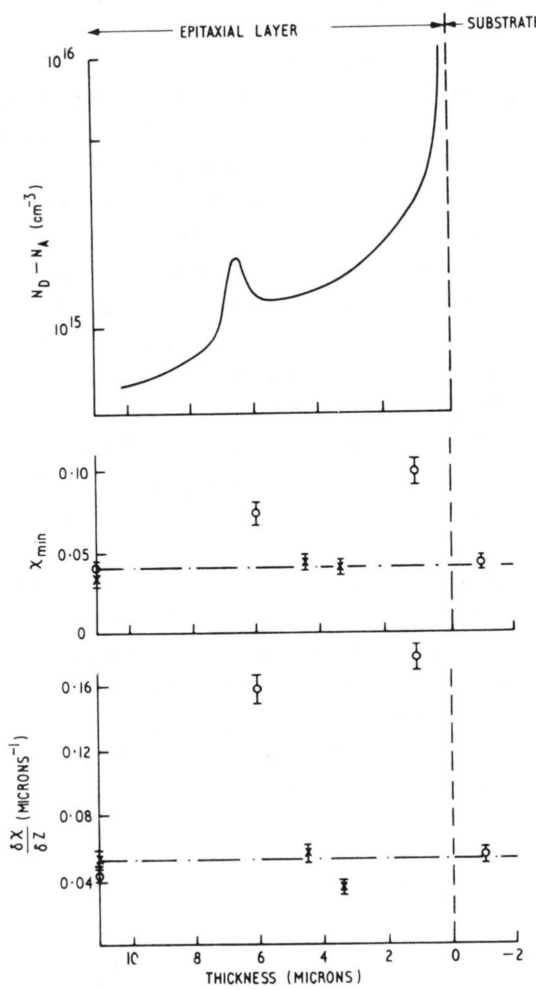

Figure 2-3/Impurity Profile and Minimum Yield in a GaAs Layer.

successively etched GaAs epitaxial layer. The comparison of measured impurity profile with the ratio aligned yield/random yield (χ_{min}) as a function of depth is seen in Figure 2-3.

Thickness Measurement

The simplest method of measuring layer thickness is by cleave and stain. The process is direct and does not need further elaboration other than to list suitable etchants. The disadvantages of the process are that it is destructive and resolution is limited to around ± 0.5 μm. The thickness measurement of thin films is thus difficult, although angle lapping can improve the resolution to some extent.

For GaAs etchants for staining n^+-n and p-n junctions are:

1./HF:HNO_3:H_2O in the ratio 1:3:4 for 5-20 sec.

2./KOH (120g./500 ml.H_2O): K $Fe(CN)_6$ (80g./500 ml.H_2O) 1:1 which can be used hot or cold

3./2ml.H_2O:8 mg.Ag NO_3:1g.CrO_3 1ml.HF at 65°C ~ 30 sec.

For InP, etch # 2 works efficiently, as does a hot solution of nitric acid in which iron has been dissolved. For non-destructive thickness measurement the most widely used technique is infra-red interference. This method is fast and accurate and the normal arrangement is composed of a spectrophotometer with I-R optics and a specular reflection attachment. The film thickness is given by:

$$t = \frac{n}{2\sqrt{n_i^2 - \sin^2\theta \cdot \nu}}$$

where n is the interference order 1/2, 3/2, 2 . . . . etc., n_i is the refractive index, and ν is the radiation frequency.

The measurement range is from 0.5 to about 15 μm, and it works well apart from low doped layers that are grown on insulating substrates. X-ray interferometry can be used to cover the thickness range up to about 0.1 μm.

Ellipsometry

The measurement of very thin films (below 5000Å) of dielectric materials is achieved by mechanical methods such as the Talystep, and a range of interferometric techniques. The most accurate instrument for the measurement of refractive index and thickness is the ellipsometer which, under optimum conditions, operates with film thickness down to 5Å. The ellipsometer measures the effect of reflection on the state of polarisation of incident polarised light. (See Figure 2-4.) Normal measurements are of two variables — the azimuth (ϕ) and the ellipticity (ψ) of the light reflected from the sample. In a single measurement, given n and K, one can determine n and d. It can be shown that:

$$\tan 2\phi = 2 (r_p/r_s) \cos \Delta/[1 - (r_p/r_s)^2]$$

and

$$\cos 2\psi \cos 2\phi = [1 - (r_p/r_s)^2]/[1 + (r_p/r_s)^2]$$

Thus, r_p/r_s and Δ can be calculated; r_p and r_s are the reflected amplitudes, differing in phase by Δ and combining as elliptically polarised light with azimuth ϕ.

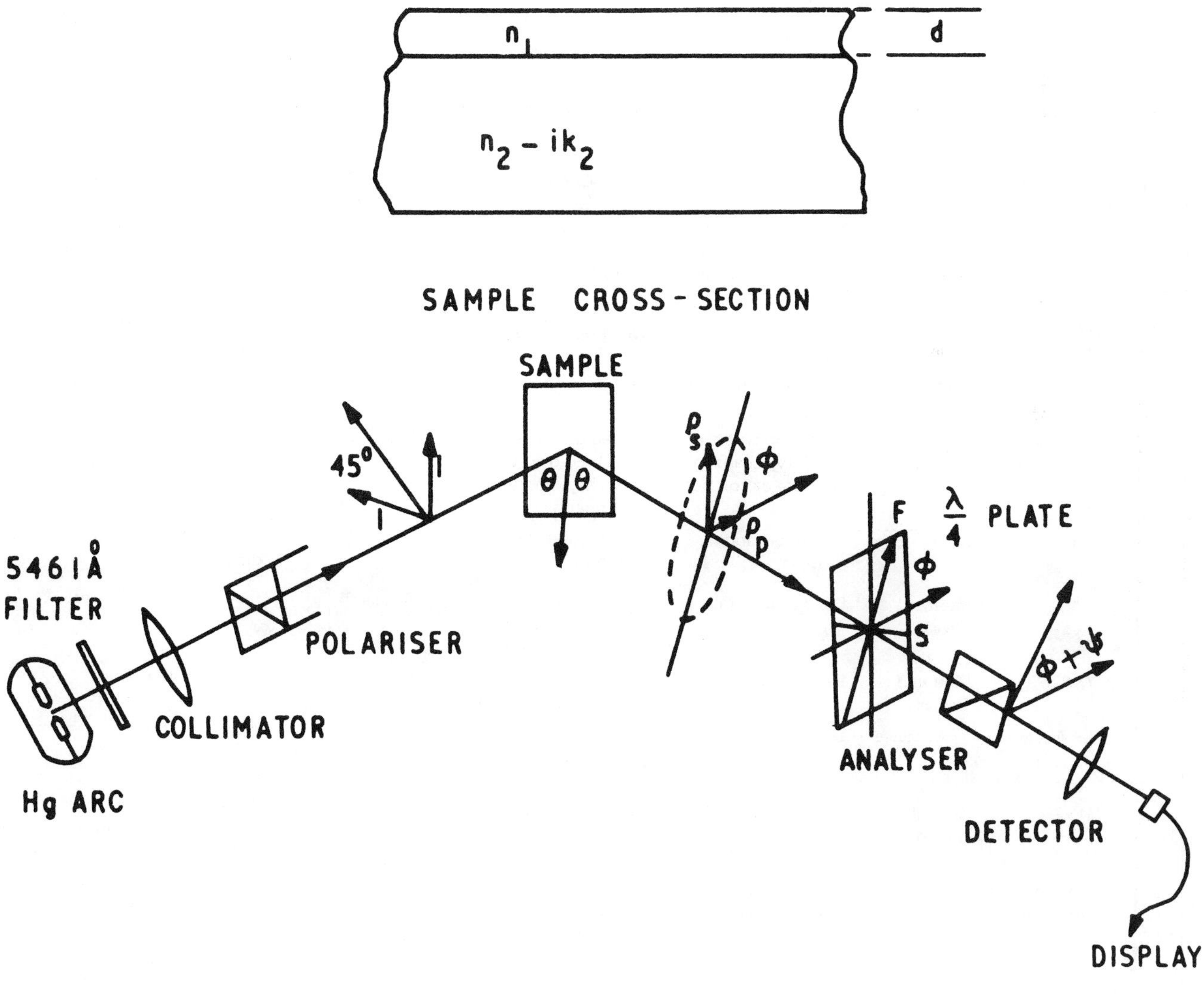

Figure 2-4/Conventional Ellipsometer

The measured quantities r_p/r_s and Δ are related to the optical properties of the sample by Fresnel reflection coefficients. They are normally complex and result in transcendental equations. Sources of error can be introduced in the measurement[29], and the computation can be quite complex. Once the equipment is properly set up it is easy to use, and measurements have been carried out on surface films on GaAs. For a film thickness greater than a few hundred angstroms, it is advantageous to employ simpler interferometric methods[30].

2.3 electrical analysis

The electrical assessment of material is a complex and wide-ranging topic, covering analysis of bulk material, thin monitor epitaxy layers (grown simultaneously with device layers), average measurements over whole wafer surfaces, determinations within a particular device volume, measurements on the completed device, and experiments aimed at defining a special property of the material.

This section covers the major techniques involved in the electrical analysis of material — not in the analysis of devices except when the

	Method	Sample	Details	Conditions	Measurement Frequency	Limitations
ROUTINE	*Van der Pauw*	Clover leaf Layer grown on Insulating substrate	Measures mobility resistivity and net carrier concentration	Room temperature and 77°K. Magnetic field ~ 2 kG	Every growth run on epitaxy	Not on device layer generally. Time consuming. Average over sample.
ROUTINE	*4-Point Probe*	Slice	Measures average resistivity	Need reliable contacts		Not reliable on GaAs or InP
ROUTINE	*Type Check*	Slice	Measures conductivity type	Hot or Cold Probe		Only works with highly doped material
ROUTINE	*Magneto-resistance*	TED device no junctions	Magnetoresistance mobility. Leads to estimate of contact resistance	Pulsed device, can be measured over temperature range and under illumination	Every device	Ideally suited to TED's. Limited with other devices
ROUTINE	*C-V*	Slice	Computes doping profile	Needs good step junction	Every slice	Limited by breakdown of diode
SPECIAL	*C-V*	Mounted device	Deep levels, traps	Different conditions of bias, temperature, time, illumination	Research	Instrument Sensitivity
SPECIAL	*TSC*	Mounted Device	Deep levels	Closely controlled temperature	Research	High quality diodes, low leakage currents
SPECIAL	*LF Noise*	TED device	Deep levels	Controlled temperature	Research	Analysis difficult Ultimate sensitivity not clear
SPECIAL	*MOS*	Special process	Surface states, MOS profile	Rigorous control of all process and measurement conditions	Research	Analysis difficult Ultimate resolution not clear
SPECIAL	*Electro-chemical*	Slice	Continuous profile with etching. Deep levels	Difficult experimentally Corrosive liquids Control	Research	Not yet fully exploited

Table 2-3/Comparison of Methods of Electrical Analysis

discussion is aimed at obtaining a material parameter. A comparison of the various methods is given in Table 2-3.

2.3.1 routine analysis

Van Der Pauw Measurement

The Van Der Pauw method of measuring the mobility and resistivity of a semiconductor sample, described in his paper[31] in 1958, has become the standard, particularly for compound semiconductors. Essentially, it is necessary only to make four small diameter contacts (A, B, C, and D) along the circumference of an arbitrarily shaped flat sample, and to measure the resistances given by V_{DC}/I_{AB} and V_{BC}/I_{AD}. (Voltage is measured between D and C (V_{DC}) with current through the contacts A and B (I_{AB}).) The sample resistivity is then given by:

$$\rho = \frac{\pi d}{2\ell n2}\left(\frac{V_{DC}}{I_{AB}} + \frac{V_{BC}}{I_{AD}}\right) f$$

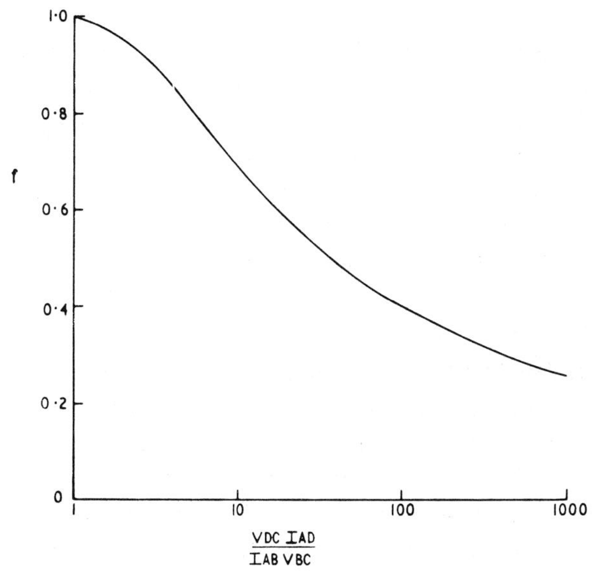

Figure 2-5/The Factor f in Van der Pauw Measurement.

where d is the sample thickness and f is a function of $V_{DC}\ I_{AD}/V_{BC}\ I_{AB}$ (the latter is a geometrical correction factor only).

When the sample is a circle or a square and the contacts are spaced at equal distances around the periphery, then f = 1. The sample is inhomogeneous if the resistance ratio is not unity. The factor f is shown plotted against the resistance ratio in Figure 2-5. The Hall mobility is determined by measuring the change of the resistance V_{AC}/I_{BD} when a magnetic field is applied in a direction perpendicular to the sample. The Hall mobility:

$$\mu_H = \frac{d}{B} \cdot \frac{\Delta V_{AC}/I_{BD}}{\rho}$$

(Note that μ_H is independent of d.)

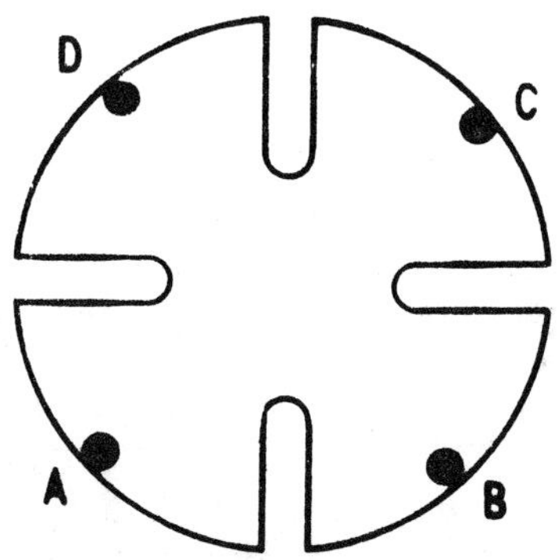

Figure 2-6/"Clover-Leaf" Hall Sample

Since errors can be introduced by the finite diameter and position of the contacts on the circumferences, it is common to sandblast the sample into a clover-leaf shape, as in Figure 2-6. Measurements are normally carried out at both 300 and 77°K in a magnetic field of 2 kG. Bulk crystals are assessed by Van der Pauw measurements taken from a wafer at each end of the as-grown crystal; however, only epitaxial layers grown on insulating substrates are suitable. Heavily conducting substrates short out effectively, the layer. For this reason, a monitor semi-insulating substrate is included in each epitaxy growth run. The technique gives only an average figure for the grown layer.

Hall measurements, a major assessment method in the rudimentary stages of these materials, have since become a way of occasionally monitoring the quality of the epitaxial reactor. Measurements taken on a monitor slice do not generally correlate with device results; thus, more direct assessment methods have been developed.

4-Point Probe

The 4-point probe is a well established technique with Si — why should it not work with compound semiconductors? The 4-point probe on thin layers gives an average resistivity of:

$$\bar{\rho} = 4.5 \frac{V}{I} \cdot X$$

where X = layer thickness in cm.

Problems occur when the point-contacts form electrical barriers to the semi-conductor, when wider band-gap materials are used. These barriers are broken down by alloying contacts or by electrical means. The result, though, is a destructive measurement.

Resistivity measurement is obtained by point contact breakdown using two or three probes and biasing one until breakdown occurs. But, all probe methods give unreliable results and are generally not used.

Thermoelectric Type Checking

When two contacts are made to a sample and held at different temperatures, a potential appears between them. The polarity of the cold probe is positive for p-type and negative for n-type material. The normal arrangement uses a small soldering iron as the hot probe, and a zero-centre galvanometer as the detector. The sensitivity is affected by blocking contacts, but the procedure has been successful on *heavily* doped ($\sim 10^{18}$) material for conductivity type checking.

Magnetoresistance

Of particular value for transferred electron devices, is magnetoresistance. It can be carried out on a completed unit and provides a measure of how processing has affected the starting material. The key paper by Larrabee et al. evaluates the method as applied to devices. In the structure of the transferred electron component, the active length is short when compared to the effective diameter; the end contacts short out the Hall Field as in the Corbino disc. Aspects of this geometry have been neglected in the study of galvanomagnetic properties of high mobility semiconductors, primarily because of the difficulty in making reliable, low resistance contacts to the material. With TED's, though, if contacts are not uniformly low resistance this information is valuable of itself.

Consider a long thin rectangular bar, the conventional Hall plate, with an applied transverse magnetic field (see Figure 2-7(b)). If the conduction is isotropic and caused by a single band of carriers, and scattering time is independent of energy, the Lorentz force is exactly balanced by the Hall field. The current then flows in the axis direction of the bar, with the net field at the Hall angle θ_H. The resistivity is the same with or without the magnetic field. In implementation, the increase in resistivity observed when the sample is in the magnetic field is known as physical magnetoresistance (PMR). It occurs when the above conditions are not fulfilled.

In the structure of Figure 2-7(c), the Hall field is now shorted out by the contacts; the Lorentz force is no longer compensated. The result is that the carriers flow at the Hall angle with respect to the electric field. A change in resistivity, known as geometrical magnetoresistance (GMR), occurs in the magnetic field even if the material exhibits no PMR. PMR is always present, but it

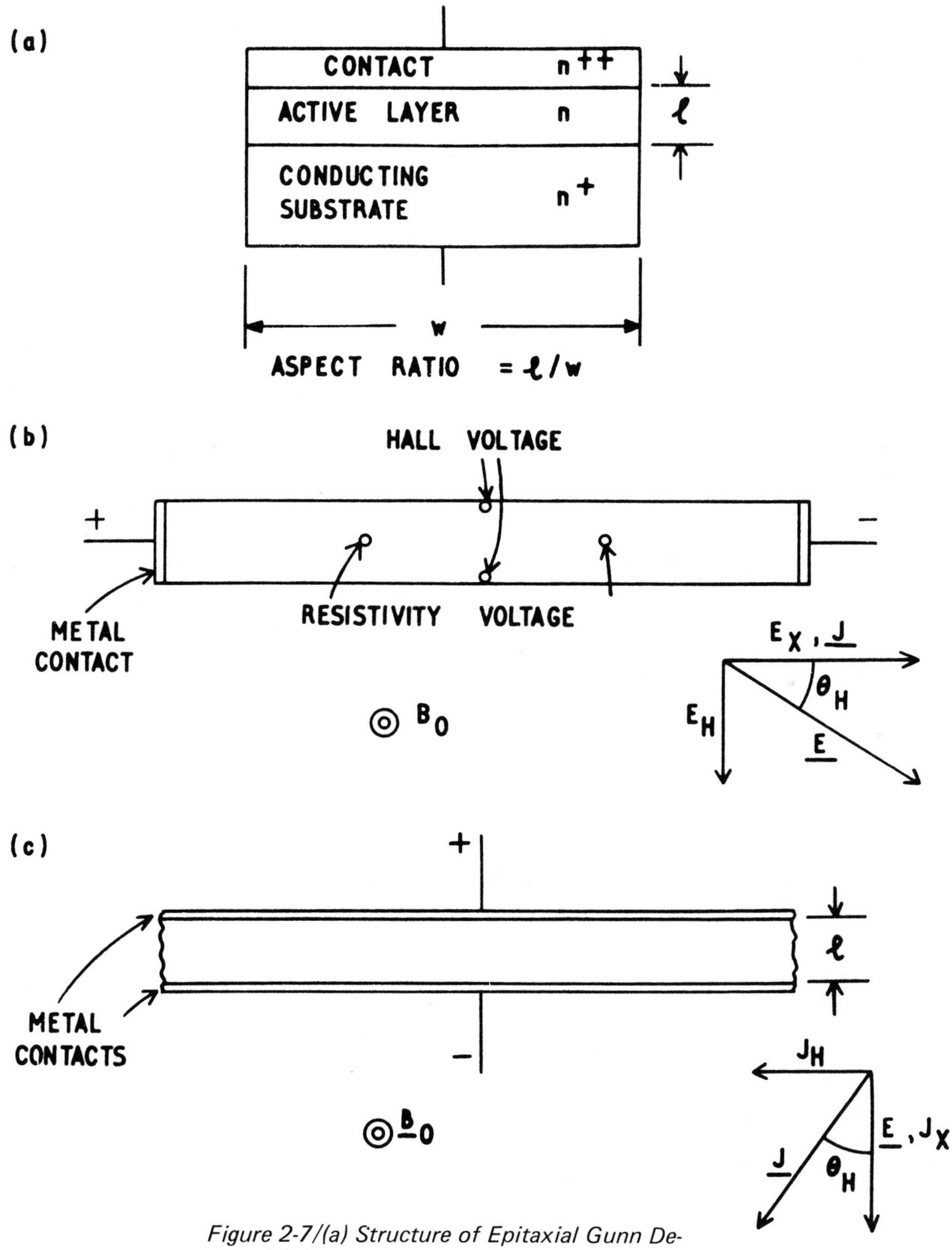

Figure 2-7/(a) Structure of Epitaxial Gunn Device; (b) Conventional Hall Plate; (c) Thin Semiconducting Plate with Plane Conducting Contacts.

is a much smaller effect than GMR. GMR is 25 times larger than PMR in GaAs at room temperature and in low electric fields. An expression for the magnetoresistance of the plate in the limit of small magnetic fields is[32]:

$$\frac{\Delta R}{R_o} = (\xi \mu_H B_o)^2$$

where μ_H is the Hall mobility and ξ is a parameter which depends on the scattering mechanisms.

Experimentally, $\Delta R/R_o = (\mu_m B_o)^2$, defining μ_m as the magnetoresistance mobility. A correction is necessary for finite values of ℓ/W such that:

$$\frac{\Delta R}{R_o} = \frac{\Delta R/R_o \text{ (measured)}}{(1-0.543\ \ell/W)}$$

The factor ξ has been determined for a number of ideal scattering mechanisms. For GaAs and InP, scattering is energy dependent; but at low electric fields, the parameter ξ is near unity. GMR appears a valid measure of mobility at low electric fields; at high fields, relative measurements can be meaningful. Figure 2-8 shows the experimental arrangement that has been used in this laboratory. Many transferred electron devices are of low impedance; to determine magnetoresistance as a function of electric field, the measurement thus is pulsed to avoid heating effects. The resulting measurement technique is more versatile, but the circuitry is more complex. The circuit consists of an astable multivibrator 1 kHz square wave generator, the output of which feeds both the pulse generator across the diode terminals and a sampling circuit. Pulse generators are SN74121 monostable multivibrators which, once fired, are independent of further transitions at the input. They are a function of the timing components only. Output pulse lengths are variable from 40 ns to 40 seconds.

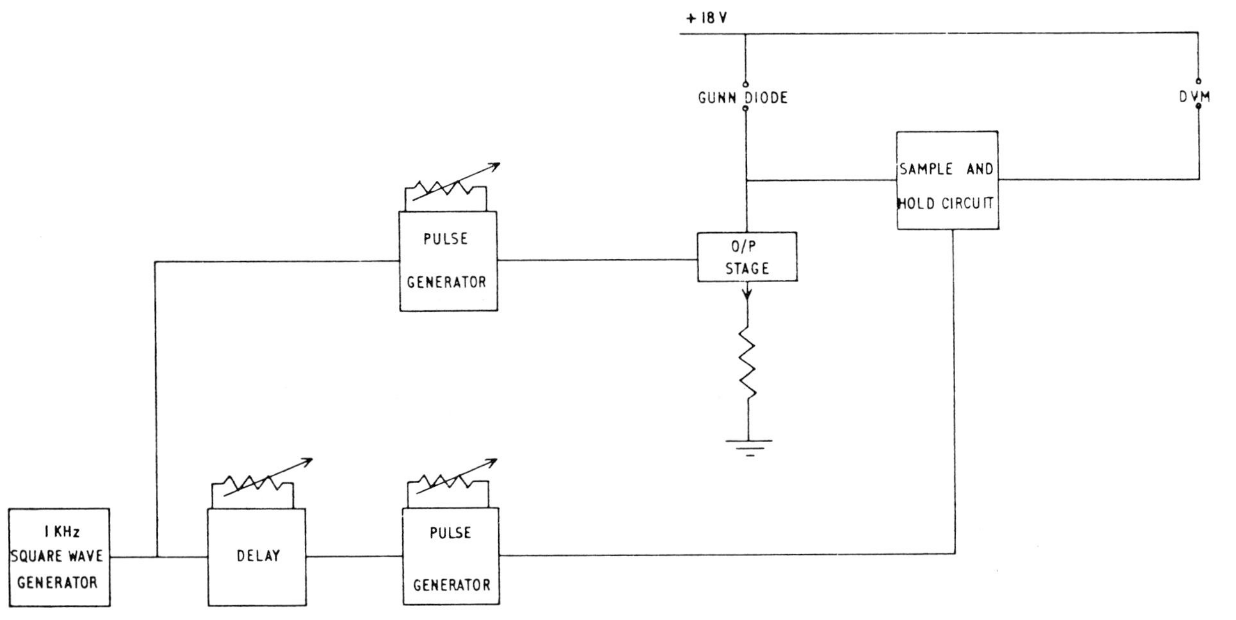

Figure 2-8/Magnetoresistance Measurement

The delay circuit is identical; it is designed to incorporate 100-600 ns delay in the sampling circuit. In this way, a 300 ns pulse is applied across the diode terminals at constant current. The voltage (sampled after the start of the pulse to avoid transient effects) then transfers to a sample and hold circuit[33] and is presented on a digital voltmeter. The procedure is to measure the voltage at a constant current with the sample first outside the magnetic field, then within. The voltage change is directly proportional to that of the resistance of the sample:

$$\mu_M = \frac{1}{B}\left(\frac{\Delta V}{V_o}\right)^{1/2}$$

The magnetic field is probed and calibrated with a Hall probe, and the linearity of ΔV with field is ensured before measurement at magnetic fields up to 10 kG. There are a number of potential sources of error:

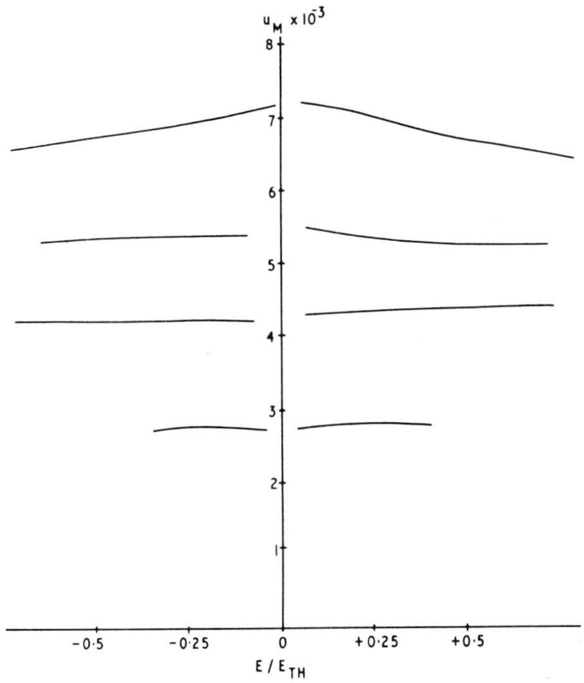

Figure 2-9/Variation of Magnetoresistance Mobility with Magnitude and Polarity of E, GaAs.

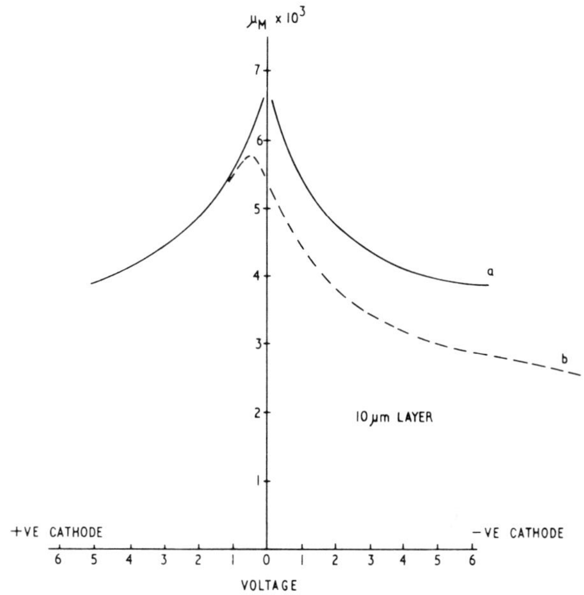

Figure 2-10/Magnetoresistance as a Function of Voltage and Polarity for InP Devices.
(a) Classical Contact, 6% Efficiency.
(b) Anomalous Contact, 9% Efficiency.

1./Geometrical effects due to a ℓ/W ratio that is too high.
2./Magnetic effects caused by diode packages with magnetic materials in their construction.
3./Temperature effects resulting from the mobility dependence on temperature.

Experiments using the magnetoresistance equipment described previously have been executed on Q-band, evaporated contact GaAs diodes, and alloy dot contacted InP diodes[34]. The electric field is reversed in polarity for measurement; thus the value of electric field varies over a suitable range.

Figure 2-9 shows the results on the GaAs diodes. The measured mobility is acceptable for 5.10^{15}

doped material; however, it falls with increasing contact resistance. Device performance is best on the highest mobility devices. (Note — the mobility curve is essentially flat and symmetrical with electric field.)

Figure 2-10 illustrates results of a similar experiment with InP devices. There are striking differences between these curves and those with GaAs.

1./The classical InP devices have a characteristic bell shaped curve. Similar curves are obtained from many different materials, both VPE and LPE.
2./Asymmetry is introduced by the "anomalous" contact behaviour.

The details of contact behaviour are discussed later, but this curve is not easy to explain.

Obvious extensions of the magnetoresistance analysis exist up to high electric fields, as functions of temperature and illumination. The analysis manifests the presence of deep levels and traps; however, the results are difficult to interpret quantitatively.

In summation, magnetoresistance is a very rapid and convenient method for assessing active devices. Extracting detailed information from the results is difficult, although an independent check of contact resistance alone is valuable. Since

$$\frac{R_c}{R_b} = \left(\frac{\mu_H}{\mu_M}\right)^2 - 1$$

where R_c is the contact resistance and R_b is the resistance of the bulk.

In conjunction with other device measurements, magnetoresistance provides essential qualitative information, and the technique can be readily adapted for on-line measurement.

C-V Measurements

To assess epitaxial layers, diffused layers, and junction structures, C-V measurements are currently the major technique used. Aspects of this procedure are described in the key paper by Hughes. The unique advantage of a C-V measurement is that it can be three-dimensional — each separate measurement is made on an element of device dimensions.

The principle of this method is dependent on the change in reverse bias capacitance of a junction when the voltage is changed. The fluctuation is a function of both the depletion of the bulk material and the particular junction properties creating the depletion. Standard semiconductor theory expresses the depletion capacitance associated with different junction structures. The asymmetrical step junction and Schottky barrier junction are the only ones dealt with here, as the majority of measurements are made to use this particular type of junction.

Practical methods of determining the net carrier density rely on two different procedures. The first uses the terms derived from Poisson's equation for the step junction:

$$X = \left(\frac{2\mathcal{E}_o \mathcal{E}_r V_t}{qN}\right)^{1/2}$$

and

$$C = \frac{\mathcal{E}_o \mathcal{E}_r}{X}$$

where X is the depletion layer width, $\mathcal{E}_r$ is the permittivity of the semiconductor, V_t is the total voltage across the junction, that is the applied voltage + built-in voltage, and C is the depletion capacitance per unit area.

It follows that

$$C = \left(\frac{q\,\mathcal{E}_o \mathcal{E}_r N}{2V_t}\right)^{1/2}$$

and

$$N = \frac{C^3}{q\mathcal{E}_o \mathcal{E}_r\,(\text{-}dc/dV)}$$

The results of the last equation are two measurable quantities, C and dC/dV, that determine the carrier density at the depletion depth, $X = \mathcal{E}_o \mathcal{E}_r / C$.

A second approach utilizes the fact that an incremental change in voltage due to an incremental change in charge ΔQ is found to the second order in ΔQ.

$$\Delta Q = \frac{I \cos(\omega t)}{\omega A}$$

for an ac current $I \sin(\omega t)$.

Then it can be shown that:[35]

$$\Delta V = \frac{I \cos(\omega t)}{\omega \mathcal{E} A} \cdot X + \frac{I^2 [\cos(2\omega t) + 1]}{4\omega^2 q \mathcal{E} A^2} \cdot \frac{1}{N(x)}$$

The first term on the right has an amplitude that is proportional to the depletion layer depth, X. The second term, the incremental term at twice the frequency of the driving current, is proportional to the reciprocal of the net carrier density.

In both methods, the requirements of the test diode and the depletion layer limitations are identical. The only differences exist in the mathematical application of step junction theory. Commercial instruments are made for both methods.

Commercial Instruments

The requirement to measure doping profiles is not new; neither is the concept of using a reverse biased junction to obtain this information. However, the use of an RF bridge to measure C at small voltage increments, the subsequent interpolation of dC/dV, and the final calculation are not only laborious, but often inaccurate (largely at the interpolation stage).

What was needed was an integrated test unit to measure C and dC/dV directly, with an analogue computer to immediately plot out the results. Such an instrument has been designed by the RRE and later produced by JAC Electronics Ltd. The complete computer and probe assem-

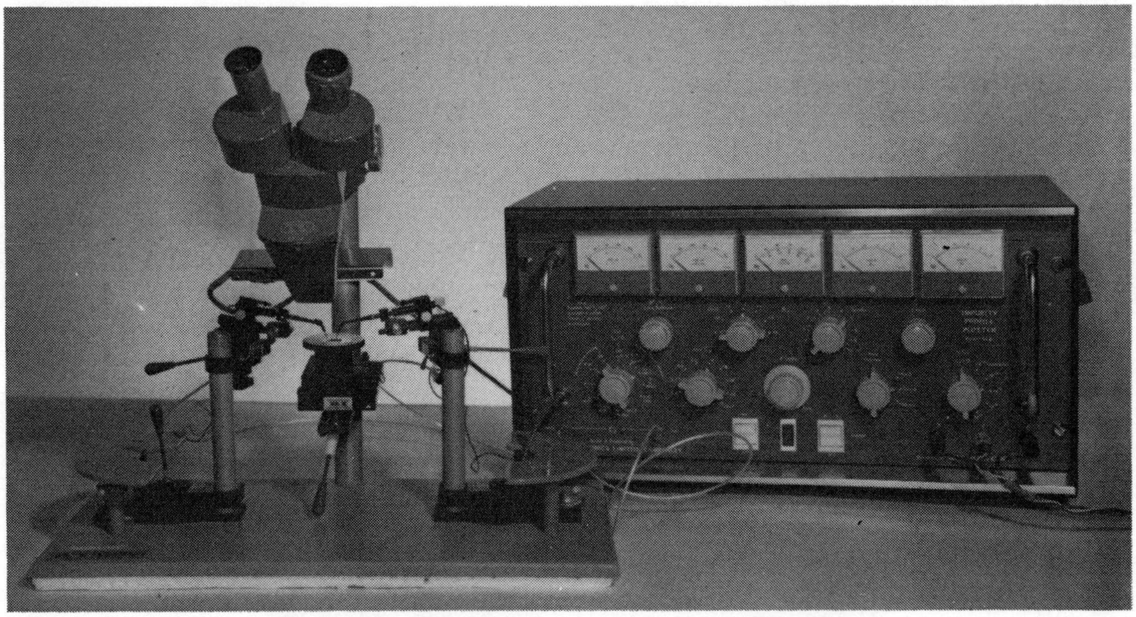

Figure 2-11/Impurity Profiler and Probe Assembly

Figure 2-12/Probes to Test Diodes on a GaAs Wafer.

bly is shown in Figure 2-11; Figure 2-12 shows a close-up of the probe to the contacts on a GaAs wafer. Cost of this instrument is approximately £1000 or $2000.

An instrument employing the second method also has been marketed in the US; however, it is more complex and expensive in its commercial form, and it is also more difficult to setup and use. The price of this instrument is around $10,000. Several commercial instruments have since appeared on the market; the semiconductor engineer can now choose a device to suit the particular requirements.

Limitations of the Measurement

All profiling methods that use reverse biased junctions are limited by the breakdown of the characteristics of the junction. For GaAs and InP, these specifics are shown in Table 2-4.

The complete curves of depletion width, breakdown voltage, and carrier concentration are illustrated in Figure 2-13. The measurement can

Doping (cm^{-3})	Zero Bias Depletion Width (μm)	Maximum dc (Volts)	Maximum Depletion Width (μm)	Depth Resolution (μm)
10^{14}	3.0	500	80	0.5
10^{15}	1.0	100	12	0.15
10^{16}	0.3	20	2	0.05
10^{17}	0.1	4	0.3	0.015

Table 2-4

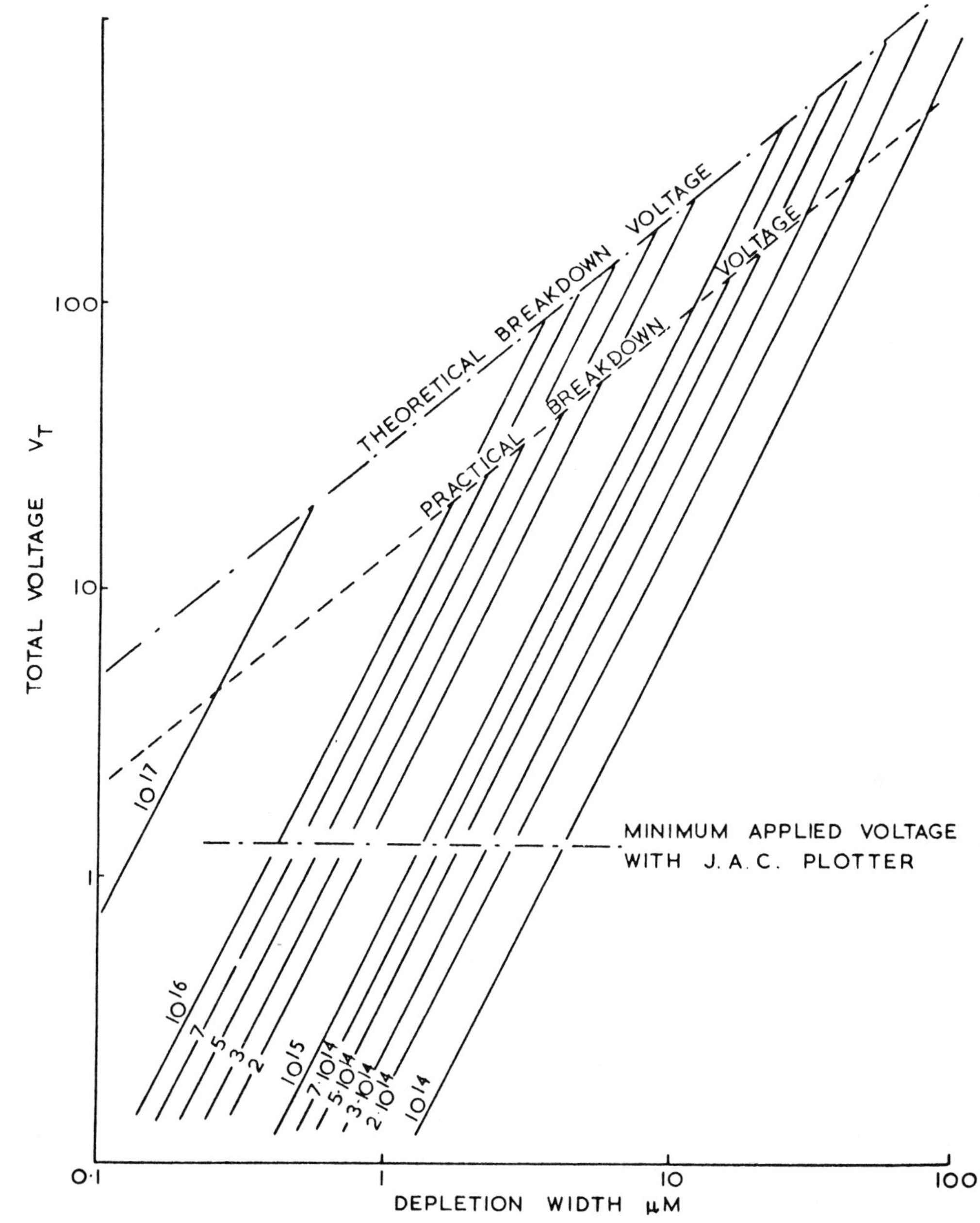

Figure 2-13/GaAs Depletion Layer Characteristics

be carried out on either p- or n-type material, providing an adequate junction can be made. In order to measure the carrier profile of layers grown on semi-insulating substrates (e.g., FET structures), it is essential to apply both ohmic and step junction contacts to the epi-layer surface. For measurements of layers thicker than the maximum depletion width, a successive etching of the wafer must be used. Otherwise C-V measurements are non-destructive.

Errors

A major source of error is the introduction of stray capacitance in the mechanics of contacting the junction or within the junction itself. The former can be overcome by careful design of the probe assembly, but the latter is inherent in the dimensions of the junction. Stray capacitance in the junction happens when the depletion layer is not cylindrical and spreads in the plane of the junction and around its edge. The error in depletion depth can be obtained from:

$$X = X_a \left[1 + b\left(\frac{X_a}{r}\right)\right]$$

where x_a is the apparent depth, b is 1.5 from numerical calculation, and r is the diode radius.

It follows that:

$$N_a = N \left[1 + b\left(\frac{X_a}{r}\right)\right]^3$$

It is therefore essential to make the diameter/depth ratio as large as possible, or to adjust the test instrument for the error introduced[37]. The magnitude of the error in doping density as a function of diode diameter and depletion depth is seen in Figure 2-14.[38]

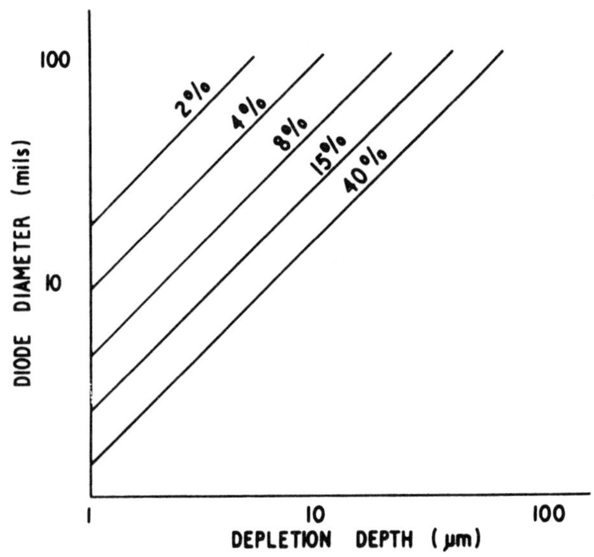

Figure 2-14/Doping Measurement Error Due to Edge Effects

Junction Formation

Metal-semiconductor Schottky barriers are relatively easy to form on GaAs, but not with InP. There are two different approaches, each with relative advantages and disadvantages. One way is to evaporate an array of metal (usually gold or nickel) barrier contacts as a matrix on the wafer. The contact areas are well-defined and the diodes are permanent, enabling the characteristics to be re-checked at any time. (Contact areas can be removed by etching, if necessary.) Processing is relatively long and sophisticated. The other method uses a liquid barrier (such as mercury), applied to the wafer surface through a fine capillary under controlled pressure. This technique has the advantage of simplicity; one can probe any part of the wafer with instant processing without further treatment. The disadvantage is that present capillary designs are apt to introduce additional stray capacitance into the measurement; this capacitance must be accounted for in the computation.

Junction Properties

The necessary junction property is that it is abrupt in order for the depletion region to extend only into the low side of the junction (on uniform material, $V \propto 1/C^2$). The diode characteristic is a sensitive function of the preparation, resulting in errors in the doping profile measurement. It is essential with Schottky barrier contacts to determine the slope factor (n) in the log J-V forward bias diode characteristic. If n is greater than 1.1, there is evidence that the C-V relationship is affected[39] and the measured carrier profile is incorrect. The control is even more important for the mercury barrier where it is much more difficult to form "ideal" barriers.

Deep levels and traps can produce misleading results but, by choosing the correct operating conditions, information on these levels is derived from the C-V measurement. For routine measurements it is customary to ensure that the wafer is in dark conditions.

Summary

The C-V measurement has been valuable to the materials technologist, enabling him to develop the growth process to give almost any desired profile in the epitaxial layer. The C-V measurement identified the interface dip in carrier concentration between the epitaxial layer and the substrate; as a result, the yield of uniform material has increased many times. The measurement is well-suited for on-line assessment, but the operator must be aware of the limitations — particularly those introduced by "non-ideal" diode characteristics.

Unfortunately small differences in profiles have not correlated well with device performance; the correlation *may* be there, but more statistical evidence and a more detailed understanding of the role of small scale doping fluctuations on device performance are needed to establish that relationship.

key paper

A Rapid Evaluation Technique For Function Gunn Diodes

R.D. Larrabee, W.A. Hicinbothem, and M.C. Steele
(RCA Laboratories, Princeton, New Jersey, US)

IEEE Transactions on Electron Devices, Vol. ED-17, No. 4 (April 1970)
Pages 271-274

Abstract—A rapid technique for estimating the carrier mobility, average and minimum carrier concentration, and contact resistance in operational Gunn diodes is presented. Although the physical nature of the functional Gunn-effect device (i.e., an n-type active layer with only two n^+ ohmic contacts) restricts what can be measured, it will be shown that information useful for estimating device performance can be obtained if some simplifying assumptions are granted. The procedure is illustrated by data obtained from five separate wafers of vapor-phase, epitaxially grown gallium arsenide.

Introduction

In 1963 J. B. Gunn announced that gallium arsenide and indium phosphide emitted microwave energy when they were subject to pulsed electric fields of several thousand volts per centimeter [1]. This announcement stimulated a large amount of research aimed at elucidating the mechanism of emission and evaluating its device potential. By late 1965, Gunn's original observations were reasonably well explained and were discussed in a special issue of the IEEE Transactions on Electron Devices [2] devoted to this and similar subjects.

Once the basic mechanism of the Gunn effect was understood [3] and applied to Gunn's original observations of transit-time oscillations, the emphasis of research work shifted toward a study of other possible modes of operation, their material requirements, and the microwave circuitry necessary for realizing them.

The electrical properties of the material required for operation in the transit-time mode originally observed by Gunn can be realized in bulk and epitaxially grown gallium arsenide, but many of the other modes are much more restrictive in the specifications on the material required, and it is often difficult to find suitable material. In some cases, devices constructed from material that was thought to be suitable, failed to operate in the desired mode.

In the past, gallium arsenide was usually characterized by its geometrical size, room temperature carrier concentration, Hall mobility, a temperature coefficient of resistivity, and a carrier mobility at 77°K. However, it is now apparent that these parameters are inadequate and a significant improvement in the characterization of material for Gunn devices is required. This paper discusses some nondestructive measurements on operational Gunn diodes that are so simple as to warrant measurement of all devices for an initial evaluation. Although the physical nature of the operational Gunn-effect device (i.e., an n-type active layer with only two n^+ ohmic contacts) restricts what can be measured, it will be shown that information useful for estimating the performance of Gunn devices can be obtained if reasonable assumptions are granted. This technique has proved useful for the initial evaluation and the rapid selection of those devices that might be suitable for high-performance Gunn oscillators, thus minimizing the need for elaborate, exact, and time-consuming evaluations such as Schottky barrier capacitance–voltage measurements on thick samples.

Estimation of the Low-Field Mobility

The mobility of the charge carriers in semiconductors is usually measured by the Hall effect. However, this requires a minimum of four separate contacts which can be either randomly spaced around the edge of a wafer [4] or, more commonly, at the ends of two mutually perpendicular lines. A current is passed through two contacts and the Hall voltage developed across the other two contacts is measured. In any event, these conventional Hall effect measurements can not be conveniently used on very small devices that have an active layer with only two ohmic contacts.

An alternative method is to measure the change in sample resistance when the device is placed in a magnetic field aligned perpendicular to the current flow direction between the contacts [5]. Let the subscript 0 refer to conditions in zero magnetic field and the subscript b refer to conditions in which the sample is subjected to a transverse magnetic field of flux density B. If the Hall electric field is short-circuited by the contacts, the effective mobility μ_b in the magnetic field is given (in MKS units) by

$$\mu_b = \frac{\mu_0}{1 + (\mu_0 B)^2}. \tag{1}$$

The most convenient way [5] to use this magnetoresistance to measure mobility is to pass a constant current through the sample and measure the voltage across the sample before the magnetic field is applied, V_0, and the change in voltage across the sample when the magnetic field is applied, ΔV. It is a simple algebraic exercise to show that

$$\mu_0 = \frac{1}{B}\sqrt{\frac{\Delta V}{V_0}}. \tag{2}$$

Clearly, $\Delta V/V_0$ must be measured precisely if accurate values of mobility are to be obtained. For example, if $\mu_0 = 0.5\ m^2/V\cdot s$ ($5000\ cm^2/V\cdot s$) and $B = 0.5\ Wb/m^2$ (5000 gauss), (2) makes $\Delta V/V_0 = 0.0625$.

Measurement of Average Carrier Concentration

Once the carrier mobility has been determined, one can use the measured, low electric-field device resistance to compute an average value for the carrier density. If one considers that the carrier density $N(x)$ may be a function of the distance between the contacts x, one has

$$\frac{1}{N_{avg}} = \frac{1}{l}\int_0^l \frac{dx}{N(x)} = \frac{e\mu_0 A}{l} R \tag{3}$$

where

l = the total length of the sample between contacts
A = the cross-sectional area of the sample
e = the electronic charge
R = the measured sample resistance.

It has been assumed that μ_0 and A are independent of position. Note that the average value so obtained is a harmonic average over the sample length and not the arithmetic average.

Measurement of the Minimum Carrier Concentration

Consider the case in which the electron density is inhomogeneously distributed over the length of the sample between contacts according to the function $N(x)$. If a subthreshold current is passed through the sample, the electric field will, in general, be nonuniform and dependent upon $N(x)$. Since no appreciable space charge (e.g., domains) can build up below the Gunn threshold, diffusion and/or space-charge effects will be small and the electric field will be a maximum at, or very near, the point where $N(x)$ has its minimum value. As the current is increased, a point is reached where either one of two things occurs: 1) the electric field near the minimum density point reaches the Gunn threshold, or 2) if this region is sufficiently thin, it will break down via tunneling, avalanching, or space-charge injection. In the latter case, further increase in the current will ultimately cause some other low $N(x)$ point to attain threshold electric field. Experimentally, the current at which the Gunn threshold electric field is first reached at some point within the sample can be observed as the onset of any of the phenomena usually associated with the Gunn effect (e.g., bias circuit oscillations, current saturation, microwave emission, discontinuous decrease in the bias current and/or increase in the bias voltage, etc.). If the electron-drift velocity, V_d, at the Gunn-effect threshold field is independently known (2.2×10^7 cm/s for electrons in GaAs) [6], one can use the measured current immediately below threshold to compute the electron density at this minimum density point in the following way:

$$N_{min} = \frac{I_{threshold}}{AeV_d}. \tag{4}$$

This equation yields a phenomenological minimum density but does not determine where within the sample this minimum point occurs. There are, in fact, many distributions $N(x)$ that are characterized by the same values of N_{min} and N_{avg} and the present technique cannot distinguish between them. The measurement of N_{min} is based on the current just below threshold and, therefore, is independent of the mode of operation or the important parameters (e.g., Nl, N/f, etc.) of the ultimate Gunn device. It is related to, but somewhat less complicated than, a figure of merit based on the decrease in current at threshold in Gunn oscillators. Comparison of N_{min} to N_{avg} gives one a crude measure of the homogeneity of the electron density within the sample. The ratio of N_{avg} to N_{min} has proved useful as a figure of merit in samples whose gross inhomogeneity is a gradation of density between contacts. Indeed, the high efficiency Gunn oscillators reported elsewhere [7]–[9] all had N_{avg}/N_{min} ratios less than about 1.4. Samples with ratios greater than this value have not operated as high-efficiency oscillators.

The voltage applied to a Gunn device immediately below threshold is a measure of the electric field at which the drift velocity of the electrons is a maximum at some point in the sample. However, one cannot simply divide this voltage by the sample length to obtain this field because the field distribution, $E(x)$, may be nonuniform, and actually will only equal the threshold value near the minimum density point. One can, however, estimate the value of the threshold field in spite of the inhomogeneity in the following way.

From (3) and (4),

$$\frac{N_{avg}}{N_{min}} = \frac{\left(\frac{v_d}{\mu_0}\right) l}{(I_{threshold}) R}.$$

If the mobility μ_0 is assumed constant from low voltages to threshold,

$$v_d/\mu_0 = E_{threshold} \quad \text{and} \quad (I_{threshold})(R) = V_{threshold}.$$

Thus,

$$E_{threshold} = \left(\frac{V_{threshold}}{l}\right)\left(\frac{N_{avg}}{N_{min}}\right). \tag{5}$$

This equation can be qualitatively explained in the

following way. Near the minimum density point the electric field is equal to the threshold value. At all other points in the sample, the density of electrons is larger than it is at the minimum point and the electric field is lower than threshold. Consequently, the net voltage across the device is lower than one would expect if the inhomogeneity were neglected. Equation (5) states that the ratio of N_{avg}/N_{min} is a measure of just how much lower this applied voltage is for the inhomogeneous sample. Looking at it another way, unusually low values of apparent threshold electric field (i.e., $V_{threshold}/l$) portend large sample inhomogeneity (i.e., N_{avg}/N_{min}).

Estimation of Contact Resistance

The above discussion has assumed that there is negligible contact resistance. If the Hall mobility is independently known, one can compare it with the mobility as measured by the magnetoresistance technique and estimate the contact resistance in the following way. If it is assumed that the contact resistance is invariant to the magnetic field, one can use (1) to show that

$$R_c/R_b = (\mu_0/\mu)^2 - 1 \tag{6}$$

where

R_c = the resistance of the contacts
R_b = the resistance of the active Gunn layer (bulk)
μ_0 = the known correct value of the carrier mobility
μ = the mobility as measured by the present technique (2).

Of course, once R_c is known, it must be subtracted from the total sample resistance before computing N_{avg} from (3). The calculation of N_{min} is independent of any contact resistance.

However, one often does not have independent Hall measurements on the material actually used in a particular device and one would like a way to estimate the contact resistance in the particular device in hand. One may know or assume the threshold electric field for the Gunn effect and, if (5) yields too high a value, attribute it to a voltage drop in the contacts. Substituting for N_{avg} (3) in (5):

$$E_T = \left(\frac{V}{lN_{min}}\right)\left(\frac{l}{e\mu A}\right)\left(\frac{1}{R}\right).$$

Letting $V = V_{total} - V_{contact}$ and $R = R_{total} - R_{contact}$ one obtains

$$\frac{R_c}{R_b} = \left(\frac{E_t^a}{E_t}\right)^2 - 1 \tag{7}$$

where

E_t^a = the actual known (or assumed) threshold field
E_t = the apparent threshold field as calculated by (5) in which contact resistance is completely ignored.

We have had considerable success in using (7) with E_t^a equal to 3.2 kV/cm, the value found by Ruch and Kino [6]. Only occasionally is a sample found that apparently has a threshold field below 3.2 kV/cm, thus leading to a calculated negative value of contact resistance. In this case we have assumed that the contact resistance is actually zero. Once the contact resistance has been found, the mobility should be corrected by use of (6) and the value of N_{avg} recomputed using the now known correct bulk resistance.

Illustration of the Evaluation Technique

Since the computations necessary to the present evaluation technique are time consuming if many samples are to be tested, a time-shared computer was programmed to calculate the desired quantities from the measured data. Table I summarizes the results obtained on two different sessions with this computer in which five samples of vapor-phase, epitaxial n^+nn^+ GaAs were evaluated. The uppermost portion of this table represents the measured data and the lower two portions of this table the computed information. The device name, length, (cross sectional) area, (low field) resistance, and onset (threshold of anything associated with the Gunn effect) voltage and current are self-explanatory. DV/V is the relative change in sample voltage (at a constant low current) when the sample is placed in a magnetic field oriented perpendicularly to the current flow. The middle portion of Table I lists the corresponding computed values of electron mobility ($cm^2/V \cdot s$), average and minimum electron concentration (in units of 10^{15} electrons/cm^3), the inhomogeneity parameter N_{avg}/N_{min} and the ratio R_c/R_b of contact to bulk (active layer) resistance, as computed from (7). The values of mobility, N_{avg}, and N_{avg}/N_{min} have been computed taking this contact resistance into account. Notice that several samples yielded a negative contact resistance. In this case the computer was programmed to arbitrarily set R_c/R_b equal to -0.001 (essentially zero) to indicate that a negative contact resistance was found but to avoid making the contact resistance corrections with the unrealistic negative contact resistance. If the measured low electric-field sample resistance is equal to or greater than V_{onset}/I_{onset}, the computer program prints JUNCTION?? as a warning that there may be a junction at the contact(s) that may invalidate the low field measurements of resistance and DV/V. These samples should have their current–voltage curves examined in greater detail before accepting this evaluation. The bottom portion of the table lists the computed values of $N_{avg}l$ product in units of 10^{12} cm^{-2}, the transit-time (TT) frequency ($10^7/l$), and the LSA frequency (i.e., the frequency where $N_{avg}/f = 5 \times 10^4 s/cm^2$).

The first two wafers in Table I were measured in a 5000-gauss magnetic field and are representative of wafers yielding very efficient Gunn oscillators. In ex-

TABLE I

Wafer No.	Length (Microns)	Area (Sq. Mils)	DV/V	Resistance (Ohms)	V Onset (Volts)	I Onset (Amperes)
360	75.0	225.0	0.0920	13.93	22.50	1.85
P-33	9.0	9.0	0.0540	10.35	5.40	0.455
448-C	90.0	93.7	0.0490	29.41	18.00	1.30
463-A	70.0	58.4	0.0647	12.80	15.70	1.34
310	25.0	88.7	0.0400	8.70	9.50	0.90

Wafer No.	Mobility (CGS)	N Avg $E+15$	N Min $E+15$	Avg/Min	R_c/R_b	
360	6063.	0.382	0.374	1.021	−0.001	JUNCTION??
P-33	6317.	2.737	2.300	1.190	0.847	
448-C	4758.	0.664	0.631	1.052	−0.001	JUNCTION??
463-A	5788.	1.755	1.044	1.681	0.120	JUNCTION??
310	6468.	1.097	0.462	2.377	1.261	

Wafer No.	Nl $E+12$	TT Frequency (GHz)	LSA Frequency (GHz)
360	2.86	1.333	7.639
P-33	2.46	11.111	54.747
448-C	5.98	1.111	13.286
463-A	12.28	1.429	35.098
310	2.74	4.000	21.945

periments by Berson [7], wafer 360 has produced pulsed devices that operated with efficiencies exceeding 30 percent. Wafer P-33 has yielded X-band pulsed devices with efficiencies exceeding 15 percent, in spite of its very high apparent contact resistance. It is our experience that samples operable at high efficiency usually have inhomogeneity parameters less than about 1.2, although samples with somewhat higher ratios have been operated with high efficiency [8], [9]. The last three wafers in Table I were measured with a magnetic field of 4650 gauss. Wafer 448-C was very inefficient compared to the first two wafers in Table I in spite of its low inhomogeneity parameter, thus illustrating that low ratios of N_{avg}/N_{min} seem to be a necessary, but not sufficient condition for high-efficiency operation. Wafer 463 yielded about 5 percent efficiency and wafer 310 could not be made to operate with efficiencies exceeding 0.1 percent. Notice that the mobilities in Table I do not correlate with the performance of the wafers (except possibly in the case of wafer 448-C). It has been our experience that the inhomogeneity parameter N_{avg}/N_{min} is a more satisfactory measure of potential performance than is the more commonly used electron mobility.

Conclusion

A technique has been presented that is capable of evaluating quickly some parameters of importance in operational Gunn effect devices. The procedure is relatively simple and applicable to functional devices, including long samples where conventional techniques may be very time consuming. The technique might be used on each Gunn device as an initial basis of evaluation and/or selection.

Acknowledgment

The authors wish to acknowledge the active cooperation of all the material preparation groups at RCA Laboratories that have been wrestling for several years with the very difficult problem of growing homogeneous gallium arsenide of the required N_{avg}. The experimental work was performed on gallium arsenide grown by R. E. Enstrom, A. Gobat, and I. Hegyi. The authors also acknowledge the assistance of V. Cannuli in preparing samples.

References

[1] J. B. Gunn, "Microwave oscillations of current in III–V semiconductors," *Solid State Commun.*, vol. 1, pp. 88–91, September 1963; also, "Instabilities of current in III–V semiconductors," *IBM J. Res. and Develop.*, vol. 8, pp. 141–159, April 1964.

[2] Special issue of *IEEE Trans. Electron Devices* devoted to semiconductor bulk-effect and transit-time devices, vol. ED-13, January 1966.

[3] Second special issue of *IEEE Trans. Electron Devices* devoted to semiconductor bulk-effect and transit-time devices, vol. ED-14, September 1967.

[4] L. J. van der Pauw, "Method of measuring specific resistivity and Hall effect of disks of arbitrary shape," *Philips Res. Repts.*, vol. 13, pp. 1–9, 1958.

[5] J. A. Copeland, private communication.

[6] J. G. Ruch and G. S. Kino, "Measurement of the velocity-field characteristic of gallium arsenide," *Appl. Phys. Letters*, vol. 10, pp. 40–43, 1967.

[7] J. F. Reynolds, B. E. Berson, and R. E. Enstrom, "High-efficiency transferred electron oscillators," *Proc. IEEE* (Letters), vol. 57, pp. 1692–1693, September 1969.

[8] F. P. Califano, "High efficiency X-band Gunn oscillators," *Proc. IEEE* (Letters), vol. 57, pp. 251–252, February 1969.

[9] M. C. Steele, F. P. Califano, and R. D. Larrabee, "High-efficiency series operation of Gunn devices," *Electron. Letters*, pp. 81–82, February 1969.

key paper

The Characterization Of Epitaxial GaAs Using Schottky Barriers

F.D. Hughes
(Mullard Research Laboratories, Redhill, Surrey, UK)

Acta Electronica, Volume 15 (1972)
Pages 43-53

The characteristics of a Schottky barrier made on a semiconductor can be used to provide highly accurate and unique data on the properties of the material. These include the carrier concentration profile, the detection of deep traps and the measurement of some of the properties of these traps.
The techniques used to make the Schottky barriers and to measure these properties are given. Some results obtained on both vapour and liquid phase epitaxial GaAs intended for Gunn devices are presented and discussed.

Les caractéristiques d'une barrière de Schottky réalisée sur un semiconducteur peuvent être utilisées pour fournir des renseignements uniques et très précis sur les propriétés du matériau. Entre autres, elles permettent de déterminer le profil de concentration des porteurs, de détecter les pièges profonds et de mesurer quelques propriétés de ces pièges. Les techniques utilisées pour réaliser les barrières de Schottky et pour mesurer ces propriétés sont exposées. Ensuite sont présentés et discutés quelques résultats concernant l'arséniure de gallium obtenu par épitaxie en phase liquide ou en phase vapeur et destiné à être utilisé dans des dispositifs Gunn.

From the properties of a Schottky barrier formed on the surface of an epitaxial layer of GaAs it is possible to obtain much information about the electrical characteristics of the material. In this context the most widely used property of the barrier is its capacity under reverse bias. This is related to the net concentration of charged centres in the depletion layer associated with the barrier. By changing either the bias on the barrier, its temperature or illumination and measuring the resulting change in capacity it is possible to obtain the free carrier profile through the layer, detect the presence of deep traps and to measure some of their properties. These measurements using Schottky barriers have the advantage of being made on material grown on a conducting substrate, that is the actual material used for the production of Gunn diodes, rather than on a separate monitor slice.

There follows a brief outline of the general theory of a Schottky barrier and how this can be used to investigate the material properties. A description is given of the techniques used to measure the carrier concentration profile together with a discussion of their limitations and some typical results. The methods of investigating the deep traps in GaAs are also described and the results on four such traps are discussed.

THEORY

A Schottky barrier is a simple metal-semiconductor junction where the Fermi levels of the two materials must be coincident at thermal equilibrium (fig. 1a). This results in the establishment of an electrostatic barrier between the metal and the semiconductor. Since the carrier concentration in the semiconductor is many orders of magnitude less than that in the metal, this barrier appears in the form of a depletion layer adjacent to the metal and extending an appreciable distance into the semiconductor. The characteristics of ideal Schottky barriers are well established in terms of the barrier height and the equivalent capacity of the space charge region [1]. Since only N-type material is of interest in connection with the Gunn effect only the results applicable to this type of material are given below.

The height of the barrier V_B, between the semiconductor and the metal, depends upon the metal work function ϕ_m and the electron affinity of the semiconductor χ. For a barrier with no surface states present, this is given by :

$$V_B = \phi_m - \chi \; . \tag{1}$$

The depletion layer in N-type material is produced by the removal of electrons resulting in a net positive

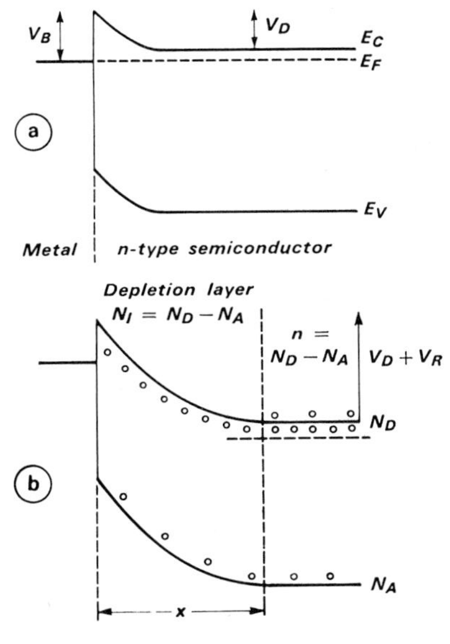

Fig. 1. *A Schottky barrier on an N-type semiconductor.*

space charge arising from the ionized donors (fig. 1b). From Schottky barrier theory the space charge Q and the associated depletion layer capacity are related to the barrier height by :

$$C = \frac{\partial Q}{\partial V_R} = \left[\frac{q\varepsilon\varepsilon_0 A^2 N_I}{2(V_D + V_R)}\right]^{\frac{1}{2}} , \tag{2}$$

where :

$$V_D = V_B - \frac{(E_C - E_F)}{q} \; . \tag{3}$$

Here N_I is the net ionized impurity concentration, A the area of the barrier, V_R the reverse bias applied to the barrier, V_D the diffusion potential and E_C and E_F the energy positions of the semiconductor conduction band and Fermi level respectively.

Clearly if the capacity of a Schottky barrier is measured as a function of reverse bias then from a plot of C^{-2} against V_R both V_D and N_I may be obtained from the intercept at $C^{-2} = 0$ and the slope respectively. For material which has only shallow donors and acceptors the net ionized impurity concentration is given by :

$$N_I = N_D(1 - P_D) - N_A \; , \tag{4}$$

where the occupation number of the donor P_D lies in the range 0 to $(1 - N_A/N_D)$, and N_D and N_A are the donor and acceptor concentrations. If N_I is constant through the material then clearly it is equal to the free electron concentration n in the conduction band and has a maximum value of $N_D - N_A$ when the donor is fully ionized. This will occur at room temperature for a sufficiently shallow donor. However if N_I is not constant through the material then the C^{-2} versus V_R plot will deviate from linearity. This change in slope may be used to measure changes in N_I through the material as the reverse bias and hence the depletion layer width are increased [2].

A much more convenient way of measuring n as a function of position becomes apparent if equation (2) for the small signal capacity is expressed in terms of the free carrier concentration at the edge of the depletion layer, $n(x)$ (see for example [3]):

$$n(x) = \frac{-C^3}{q\varepsilon\varepsilon_0 A^2}\left(\frac{\mathrm{d}C}{\mathrm{d}V}\right)^{-1} . \tag{5}$$

By using equation (5), in conjunction with the traditional expression for the capacity of a Schottky barrier [1] :

$$C = \frac{\varepsilon\varepsilon_0 A}{x} , \tag{6}$$

it is possible to measure $n(x)$. Here x is the width of

the depletion layer. These equations form the basis of the differential capacitance technique for measuring the free carrier concentration profile and will be discussed in more detail later.

In the above discussion it has been assumed that only shallow donors and acceptors are present in the depletion layer, however if the material contains deep lying traps which are ionized then they also contribute to the net space charge. Equation (4) is readily extended to include both donor- and acceptor-like traps in the net ionized impurity concentration N_I. For example for a material with two deep traps, one donor-like and one acceptor-like with concentrations N_{T_1} and N_{T_2} respectively, N_I is given by :

$$N_I = N_D(1-P_D) - N_A + N_{T_1}(1-P_1) - N_{T_2}P_2 \ , \quad (7)$$

where P_1 and P_2 are the occupation numbers of the traps. Equation (2) may also be rearranged to express the net ionized impurity concentration in the depletion layer explicitly in terms of its capacity :

$$N_I = \frac{2C^2(V_D+V_R)}{q\varepsilon\varepsilon_0 A^2} \ . \quad (8)$$

where N_I is now given by equation (7).
Through equations (7) and (8) it may be seen that if the occupation of one or both of the traps in the depletion layer is changed then this is reflected in a change in the barrier capacity. Therefore using such a change in capacity as a measure of the change in occupation of a trap it is possible to detect the presence of trapping centres in the depletion layer. The occupation may be changed by varying the temperature or illumination of the material and, as will be shown later, the detailed dependence of the capacity on both these parameters can be used to obtain information about the properties of the traps.
These then are the basic properties of a Schottky barrier which we will use in the following discussion on their application to the electrical characterization of epitaxial GaAs for Gunn diodes.

CARRIER CONCENTRATION PROFILES

Methods of measurement

The carrier concentration profile, that is the carrier concentration as a function of position through the epitaxial GaAs layer, has been shown to be an important parameter in the performance of a Gunn diode [4] and this can be conveniently measured using the differential capacitance technique. Essentially the capacity of a reverse biased Schottky barrier formed on the epitaxial surface is measured as a function of bias and equations (5) and (6) are solved to obtain the profile.

Traditionally the capacity has been measured point-by-point using a conventional capacitance bridge [5]. However this has the disadvantage of not giving a continuous record of the profile and is very time consuming. Recently several techniques have been proposed to automate the measurements and these have reduced the measurement time from a few hours to a few minutes.
One such method, developed at the Royal Radar Establishment (RRE) [6], drives a 100 kHz current through the Schottky barrier which produces a 100 kHz voltage proportional to $C(x)$. The current is amplitude modulated at 1 kHz to give a signal proportional to $\mathrm{d}C(x)/\mathrm{d}V$ and the equipment electronically computes $n(x)$ and x from equations (5) and (6). The results are presented in the form of D.C. voltages proportional to lg $[n(x)]$ and x which can be displayed on an X-Y recorder.
A second method proposed by J.A. Copeland [7] utilises the non linear behaviour of a Schottky barrier to generate harmonic voltages when driven by a constant R.F. current. It may be shown that the voltage generated at the fundamental frequency is proportional to the depth of the depletion layer and that at the second harmonic is proportional to $[n(x)]^{-1}$. These signals are separated and transformed to D.C. voltages which again can be displayed on an X-Y recorder. In both these techniques the profile is obtained directly by slowly sweeping the reverse bias, giving a continuous record of the profile in a few minutes once a Schottky barrier has been formed. We have measured profiles on both Copeland [7] and RRE [6] plotters in our work and obtained excellent agreement between them.
Other techniques have been demonstrated which are analogous to the point-by-point technique in that the capacity is measured directly and this is then processed electronically to give the profile [8, 9, 10]. All these methods are useful for rapidly giving a continuous profile measurement.

The Schottky barriers on epitaxial GaAs may be conveniently made by the evaporation of gold or nickel onto the epitaxial surface [11], [12]. The metal is deposited onto the chemically cleaned surface either through an incontact mask *peppered* with holes or over the whole surface when the barrier areas are defined by photolithographic techniques.
An ohmic contact must be made to the conducting substrate and this is readily provided by alloying an evaporated gold-4 % tin mixture in a reducing atmosphere. The barriers are then mounted for use with the profiling machine. Unfortunately this technique of making Schottky barriers is time consuming in comparison with the time taken to measure the profile.

A much more rapid process has been developed which uses mercury as the barrier metal. A similar technique using mercury has been employed for measurements on MOS structures on silicon [10]. In figure 2 is shown a simple mount for making mercury Schottky barriers which is suitable for use with a Copeland plotter [7] where it is desirable to

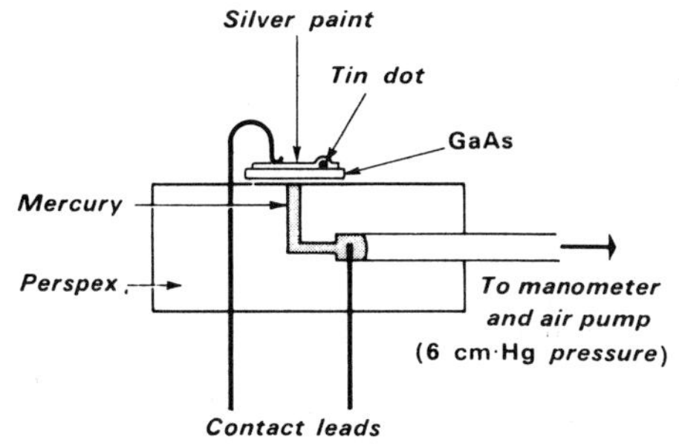

Fig. 2. *A simple mercury Schottky barrier mount suitable for use with a Copeland plotter* [7].

maintain the stray capacity at a minimum. A perspex block is drilled with a small hole and is filled with mercury. The cleaned epitaxial GaAs is held against the end of this tube and an air pressure is applied to the thread of mercury to force it up against the epitaxial surface. A pressure of 6 cm of mercury (7.98×10^3 $N \cdot m^{-2}$), is sufficient to overcome the meniscus ensuring that the end of the tube is filled thus giving a barrier of reproducible area. The ohmic contact to the substrate may be made using an alloyed tin bead.

With an RRE plotter [6], where, as will be shown later, it is possible to compensate for the effects of the stray capacity [13], a much more versatile mercury probe may be used. One such probe [13], based on a design by R. Hammer [14], is shown schematically in figure 3. The epitaxial GaAs is held against a vacuum chuck and the probe tip, mounted on the end of a flexible U-tube, is brought up against the surface with an X-Y-Z micropositioner. The U-tube is filled with mercury and pressure is accurately applied to the barrier by raising one side which is mounted on a rack and pinion slide. In this design the probe tip is easily changed which allows barriers of different sizes to be made. The area is chosen so that for a particular carrier concentration the capacity falls within the range of the instrument, 30 pF to 0.03 pF. Through the X-Y micropositioner it is possible to measure the profile at any point on the epitaxial surface.

The time taken to make Schottky barriers is therefore greatly reduced by using either of these mercury systems and also has the added advantage of being non-destructive. This increase of speed of profiling is very useful for the rapid feedback of information on growth experiments and for profiling large numbers of layers for device production.

It is an important requirement for the differential capacitance technique that the contribution to the capacitance from the semiconductor surface states is small [12]. This can be checked by measuring the forward current voltage characteristics of a barrier and fitting it to the equation for an ideal Schottky barrier [11] :

$$I = I_s \left[\exp\left(\frac{qV}{\beta kT}\right) - 1 \right] . \tag{9}$$

If the surface state contribution is small then the numerical constant β should be close to unity. For evaporated metal and mercury barriers prepared as described above a value of $\beta = 1.05$ is typically obtained over several orders of magnitude of current.

Before discussing some of the results obtained from profile measurement it is worthwhile examining some of the limitations and possible errors in the technique. Firstly it should be emphasized that the technique measures the free carrier concentration and not the impurity concentration. D.P. Kennedy et al. [15] have shown that only in the case of charge neutrality in the semiconductor will the carrier profile and the impurity profile be identical. In general the impurity concentration will not be constant, giving rise to carrier diffusion, and it is this diffused profile which is measured. Also since all the automatic profiling techniques involve measurements at high frequencies then only the thermally ionized carriers from the shallow donors will be measured, since electrons in the deeper traps will not be able to respond to the test signal. A further limitation is avalanche breakdown of the Schottky barrier at large reverse bias which limits the depth to which measurements can be made. S.M. Sze and G. Gibbons [16] have calculated the breakdown voltage of an abrupt junction in GaAs as a function of carrier concentration. They give the maximum depth of the depletion layer at breakdown as 20 μm and 3 μm for material with carrier concentrations of 1×10^{15} cm^{-3} and 1×10^{16} cm^{-3} respectively. Therefore for epitaxial GaAs intended for X-band Gunn diodes with a thickness of 10 μm and a carrier concentration of 2×10^{15} cm^{-3} it is possible to profile to the layer-substrate interface with a good quality barrier.

From equation (5) it is clear that any error in the capacity or barrier area has a considerable effect

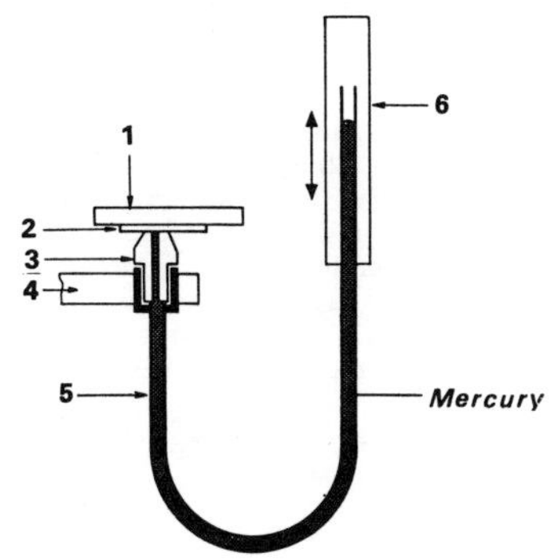

Fig. 3. *Schematic diagram of an adjustable mercury Schottky barrier probe for use with an RRE plotter* [6].

1. vacuum
2. chuck
3. probe tip
4. X-Y-Z micropositioner
5. flexible U-tube
6. rack and pinion slide.

on the value obtained for $n(x)$. Therefore it is necessary either to measure the area of each barrier or to reproduce it accurately. For the evaporated metal barriers the area can easily be measured, however care should be taken to eliminate any penumbra produced during evaporation either by etching or using photolithographic techniques. It is not possible to measure the diameter of each mercury barrier but it has been found that with careful regulation of the pressure on the mercury a contact diameter reproducible to better than 1 % can be achieved [13, 14]. I. Amron [15] has discussed in detail the effects of these errors and showed that for a total error of 10 % in $n(x)$ the capacity and area must be measured to within 0.5 % and 1.0 % respectively.
Unavoidable errors in the capacity arise also from the external stray capacitances and from the edge capacity of the diode itself [18]. This latter error may be minimized by using a large diameter barrier but J.A. Copeland [19] has shown that for a barrier diameter of 1 mm the error in $n(x)$ can still be as great as 10 % at 10 μm depth.

However if the total stray capacity C_s is known then the real profile may be easily calculated from the measured profile through the equations [13] :

$$x_R = x_M \left(1 - \frac{x_M C_s}{\varepsilon\varepsilon_0 A}\right)^{-1}$$

and

$$n_R(x_R) = n_M(x_M)\left(1 + \frac{x_R C_s}{\varepsilon\varepsilon_0 A}\right)^{-3},$$

where the subscripts R and M refer to the real and measured quantities. C_s may be calculated initially through these equations from the profile measured on a sample which is known, from step etching and reprofiling, to have a constant carrier concentration. See, for example, figure 4 where the solid lines are the measured profile of a step etched layer before being corrected for the stray capacity. Recently a modification to the RRE plotter has been demonstrated by F.D. Hughes et al. [13] which automatically compensates the measurements for these stray capacities. This has allowed the versatile mercury Schottky barrier probe, as described earlier, to be used. Also in figure 4 is shown the compensated profile, the broken lines, measured using this modification. The total stray capacity in this case was 0.57 pF for a barrier of 473 μm diameter. J.A. Copeland [19] has calculated the edge capacity for this size of barrier to be approximately 0.12 pF, the extra capacity being the stray introduced by the probe design.
From the last etch step of the profile in figure 4 the compensation is clearly demonstrated where both $n(x)$ is reduced and x increased.

Results

Considerable variations have been observed in the types of profile measured on epitaxial GaAs. Four typical profiles are shown in figures 4 and 5 for vapour phase material grown on N^+ substrates. The substrates have a carrier concentration in the range 10^{17} to 10^{18} cm^{-3} and at the interface the profile exhibits a sharp rise. The two flat profiles in figure 5 show near ideal material for X and Q band Gunn devices respectively. The third profile illustrates a common fault in having a large dip in the carrier concentration near the layer-substrate interface. This type of profile has been reported for material grown at many laboratories throughout the world and the origin of the dip has received considerable attention. There is some suggestion that it may be due to a region with a net acceptor concentration, [20, 21], although this has yet to be proved. Other profiles have been measured which show bumps or notches in the carrier concentration. These can often be related to fluctuations in the growth conditions such as the source temperature or the temperature of the $AsCl_3$.

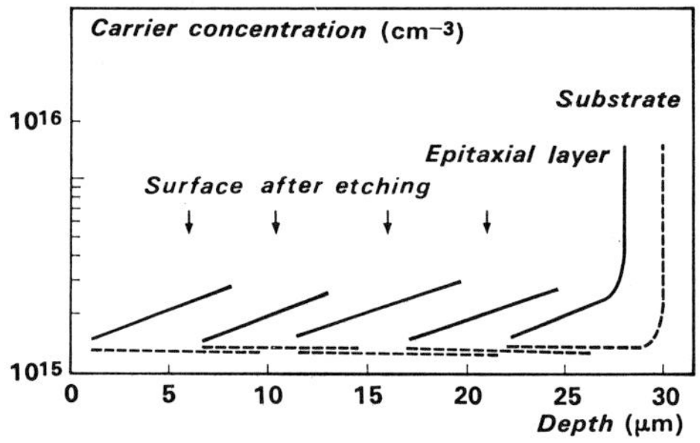

Fig. 4. *Carrier profile of an epitaxial GaAs layer ; full line, uncompensated profile, broken line, compensated profile.*

Generally only one or two profiles are measured to characterize a full layer. Profile maps have been measured on complete slices and these indicate that the material is usually homogenous across the layer. However occasionally a map has indicated a gradually changing profile along a substrate and this change usually correlates with the direction of the gas flow during growth. The flatter profiles occurring at the upstream end of the substrate. However provided the profile is measured away from the edge of the slice it is usually representative of the whole layer.

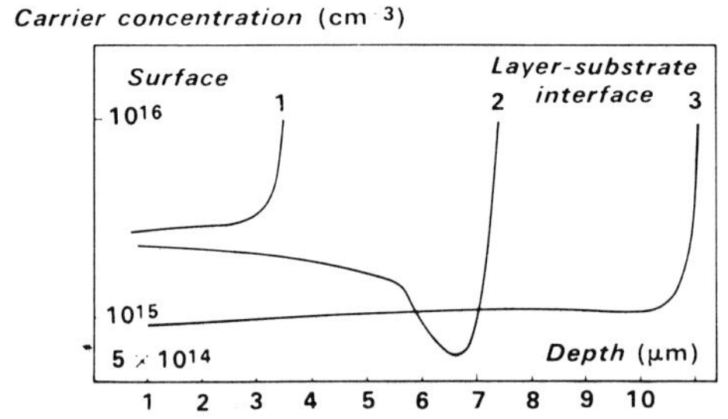

Fig. 5. *Three typical epitaxial GaAs profiles : 1 and 3 are good Q and X band layers respectively, 2 demonstrates a common fault.*

A further interesting application of the profile technique has been the removal of the substrate and the measurement of the carrier profile beneath an alloyed ohmic contact. J.S. Harris et al. [22] have shown an increase of the carrier concentration arising from an alloyed Au-Ge-Ni contact from the surface to a depth of 3 μm. This obviously could have a considerable influence on the performance of the Gunn device, and is worthy of further investigation.

DEEP TRAPS

General methods of detection

It has been suggested that the presence of trapping centres in GaAs can affect its operation as Gunn device material [23], and as indicated previously a Schottky barrier can be used for detecting the presence of such centres. Through equations (7) and (8) it can be seen that changes in the occupation of a deep trap are reflected in a change in the capacity of the barrier at constant reverse bias. A number of techniques for changing the occupation have been reported. These include instantaneously applying the reverse bias [24, 25], illuminating the depletion layer [24, 26], and thermally stimulated variations in capacity [27]. A summary covering all these techniques in relation to abrupt P-N junctions in Si has been given by C.T. Sah et al. [28].

These methods are easily understood in terms of a material with a single deep donor-like trap (fig. 6a) with a net ionized impurity concentration given by :

$$N_I = N_D(1-P_D) - N_A + N_{T_1}(1-P_1) . \qquad (10)$$

Suppose the trap is more than a few kT below the Fermi level in the bulk of the material, then its occupation number will be close to unity. If a large reverse bias is instantaneously applied to a Schottky barrier formed on the surface of this material, a depletion layer will be formed whose width is determined solely by the space charge arising from the uncompensated shallow donors. There will be no contribution from the deep trap since only the free electrons can respond rapidly to the change of field. Therefore the occupation of the deep trap at this instant is still unity and the capacity is proportional to :

$$N_D(1-P_D) - N_A .$$

However in the high field region of the depletion layer the deep traps will slowly empty and the capacity will relax to an equilibrium value. It is easy to show that the rate at which the capacity relaxes is (e_n+e_p) and that the equilibrium occupation number of the trap is $\frac{e_p}{e_n+e_p}$, where e_n and e_p are the electron and hole thermal emission rates respectively (fig. 6b). Therefore the equilibrium capacity is proportional to : $N_D(1-P_D)-N_A+N_{T_1}(1-P_1)$ and from the initial and equilibrium capacities, C_0 and C_∞, we get, through equations (7) and (8) :

$$N_{T_1}\frac{e_n}{e_n+e_p} = \frac{2(C_\infty^2 - C_0^2)(V_D+V_R)}{q\varepsilon\varepsilon_0 A^2} . \qquad (11)$$

Also (e_n+e_p) is obtained from the capacity relaxation rate. From these measurements the concentration of the trap and the termal emission rates may be determined if e_n/e_p, e_n or e_p can be measured independently or if the conditions of the experiment can be adjusted such that $e_n \gg e_p$.

The electron emission rate is related to the capture cross section σ_n through the principle of detailed balance :

$$e_n = \sigma_n \langle v \rangle N_C \exp -\left[\frac{E_C - E_T}{kT}\right] , \qquad (12)$$

where $\langle v \rangle$ and N_c are the thermally averaged electron velocity and the density of states in the conduction band, and E_T is the energy position of the trap. A similar expression holds for the hole emission rate and as may be seen e_n and e_p are highly temperature dependent. Therefore it is often possible to find a temperature range such that $e_n \gg e_p$. This condition may be verified either through measuring $N_T e_n/(e_n+e_p)$ as a funtion of temperature, where a constant value would indicate $e_n \gg e_p$, or through the optical experiment described below. In addition from the temperature dependence of e_n a thermal activation energy may be obtained.

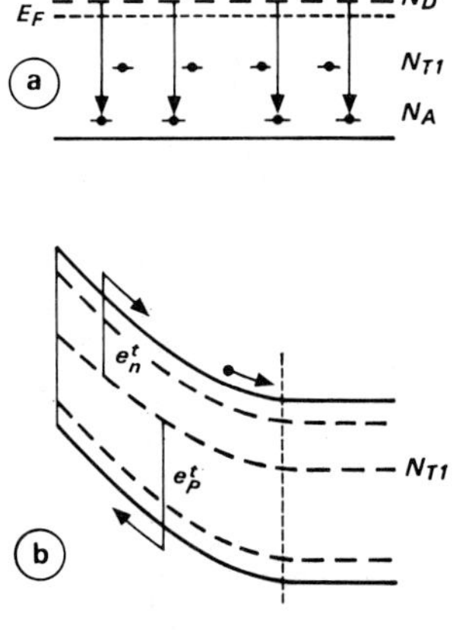

Fig. 6. *Semiconductor containing a deep donor-like trap.*

Again consider figure 6 for a material with a single deep donor-like trap. If the material is illuminated with monochromatic light of energy greater than the separation of the trap from the nearest band edge and the previous experiment is repeated, then the capacity will again relax but at a faster rate. This new relaxation rate is $e_n+e_p+e_n^0+e_p^0$ where e_n^0 and e_p^0 are the optically induced emission rates of the electrons and holes respectively. Depending upon the energy of the light relative to the position the trap either e_n^0 or e_p^0 may be zero. The equilibrium occupation number of the trap will also be changed by the illumination to $\frac{e_p+e_p^0}{e_n+e_p+e_n^0+e_p^0}$.

If the trap is closer to the conduction band than the valence band then illumination with light equal to half the band gap will increase the relaxation rate ot $e_n+e_p+e_n^0$ and the initial and equilibrium capacities will give $N_{T_1}\frac{e_n+e_n^0}{e_n+e_p+e_n^0}$ in equation (11). This will represent an increase in capacity from the equilibrium capacity in the dark unless $e_n \gg e_p$ when the trap will already be fully emptied thermally. However if the trap is nearer to the valence band then illumination with the same wavelength light will again increase the relaxation rate to $e_n+e_p+e_p^0$ but will reduce the equilibrium capacity to give $N_{T_1}\frac{e_n}{e_n+e_p+e_p^0}$.

Thus by measuring the threshold energy of the light for producing a change in relaxation rate and by observing the equilibrium capacity it is possible to obtain the energy position of the trap unambiguously and, provided the trap is in the upper half of the band gap, to again demonstrate if $e_n \gg e_p$ in the thermal relaxation experiment.
The thermal activation energy and (E_c-E_T) given by the optical threshold will in general differ since, as may be seen from equation (12), no allowance has been made for the temperature dependence of the capture cross section. A general calculation of the capture cross section for attractive centres in polar semiconductors [29] indicates that σ_n can have a large temperature dependence. However having measured e_n and (E_c-E_T) the capture cross section may be obtained through equation (12).
By combining these experiments it is possible to measure the energy position, concentration and capture cross section of a deep trap.

The preceeding discussion has been presented in terms of a single donor-like trap, a similar argument leads to the same conclusions for an acceptor-like trap. In GaAs the real situation is considerably more complex since several traps will be present together. However, because the emission rates are highly temperature dependent it is often possible to separate the contributions from the different traps by a suitable choice of temperature and experiment.

Experimental details and results

The gold Schottky barriers used in these experimente are prepared by exactly the same method as for thd profile measurements. Their quality is again checkes by measuring the forward current-voltage characteristic before mounting in an optical cryostat. The depletion layer can be illuminated, through the substrate, with light from a Bausch and Lomb monochromator and the capacity measured using a Boonton 74D capacitance bridge. This has an output proportional to the capacity which is used to record the capacity relaxations as a function of temperature and illumination.

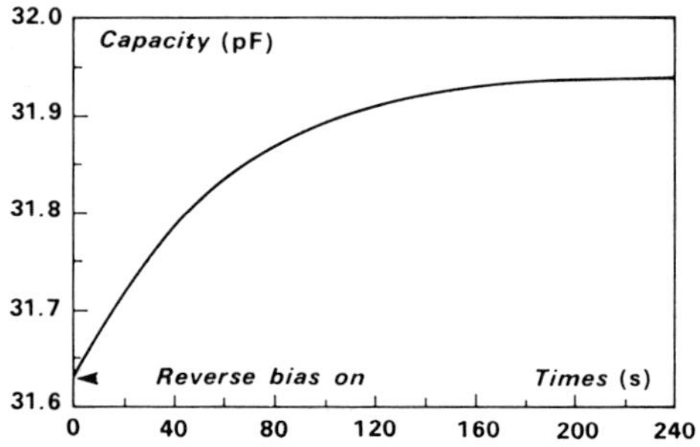

Fig. 7. *Capacity relaxation at 295 °K of a Schottky barrier on epitaxial GaAs.*

Measurements of the capacity relaxation at room temperature of Schottky barriers formed on both vapour and liquid phase material exhibit a single exponential increase of capacity (fig. 7). Similar results have been reported by a number of authors on both bulk and epitaxial GaAs samples [24 to 27, 30, 31]. The thermal activation energy we obtained from the temperature dependence of this relaxation rate is 0.89 eV and is in good agreement with values given by [25, 30, 31]. A more reliable estimate of the trap position was obtained by measuring the photon threshold energy for a change in the emission rate. Measurements of this kind on the same samples used in the relaxation experiments located the trap at 0.82 eV below the conduction band at 295 °K. The discrepancy between the thermal activation energy and the optical threshold is accounted for by the temperature dependence of the density of states, electron velocity and the capture cross section in equation (12).

R.R. Senechal and J. Basinski [25] found that $N_T e_n/(e_n+e_p)$, obtained from C_0 and C_∞ through equation (11), varied by about 3 % over a temperature range of 70 °K and measurements on the 0.82 eV trap in our material gave similar results. These suggest that $e_n \gg e_p$ for this trap at room temperature and hence from C_0 and C_∞ the total trap concentration may be obtained. Some typical results are given in table I for the concentration in various types of epitaxial material. Previous authors [24 to 27] have reported typical concentrations for this trap of 4×10^{15} cm^{-3} to 4×10^{16} cm^{-3} in bulk grown GaAs.

TABLE I

Relative concentrations of trap 0.82 eV below conduction band at 295 °K.

Material	N_T (cm^{-3})
Vapour Epitaxy (Reactor 1)	5.5×10^{13}
Vapour Epitaxy (Reactor 1)	4.2×10^{13}
Vapour Epitaxy (Reactor 1)	1.0×10^{14}
Vapour Epitaxy (Reactor 2)	1.3×10^{14}
Vapour Epitaxy (Reactor 2)	5.9×10^{13}
Liquid Epitaxy	2.8×10^{12}
Liquid Epitaxy	2.1×10^{12}

As may be seen the concentration is an order of magnitude greater in vapour phase than liquid phase GaAs and the results of R.R. Senechal and J. Basinski [25] suggest that in bulk GaAs it is an order of magnitude greater still. Also because $c_n \gg e_p$ the capacity relaxation rate is a direct measure of the electron thermal emission rate and is typically 0.026 s^{-1} at 295 °K giving, through equation (12), an electron capture cross section of 1.4×10^{-12} cm^2.

A reduction in the temperature of the material from room temperature reduces the relaxation rate until, by about 270 °K, no capacity relaxation is observable. With a further reduction to 197 °K a second thermal relaxation is measurable with a threshold photon energy of 0.47 eV for an increase in the relaxation rate. However in this case there is no change in the equilibrium capacity from that in the dark. This indicates that the trap is 0.47 eV below the conduction band and that $e_n \gg e_p$ since the trap is fully emptied thermally. Using the initial and final capacities in equation (11) gives the concentration of the trap as typically 3×10^{13} cm^{-3}. Again the capacity relaxation rate is a direct measure of the electron emission rate giving 0.056 s^{-1} at 197 °K. These can be used in equation (12) to calculate the electron capture cross section as 1.1×10^{-14} cm^2. The theoretical value for capture at an attractive centre is GaAs has been calculated by H.I. Ralph and F.D. Hughes [29] to be approximately 1×10^{-14} cm^2 at 197 °K and this close agreement indicates that the trap is donor-like. The properties of this trap are summarized in table II.

TABLE II

Summary of the properties of the 0.47 eV trap at 197 °K in GaAs.

$(E_C - E_T)_{197\,°K}$	0.47 eV
$(E_C - E_T)_{77\,°K}$	0.46 eV
Concentration	typically 3×10^{13} cm^{-3}
e_n	0.056 s^{-1}
σ_n	1.1×10^{-14} cm^2

J.C. Carballès and J. Lebailly [27] also report a trap situated at approximately 0.4 eV in bulk GaAs. They found this using the thermally stimulated variations of capacity technique, which is similar to the well known thermally stimulated current method [32].

If the reverse bias is applied at 77 °K to a Schottky barrier made on epitaxial GaAs then no capacity relaxation is observable. That is all the deep traps remain filled with an occupation number of unity. However by illuminating the depletion layer with monochromatic light of suitable wavelength it is possible to at least partially empty the traps. Therefore a plot of the capacity as a function of photon energy at 77 °K will reveal the presence of a number of traps as an increase in the capacity. Figure 8 shows such a plot made for a sample of vapour phase epitaxial GaAs. Four increases in capacity may be seen indicating four traps.

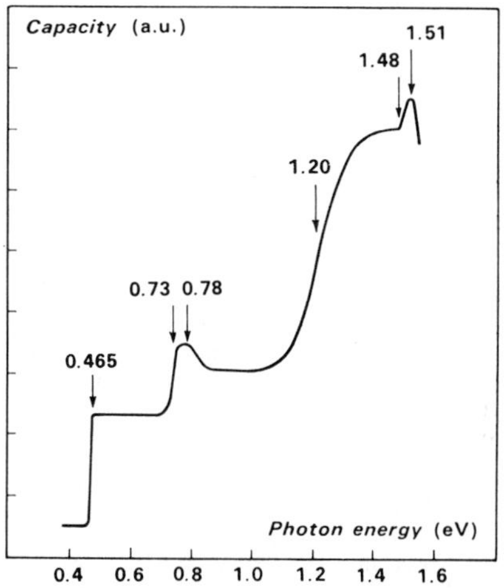

Fig. 8. *Capacity as a function of incident photon energy at 77 °K for vapour phase epitaxial GaAs.*

The first trap is situated at 0.46 eV below the conduction band and is the same one as is observed at 197 °K and summarized in table II. A second trap empties at 0.73 eV below the conduction band and this has been identified as the trap which relaxes thermally at room temperature, by measuring the energy position at intermediate temperatures. Since the band gap at 77 °K is 1.51 eV then the complementary energy for the 0.73 eV trap is 0.78 eV and photons of this energy will partially refill the trap with electrons from the valence band. The corresponding decrease in capacity can be seen at this energy. Y. Furukawa [33] also observes at 77 °K a similar increase in capacity at 0.7 eV.

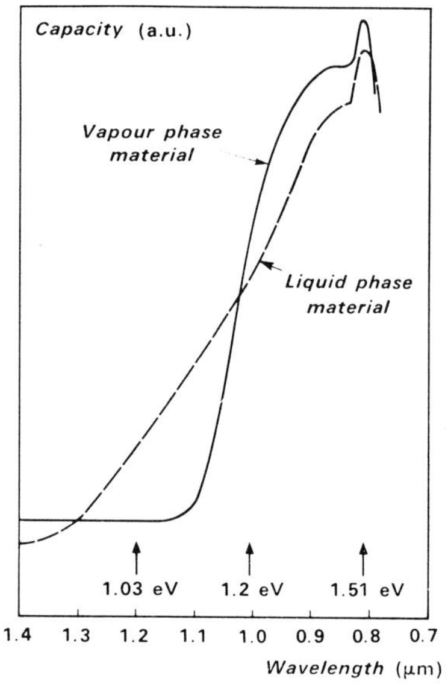

Fig. 9. *Comparison of liquid and vapour phase GaAs measured at 77 °K.*

The third increase of capacity in figure 8 has a broad energy response centred at approximately 1.2 eV below the conduction band. This trap has been observed by many investigators [33, 34], with the total change in capacity typically giving a concentration of 2×10^{14} cm^{-3} for the traps emptied optically. However this must be regarded as a lower limit for its concentration since it is closer to the valence band than the conduction band and both electron and hole emission must be stimulated by the photons. A peak in the cathodoluminescence spectrum at 1.25 eV has been observed by P.M. Williams [35] on the same epitaxial GaAs on which these capacity measurements were made. This is presumably the same peak as reported in photoluminescence by E.W. Williams [36] of bulk GaAs which he attributed to a donor-vacancy complex.

It is interesting to note a difference between vapour and liquid phase GaAs with respect to the response at this trap. Figure 9 shows the capacity-energy plot at 77 °K for both types of material. The capacity response for the liquid phase GaAs extends over a much wider energy range and it may even be possible to resolve the curve into the emptying of two separate traps.
Finally returning to figure 8 a fourth trap is indicated by a small peak with a threshold of 1.48 eV. Since the depletion layer is illuminated through the substrate the decrease in capacity is due to the absorption of the light by the band edge absorption at 1.513 eV [37]. The increase of capacity is therefore due to a trap 0.030 eV above the valence band. This is consistent with silicon acting as a shallow acceptor where an activation energy of 0.029 eV is given by E.H. Bogardus and H.B. Bebb [38]. Though other shallow acceptors such as zinc have closely similar activation energies.

Because of the proximity of the trap to the valence band it is not possible to obtain an estimate of its concentration. However E. Fabre [39] has shown a variation in this peak height at various points through a layer giving the relative concentrations of this trap. These results show an increased trap concentration near the layer-substrate interface for material which has a large dip in the profile. This suggests that the dip may therefore be associated with an increased acceptor concentration at the interface.

CONCLUSION

Schottky barriers are very useful in enabling the free carrier profile of epitaxial Gunn layers to be measured. In the past few years with the development of rapid measuring techniques the carrier profile has become an important parameter in investigating the effects of growth conditions on the epitaxial layer. In particular a considerable amount of effort has been devoted to investigating the dip near the layer-substrate interface which appears in many profiles. A few authors have reported a correlation between the performance of a Gunn device and the carrier profile [4, 40], which agree with computer simulations of domain dynamics and suggest that material with a flat profile gives the highest efficiencies.

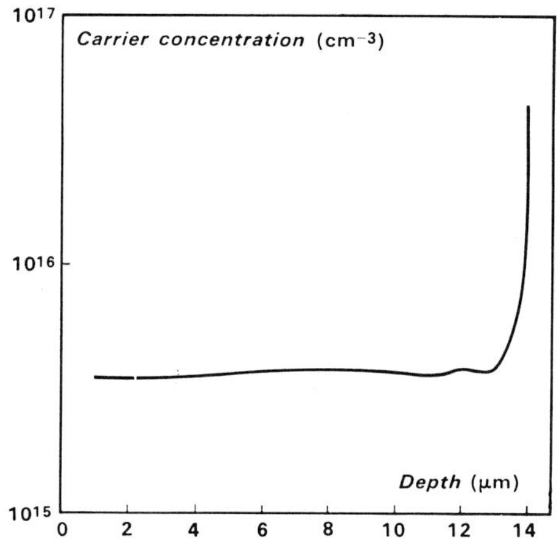

Fig. 10. *Carrier concentration profile of liquid phase GaAs.*

As indicated in figure 5 it is possible to grow vapour phase epitaxial GaAs with such a profile and recent results on liquid phase GaAs have also given flat profiles as shown in figure 10. This latter material gave between 50-60 mW power at 9.8 GHz with 3.5 % efficiency when made into conventionally mounted C.W. devices (i.e. not inverted).

J.S. Heeks [41] has demonstrated that the presence of field dependent trapping centres can in fact lead to the disappearance of Gunn oscillations after the first few cycles and Schottky barriers are valuable for detecting the presence of these deep traps. At least four traps have been shown to be present in epitaxial GaAs in concentrations close to the free carrier concentration. However it has not yet been possible unambiguously to identify the impurities or defects which give rise to all of these traps. Some identifications have been reported but discrepancies arise between different authors. For example it has been suggested that oxygen is responsible for traps covering the range 0.65 eV to 0.76 eV. However it appears to be widely accepted that the trap located 1.2 eV below the conduction band is related to a vacancy and that the shallow acceptor 0.030 eV above the valence band is probably silicon [21]. This latter identification is not unlikely since epitaxial GaAs is usually grown in a silica glass reaction tube.

ACKNOWLEDGMENTS

The author wishes to thank Drs H.I. Ralph, P. Blood and J.W. Orton for many helpful discussions and T.J. Harris for assistance with some of the measurements. He would also like to thank the GaAs growth group at M.R.L., B.C. Easton, Miss C.A. Fisher, M.J. King and J.A. Wilkinson for the helpful provision of material used in this work.

Bibliography

1 HENISCH (H.K.). — **Rectifying semiconductor contacts.** Oxford University Press, 1957

2 VAN OPDORP (C.). — **Evaluation of doping profiles from capacitance measurements** Solid-State Electron., GB, vol. 11, 1968, pp. 397-406.

3 AMRON (I.). — **A slide rule for computing dopant profiles in epitaxial semiconductor films.** Electrochem. Technol., USA, vol. 2, 1964, pp. 327-333.

4 COLLIVER (D.J.), GIBBS (S.E.) and TAYLOR (B.C.). — **Material selection for coefficient transferred-electron devices at Q-band.** Electron. Lett., vol. 6, 1970, pp. 353-355.

5 DECKER (D.R.). — **Measurement of epitaxial doping density vs. depth.** J. electrochem. Soc., USA, vol. 115, 1968, pp. 1085-1089.

6 BAXANDALL (P.J.), COLLIVER (D.J.) and FREY (A.F.). — **An instrument for the rapid determination of semiconductor impurity profiles.** J. Phys. E, vol. 4, 1971, pp. 213-221.

7 COPELAND (J.A.). — **A technique for directly plotting the inverse doping profile of semiconductor wafers.** IEEE Trans. Electron. Devices, USA, vol. ED-16, 1969, pp. 445-449.

8 CALIFANO (F.P.) and LUCCIANO (A.). — **An automatic test set for measuring the doping profile of semiconductor epitaxial layers.** Rev. sci. Instrum., USA, vol. 41, 1970, pp. 865-869.

9 SALAMA (C.A.T.) and HOLMES (F.). — **A technique for directly plotting the doping profile of semiconductor wafers.** Solid-State Electron., GB, vol. 14, 1971, pp. 178-181.

10 SEVERIN (P.J.) and POODT (O.). — **Automated capacitance-voltage measurements on a mercury-silicon diode.** Philips Res. Lab. Rep., n° 4522, 1970.

11 SMITH (B.L.). — **The effect of surface treatment on gallium arsenide Schottky barrier diodes.** Solid-State Electron., GB, vol. 11, 1968, pp. 502-504.

12 SMITH (B.L.) and RHODERICK (E.H.). — **Possible sources of error in the deduction of semiconductor impurity concentrations from Schottky barriers (CV) characteristics.** J. Phys. D, vol. 2, 1969, pp. 465-467.

13 HUGHES (F.D.), HEADON (R.F.), and WILSON (M.). — **On the measurement of semiconductor carrier concentration profiles.** J. Phys. E. : Scientific Inst., vol. 5, 1972, to be published.

14 HAMMER (R.). — **A mercury contact probe for MOS measurements on oxidized silicon.** Rev. sci. Instrum., USA, vol. 40, 1970, pp. 292-293.

15 KENNEDY (D.P.), MORLEY (P.C.) and KLEINFELDER (W.). — **On the measurement of impurity atom distribution in silicon by the differential capacitance technique.** I.B.M. J. Res. Develop., USA, vol. 12, 1968, pp. 399-409.

16 SZE (S.M.) and GIBBONS (G.). — **Avalanche breakdown voltages of abrupt and linearly graded p-n junctions in Ge, Si, GaAs, and GaP.** Appl. Phys. Letters, USA, vol. 8, 1966, pp. 111-113.

17 AMRON (I.). — **Errors in dopant concentration profiles determined by differential capacitance measurements.** Electrochem. Technol., USA, vol. 5, 1967, pp. 94-97.

18 GOODMAN (A.M.). — **Metal-semiconductor barrier height measurements by the differential capacitance method - One carrier system.** J. appl. Phys., USA, vol. 34, 1963, pp. 329-338.

19 COPELAND (J.A.). — **Diode edge effect on doping profile measurements.** IEEE Trans. electron. Devices, USA, vol. ED-17, 1970, pp. 404-407.

20 BLOCKER (T.G.), COX (R.H.) and HASTY (T.E.). — **Interpretation of anomalous layer GaAs n^+ - n step junctions.** Solid State Communic., GB, vol. 8, 1970, pp. 1313-1315.

21 DI LORENZO (J.V.), MARCUS (R.B.) and LEWIS (R.). — **Analysis of impurity distribution in homoepitaxial non n^+ films of GaAs which contain high resistivity regions.** J. appl. Phys., USA, vol. 42, 1971, pp. 729-739.

22 HARRIS (J.S.), NANNICHI (Y.) and PEARSON (G.L.). — **Ohmic contacts to solution - grown gallium arsenide.** J. appl. Phys., USA, vol. 40, 1969, pp. 4575-4581.

23 SHUR (M.S.). — **Influence of the field-dependant trapping of carriers on the Gunn effect.** Soviet Phys. Semicond., vol. 4, 1970, pp. 1022-1023.

24 WILLIAMS (R.). — **Determination of deep centres in conducting gallium arsenide.** J. appl. Phys., USA, vol. 37, 1966, pp. 3411-3416.

25 SENECHAL (R.R.) and BASINSKI (J.). — **Capacitance measurements on Au-GaAs Schottky barriers.** J. appl. Phys., USA, vol. 39, 1968, pp. 4581-4589.

26 FURUKAWA (Y.) and ISHIBASHI (Y.). — **Trapping effects in Au-n-type GaAs Schottky barrier diodes.** Jap. J. appl. Phys., vol. 6, 1967, pp. 503-508.

27 CARBALLÈS (J.C.) and LEBAILLY (J.). — **Trapping analysis in gallium arsenide.** Solid State Communic., GB, vol. 6, 1968, pp. 167-171.

28 SAH (C.T.), FORBES (L.), ROSIER (L.L.) and TASCH Jr (A.F.). — **Thermal and optical emission and capture rates and cross sections of electrons and holes at imperfection centres in semiconductors from photo and dark junction current and capacitance experiments.** Solid-State Electron., GB, vol. 13, 1970, pp. 759-788.

29 RALPH (H.I.) and HUGHES (F.D.). — **Capture cross section of trapping centres i polar semiconductors.** Solid State Communic., GB, vol. 9, 1971, pp. 1477-1480.

30 GLOVER (G.H.). — **Measurement of the doping profile of semiconductors having deep impurity centres.** General Electric Company, Report No. 71-C-057, 1971.

31 HASEGAWA (F.). — **Deep energy levels in the high resistance region at GaAs vapour epitaxial film-substrate interface.** Jap. J. app. Phys., vol. 9, 1970, pp. 638-646.

32 BUBE (R.H.). — **Photoconductibility of solids.** New York, Wiley, 1960.

33 FURUKAWA (Y.). — **Trap levels in gallium arsenide.** Jap. J. appl. Phys., vol. 6, 1967, pp. 675-679.

34 CARBALLÈS (J.C.), DIGUET (D.) and LEBAILLY (J.). — **Electrical properties and solution grown GaAs layers.** Proc. 2nd. Int. Symp. on gallium arsenide : Inst. Phys. Soc. Conf. Ser. n° 7, 1968, pp. 28-35.

35 WILLIAMS (P.M.). — 1970. Private Communication.

36 WILLIAMS (E.W.). — **Evidence for self-activated luminescence in GaAs : the gallium vacancy-donor centre.** Phys. Rev., USA, vol. 168, 1968, pp. 922-928.

37 PARISH (M.B.) and CASEY (H.C.). — **Temperature dependance of the energy gap in GaAs and GaP.** J. appl. Phys., USA, vol. 40, 1969, pp. 163-167.

38 BOGARDUS (E.H.) and BEBB (H.B.). — **Bound-exciton, free-exciton, band-acceptor, on Auger recombination in GaAs.** Phys. Rev., USA, vol. 176, 1968, pp. 993-1002.

39 FABRE (E.). — **Caractérisation de centres profonds dans l'arséniure de gallium par des méthodes photocapacitives.** C.R. Acad. Sci., Fr., vol. 270, 1970, pp. 848-856.

40 WOLFE (C.M.), FOYT (A.G.) and LINDLEY (W.T.). — **Epitaxial gallium arsenide for high efficiency Gunn oscillators.** Electrochem. Technol., USA, vol. 6, 1968, pp. 208-214.

41 HEEKS (J.S.). — **Some properties of the moving high-field domain in Gunn effect devices.** IEEE Trans. electron. Devices, USA, vol. ED-13, 1966, pp. 68-79.

2.3.2 special techniques

There are numerous specialised methods directed at particular characteristics of device material. The majority of these techniques depend on junction structures and provide information on deep centres.

C-V Measurements

Apart from the determination of the shallow level doping profiles of semiconductors, C-V measurements can reveal the slowly varying component of depletion layer capacity (when the reverse bias is abruptly incremented) due to the presence of impurities with energies many kT below the Fermi level. These deep levels have been proven to degrade the high field performance by causing avalanching at voltages near threshold in TED's[40].

A particular level is characterised by its energy, charge state, thermal and optical capture rate, thermal and optical emission rate, capture cross section (and the dependence of these factors on sample temperature), electric field, and photon energy. Experiments have been devised to determine each of these parameters; a comprehensive paper by Sah et al.[41] discusses these areas.

Papers by Ikoma et al.[42] and the key paper by Glover, describe the determination of deep level densities and energies by switched C-V measurements. Both have shown interesting results on GaAs. Essentially, the difference in capacitance is measured between the reverse biased steady state and the state after the voltage is switched from zero (or forward bias) to the same reverse voltage. Also of significance is the time dependence of that capacitance. The experimental arrangement must be capable of measuring capacitance changes as small as 5×10^{-4} pF.

Similar measurements have been executed in the RRE laboratory on high quality epitaxial GaAs. Both forward bias injection and optical injection are maintained at 300°K and 77°K, without positive results (due to the poor sensitivity of the instrumentation at that time). Recently, a bridge, modified to the necessary sensitivity, has enabled the investigation of InP p^+-n junction diodes. The observed capacitance changes were small and ambiguous, but the indication was the deep level densities were no larger than 10^{12} cm^{-3}. No information on the energy could be obtained. The implication from published results and personal experience is that sensitivities are unlikely to be sufficient to record evidence of deep levels in material of reasonable quality. But the results at present have been interesting and the limitations of the procedure should be determined.

Thermally Stimulated Current (TSC)

TSC is executed on a reverse bias junction[43]. The temperature is lowered to the freeze out of the traps, which are then filled by light excitation or by zero biasing of the diode. The temperature is then increased at a linear rate to room temperature. The reverse leakage current is continuously monitored as a function of temperature; peaks appear in the current-temperature characteristic caused by the emptying of traps.

The temperature at which the peak appears is proportional to the activation energy of the trap. The number contributing to the stimulated peak is obtained from the amount of charge released. The trap energy is then[44]:

$$E_t \approx 24\ KT m$$

and trap density[45]:

$$N_{tt} = \frac{4Q}{qAW}$$

where Q is the total charge released and AW is the volume of depleted device.

Obviously, very low leakage diodes are required — GaAs Schottky barriers easily satisfy the need, but with InP at present, the task is much more difficult. However, no results using TSC measurements on these materials have been reported. The method is interesting and requires less sophisticated instrumentation, although the analysis is more complicated than with the simple expressions given above when dealing with practical specimens.

Low Frequency Noise

The measurement of low frequency noise can be carried out directly on TED's or any linear device configuration. The theory of generation-recombination of electrons from deep levels is discussed later[46,47]. Generation-recombination is not the only source of noise in the semiconductor; the limitation of the process is that noise at deep levels must be distinguishable from other noise sources.

The experimental arrangement established in the RRE laboratory is illustrated in Figure 2-15. The device under test is mounted in a holder which is placed in a temperature controlled box. A dc bias is fed in and the output is passed through a wideband impedance transformer to a low noise amplifier. The noise signal is then measured with a Quan Tech wave analyser (model 303). The gain of the system is checked by applying a small ac current across the device terminals to measure the output.

A complete spectrum of the noise can then be plotted in terms of current per unit bandwidth as a function of frequency out to 100 kHz. Details of the measurement, results, and analysis are given in the key paper by Copeland. It can be shown that the noise from impurities with a well-defined time constant (τ) has the shape described by[48]:

$$\frac{<i^2>}{\Delta f} = \frac{A}{1 + (f/f_o)^2}$$

where $f_o = 1/\tau$.

If the noise has a constant value, A, at low frequencies and falls as $1/f^2$ at high frequencies, f_o can be determined by extrapolating the high and low frequency parts of the log plot, with f_o being the intersection of these extrapolations.

In order to determine the density and energy of the level, A and τ must be determined over a range of temperature.

$$\text{Then } A = \frac{2 I^2 \tau D}{n^2 V} \frac{(BN_c/n) \exp(\epsilon/kT)}{1 + [(BN_c/n) \exp(\epsilon/kT)]^2}$$

where D is the deep level density at a level ϵ below the conduction band, V is the volume of the device, n is the carrier density, B is the spin

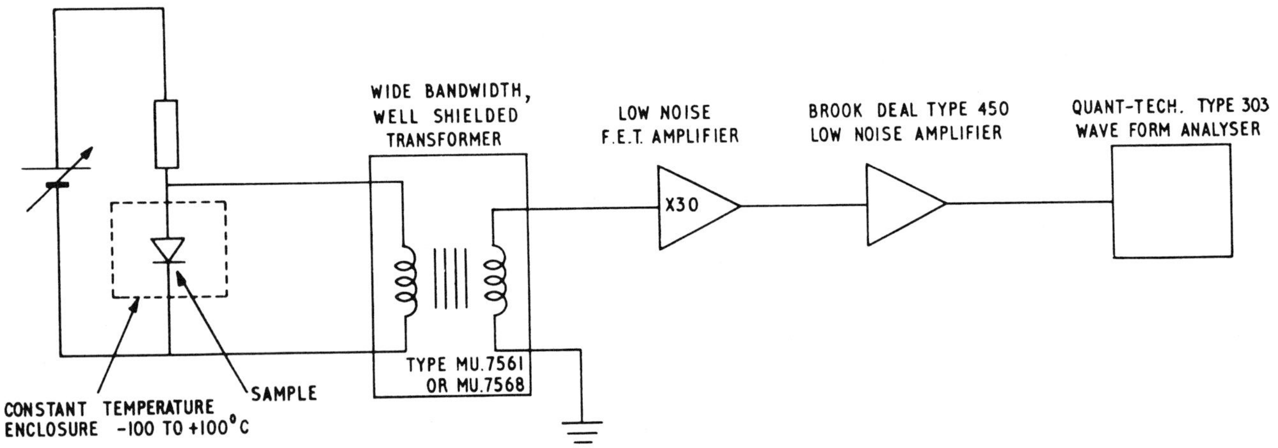

Figure 2-15/Low Frequency Noise Measurement Apparatus.

degeneracy of the level, and N_c is the effective density of states in the conduction band.

Now let the noise function $F(T) = An^2/\tau$.

Then $F(\theta) = (2I^2 D/V)/[\exp(\theta) + 2 + \exp(-\theta)]$

$$\theta(T) = -\epsilon/kT + \log_e(n/BN_c') - 1.5 \log_e(T/T')$$

where N_c' is the value of N_c at T'.

A plot of $\log_e(F)$ versus $1/T$ yields a value for D and the relationship between ϵ and B.

ϵ can then be calculated from

$$\frac{d[\log_e(F)]}{d(1/\tau)} = \pm(-\epsilon + 1.5\,kT)/k$$

where 1.5 kT is due to the temperature dependence of the effective density of states.

The emission time constant τ_e and capture cross section S are given by

$\tau_e = \tau[1 + \exp(\theta)]$

and

$S = 1/nV\tau\ [1 + \exp(\theta)]$

This analysis applies only to one level; for many levels the noise spectrum can be written in the form:

$$\frac{\langle i^2 \rangle}{\Delta f} = \frac{A}{f} + \frac{B}{1 + (f/f_1)^2} + \frac{C}{1 + (f/f_2)^2} + \dots$$

The 1/f term is incorporated to account for flicker noise, present in many of the observed spectra.

Both Copeland and Magarshak[43] present results on GaAs. Magarshak extends the measurement to correlate it with the FM and AM noise of a device that is biased above threshold. Research at RRE with InP has produced spectra of the form shown in Figure 2-16. These spectra apparently are of two distinct levels; however, the analysis of the curve is not straightforward. Experimental data must be fitted to the curve:

$$\frac{\langle i^2 \rangle}{\Delta f} = \frac{A}{f} + \frac{B}{1 + (f/f_1)^2} + \frac{C}{1 + (f/f_2)^2} + \dots$$

in order to obtain values for f_1, f_2, etc. and A, B, C, etc. at various measurement temperatures. The problem is resolved to one of computation and curve fitting to the data.

The measurement is quite complicated, but has the advantage that it can be carried out directly on a completed TED. Once deep levels have been identified, the spectra should be easily interpreted. The absolute limitation of the method on high quality materials is unknown, but, Copeland certainly had some GaAs that had a noise performance below the detectable limit of his apparatus (which was 3×10^{-21} A^2/Hz). He did not state any further information as to whether this particular material showed superior performance in any other respect.

MOS Measurements

Metal-oxide-semiconductor measurements have long been used in determining the properties of the oxided surfaces of semiconductors[49]. The application of this process to compound semiconductors has not been made to any great extent, although traps have been identified at the surface region of GaAs[50] using MOS devices (which behaved very differently to those made from silicon). The key paper by Valachas et al. discusses MOS measurements on GaAs.

It is difficult to form Schottky barriers on the surface of InP; thus, the MOS structure has been used to perform C-V measurements and compute the impurity profile. The oxide is pyrolytically deposited and the C-V measurement is carried out with a pulsed dc bias to prevent surface inversion during the measurement. Although the computed profile is much as was expected, the method is disadvantageous in terms of the need to interpolate values of dC/dV. The measurement and computation are also tedious.

There exist no MOS practical devices in GaAs or InP; therefore, information obtained from the oxide-semiconductor interface states is largely of academic interest. It is preferable to use a field effect plate held above a "real" surface to determine the surface state density and distribu-

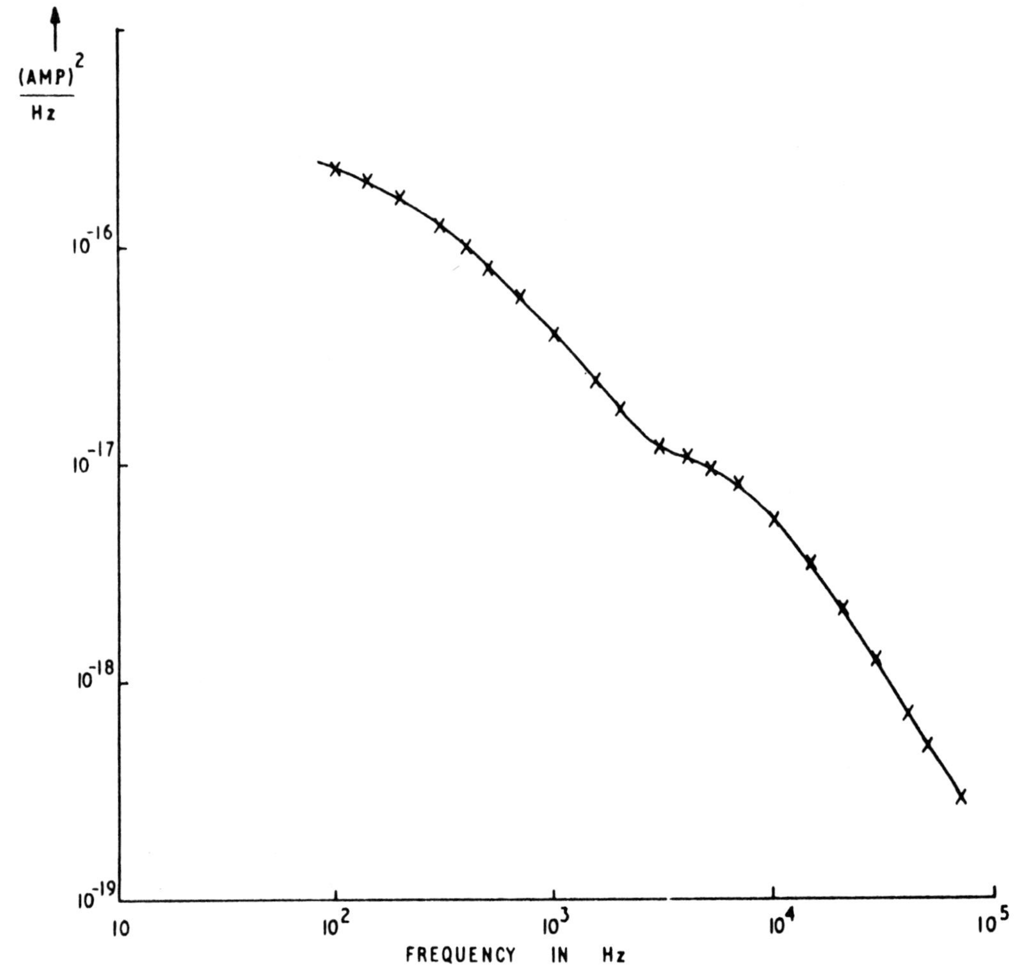

Figure 2-16/Noise Spectrum of InP Device[63]

tion. Measurement and computation are extremely complex, but the results could help to explain the behaviour of contacts to InP, both for Schottky barriers and ohmic contacts.

Electrochemical Characterisation

A new procedure of interest[51] is one using electrochemical methods which exploit the fact that, under certain conditions, the electrolyte-semiconductor interface behaves like a Schottky barrier. There are two added advantages:

1./Controlled electrochemical dissolution of the semiconductor can be performed *in situ.*
2./Illumination of the semiconductor surface is easily achieved.

The doping profile is derived as a function of depth while etching continuously, and the effect of illumination can be studied directly. This process has not been exploited fully but is certain to receive attention in the future.

key paper

Determination Of Deep Levels In Semiconductors From C-V Measurements

G.H. Glover
(Solid-State Physics Laboratory, General Electric Research and Development Center, Schenectady, New York, US)

IEEE Transactions on Electron Devices, Vol. ED-19, No. 2 (February 1972)
Pages 138-143

***Abstract*—A method is presented that allows deep level density and energy of impurities in semiconductors to be routinely determined from simple *C-V* measurements of Schottky barriers. The method is particularly useful for material having levels with very long time constants that are not accessible by noise spectra measurements. The technique is illustrated by measurements of epitaxial n-GaAs which show a donor level at 0.83 eV below the conduction band.**

I. Introduction

Capacitance-voltage (C-V) measurements on Schottky diodes have long been used to determine the shallow-level compensated doping profiles of semiconductors [1]. However, the presence of impurities with energies many kT below the Fermi level leads to a slowly varying component of depletion layer capacity when the reverse bias is suddenly incremented [2]-[4]. This in turn results in a time- and depth-dependent error in the measured profile [5]

Deep levels do not normally affect the low-field electrical transport properties of the host material. However, in GaAs bulk effect devices, such impurities have been found to degrade the high-field performance by causing avalanching at values near threshold [6]. Moreover, it is well known that deep impurities can function as generation-recombination centers [7], giving rise to noise and associate problems in solid-state devices.

Recently several techniques have been proposed for evaluating deep levels in semiconductors. Zohta [5] pointed out that the Copeland profiler [8] does not respond to deep levels, whereas standard C-V measurements do. The deep-level density was therefore inferred by combining the two measurements. Copeland described a technique [9] for evaluating the density, degeneracy, and energy of deep levels which capitalizes on the noise emission of the deep-level electrons. Implementation of this method for very deep levels (e.g., ≥ 0.5 eV), however, is difficult since the time constant of such levels becomes long (on the order of seconds or minutes). Sah *et al.*, in a very comprehensive paper [10], described transient capacitance experiments using silicon p-n junctions. Comparison of measurements made on separate samples with and without gold doping enabled determination of the gold impurity level and density.

This paper gives a method of measuring deep-level density and energy from C-V measurements on Schottky diodes. The method is based on the model of Senechal and Basinski [4] (SB), and is extended to include nonuniform distributions of shallow and deep impurities. A simple working formula is derived that requires only three measured values of capacitance to evaluate the ratio of the average density of the deep levels to the shallow level doping. In addition, conditions are given under which the shallow-level profile may be measured by C-V techniques in the presence of deep levels. For convenience, we shall consider here only deep donors. The extension to acceptors should be obvious.

Manuscript received April 4, 1971; revised September 1, 1971.

II. The Model

We shall consider a Schottky diode fabricated from semiconducting material having n_0 compensated shallow donors/cm^3 and n_t deep donors/cm^3, the latter with energy ϕ_T eV below the Fermi level in the bulk. Both densities may be a function of position x with respect to the surface. The energy diagram for this system is illustrated in Fig. 1.

We assume that at time $t=0^-$ the diode is in steady state with a bias V_1 applied and corresponding depletion layer width $x_1(\infty)$.[1] In the region $0 \leq x \leq x_1 - \lambda(x_1)$, the positive space charge is $e(n_0 - n_t)$, while for $x_1 - \lambda(x_1) \leq x \leq x_1$, the deep centers are assumed to be neutral, so that the space charge is just $e\,n_0$. Thus $x_1 - \lambda(x_1)$ defines the point where the deep level crosses the bulk Fermi level. At $t=0^+$ the bias is switched to a more negative value V_2, which increases the depletion layer width to $x_2(0)$. Following [4], we assume that the deep donors newly introduced to the depletion region become ionized in time according to $n_t(1-e^{-t/\tau})$, where τ is a constant. It shall be assumed that the capacitance is measured at a frequency high enough so that the deep levels cannot respond to the ac perturbation, but low enough that the shallow levels respond fully. We desire to obtain the response for $t \geq 0^+$.

After two integrations of Poisson's equation, one obtains

[1] The argument denotes steady state, i.e., $t=\infty$. Hereafter if t is not explicitly given we shall assume $t=\infty$.

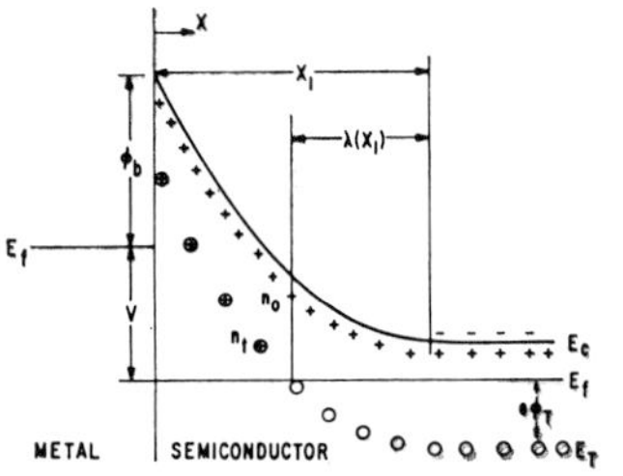

Fig. 1. Schottky barrier energy diagram showing one deep level ϕ_T eV below the Fermi level. The point $x_1-\lambda(x_1)$ is defined by the crossover of the deep level with the Fermi level.

$$
\begin{aligned}
V_2 &= V(x_1, x_2, t) + \phi_b \\
&= \frac{e}{\epsilon}\int_0^{x_2(t)} x n_0(x)\,dx + \frac{e}{\epsilon}\int_0^{\alpha(x_2(t))} x n_t(x)\,dx \\
&\quad - \frac{e}{\epsilon} e^{-t/\tau}\int_{\alpha(x_1)}^{\alpha(x_2(t))} x n_t(x)\,dx \qquad (1a)
\end{aligned}
$$

where the function $\alpha(x)$ is given by

$$\alpha(x) = [x - \lambda(x)]u(x - \lambda(x)) \qquad (1b)$$

where $u(z)$ is the unit step function, the parameter $\lambda(x)$ is given by

$$\phi_T = \int_0^{\lambda(x)} y n_0(x-y)\,dy, \qquad (2)$$

e and ϵ are the electronic charge and lattice dielectric constant, respectively, and ϕ_b is the barrier height. If n_0 may be considered constant in the region $x_1-\lambda(x_1) \leq x \leq x_1$, we have

$$\lambda^2(x_1) \simeq \frac{2}{e}\frac{\epsilon}{n_0(x_1)}\phi_T. \qquad (3)$$

For $x_1, x_2 \leq \lambda(x_1)$, we have simply

$$V_2 = \frac{e}{\epsilon}\int_0^{x_2} x n_0(x)\,dx \qquad (4)$$

since there are then no deep impurities with energy above the Fermi level.

The depletion layer capacitance for a diode with area A is

$$C(x_2) = \epsilon A/x_2(t), \qquad (5)$$

so that this value is a function of time through $x_2(t)$. Using this model we now consider the measured doping profile.

III. Measured Doping Profile

In the standard technique for obtaining doping profiles, the Schottky diode capacitance is measured as a function of reverse bias. The measured compensated donor density $n_m(x)$ is then calculated from the formula [1]

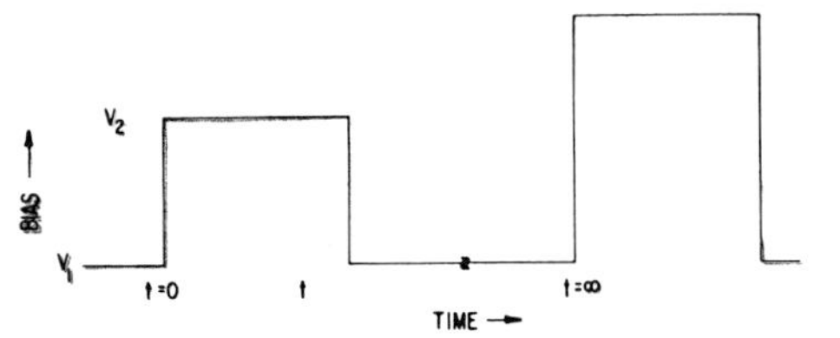

Fig. 2. Bias voltage waveform used for all measurements. Capacitance reading obtained at time t; $t=\infty$ corresponds to $t=0$ for the next bias value.

$$n_m(x) = -\frac{dV}{dC}\frac{C^3}{e\epsilon A^2}, \qquad (6)$$

where x is obtained from (5). During the measurement, the bias is normally increased in some monotonic fashion while capacitance readings are obtained, until breakdown is reached.

For samples having deep centers, however, it is advantageous to use a nonmonotonic biasing program. In particular, let us assume that after a bias voltage V_2 is applied and capacitance $C(V_2)$ measured at a time t later, the voltage is always returned to the same value V_1 for a time sufficiently long to reach steady state, before a new value of V_2 is applied (see Fig. 2.) In this manner x_1, the $t=0^-$ depletion width, is held constant. One may then write, using (1) and (4),

$$
\begin{aligned}
\frac{\epsilon}{e}\frac{dV_2}{dx_2} &= x_2 n_0(x_2) \\
&\quad + [x_2 - \lambda(x_2)]n_t(x_2 - \lambda(x_2)) \\
&\quad \cdot (1 - e^{-t/\tau})u(x_2 - \lambda(x_2)). \qquad (7)
\end{aligned}
$$

The second-order term proportional to $d\lambda/dx_2$ is ignored. Thus from (5)–(7),

$$
\begin{aligned}
n_m(x_2(t), t) &= n_0(x_2) \\
&\quad + n_t(x_2 - \lambda(x_2))\left(1 - \frac{\lambda(x_2)}{x_2}\right) \\
&\quad \cdot (1 - e^{-t/\tau})u(x_2 - \lambda(x_2)). \qquad (8)
\end{aligned}
$$

Equation (8) gives the concentration which is measured as a function of depth and time when the C-V data are analyzed with (5) and (6) in the usual fashion. Let us consider explicitly two limiting cases. At $t=0^+$,

$$n^0 \equiv n_m(x_2, 0) = n_0(x_2(0)). \qquad (9)$$

This result indicates that the shallow doping profile may be measured directly even in the presence of deep levels if the capacitance data are obtained immediately after the bias is stepped from V_1 to V_2. This follows since the deep-level impurities have not had time to emit their electrons and thus do not contribute to the posi-

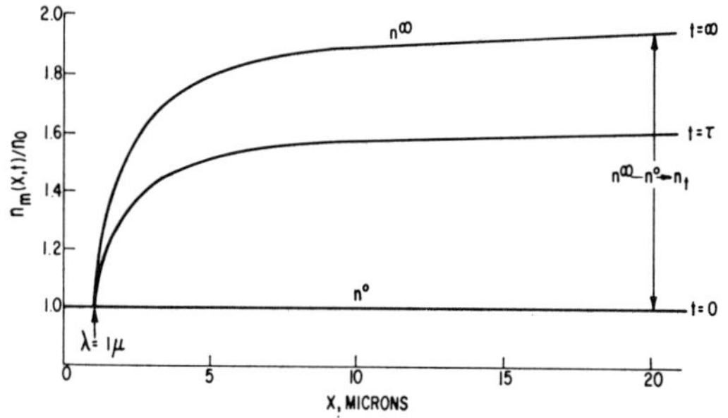

Fig. 3. Profiles which would be measured on a diode with uniform impurity distributions according to (8). $n_t = n_0$; $\lambda = 1\ \mu$.

tive space charge. On the other hand, at $t=\infty$ one has

$$n^\infty \equiv n_m(x_2,\infty)$$

$$= n_0(x_2) + \left(1 - \frac{\lambda(x_2)}{x_2}\right) n_t(x_2 - \lambda(x_1)) u(x_2 - \lambda(x_1)). \quad (10)$$

Thus the contribution from deep centers is maximized in steady state.

These two cases, as well as an intermediate time step, are shown in Fig. 3 for a diode with uniform n_0 and n_t. It is important to note that data obtained in the normal manner with monotonic bias application generally follows a curve similar to n^∞ if the rate of voltage increase is slow compared with the emission time constant τ.[2] As may be seen, the measured profile can thus be misleading when deep levels are present due to their time- and depth-dependent contribution.

Equations (9) and (10) suggest one way of evaluating the deep-level density n_t. At large values of x_2, the limiting value of $n^\infty - n^0$ becomes simply n_t, when the difference $n_0(x_2(\infty)) - n_0(x_2(0))$ may be neglected. In practice this method is not convenient, however, since two sets of data must be obtained. We turn now to a method of evaluating n_t and ϕ_T from the time dependence of the capacitance.

IV. Time Variation of Capacitance

When a bias V_2 is applied to the diode, the uncovered positive charge is an increasing function of time as the deep centers give up their electrons. Since the voltage is constant, the depletion layer capacitance must therefore increase with time. Mathematically, this requirement is given by the solution of

$$\frac{dV_2}{dt} = 0, \quad (11)$$

where V_2 is defined in (1).

Since we will be interested in $x_2(t)$ in terms of $x_2(0)$, (11) may be implemented by setting $V_2(x_1, x_2(t), t) = V_2(x_1, x_2(0), 0)$. Then one obtains from (1)

[2] The measured curve in this case is not exactly n^∞ unless the voltage increase rate is infinitely slow, since x_1 is not constant.

$$\int_{x_0}^{x(t)} x n_0(x)\, dx + (1 - e^{-t/\tau}) \int_{\alpha(x_1)}^{\alpha(x(t))} x n_t(x)\, dx = 0, \quad (12)$$

where for simplicity we have written $x_0 = x_2(0)$ and $x(t) = x_2(t)$. Further manipulation of (12) requires knowledge of $n_0(x)$ and $n_t(x)$, as well as the bias conditions V_1 and V_2. Significant simplifications result if the initial bias voltage V_1 is chosen so that $x_1 \leq \lambda\ (x_1)$. In practice this is achieved if V_1 is held at zero or slightly forward biased. Although the initial capacitance C_1 is a rapidly varying function of V_1, the determination of n_t is relatively insensitive to C_1. Thus the exact value of V_1 is not critical, so long as $x_1 \leq \lambda\ (x_1)$. This condition may be ascertained, for example, by observing negligible transient photoresponse. As a second condition, V_2 is made large enough that $x(t) \geq \lambda(x(t))$. Then by (1b), $\alpha(x_1) = 0$ and $\alpha(x(t)) = x(t) - \lambda(x(t))$. Next, we assume $n_0(x)$ may be replaced by $n_0(x_0)$ over the interval $x(t) \leq x \leq x_0$. This assumption is reasonable either when the profile $n_0(x)$ is flat, or when the interval $x_2 \leq x \leq x_0$ is small. The latter condition obtains when the deep level density is comparable or smaller in magnitude to n_0. Finally, we define an *average* deep center density $\bar{n}_t$ by

$$\bar{n}_t = \frac{2}{[x(t) - \lambda(x(t))]^2} \int_0^{x(t)-\lambda(x(t))} x n_t(x)\, dx. \quad (13)$$

Note that $\bar{n}_t$ is thus a function of V_2. Then, (12) becomes

$$[x_0^2 - x(t)^2] n_0 = (1 - e^{-t/\tau})[x(t) - \lambda(x(t))]^2 \bar{n}_t,$$

where the argument x_0 is understood for n_0.

Using (5) with obvious notation for C,

$$\frac{C^2(t)}{C_0^2} - 1 = (1 - e^{-t}/\tau)\left[1 - \frac{\lambda(x(t))}{x(t)}\right]^2 \frac{\bar{n}_t}{n_0}. \quad (14)$$

Let us first consider $t = \infty$. Then,

$$\frac{\bar{n}_t}{n_0} = \frac{\dfrac{C_\infty^2}{C_0^2} - 1}{\left[1 - \dfrac{\lambda(x_2)}{x_2}\right]^2}. \quad (15)$$

For large values of reverse bias V_2 such that $\lambda(x_2) \ll x_2$, little error is introduced by setting $\lambda(x_2) \simeq \lambda(x_1) \simeq x_1$. From (15) we obtain a working formula valid when $C_\infty/C_1 \leq 0.1$ as

$$\frac{\bar{n}_t}{n_0(x_2)} \simeq \frac{\left(\dfrac{C_\infty}{C_0}\right)^2 - 1}{1 - \dfrac{2C_\infty}{C_1}} \quad (16)$$

where

C_1	Capacitance at $t=\infty$,	$V = V_1 \simeq 0$
C_0	Capacitance at $t = 0^+$,	$V = V_2 \gg V_1$
C_∞	Capacitance at $t=\infty$,	$V = V_2$. (17)

Equation (16) is convenient for use in routine evaluation of average deep donor density since only three capacitance readings need be made. Using (10) and (16), a companion formula may be written for $\bar{n}_t$ in terms of n_∞ as

$$\frac{\bar{n}_t}{n_\infty(x_2)} \simeq \frac{\left(\frac{C_\infty}{C_0}\right)^2 - 1}{\left(\frac{C_\infty}{C_0}\right)^2\left(1 - \frac{C_\infty}{C_1}\right) - \frac{C_\infty}{C_1}}. \quad (18)$$

The lower limit on, as well as the accuracy of, $\bar{n}_t/n_0$ that may be measured is a direct function of the accuracy in determining small *changes* in capacitance (16). It was found that $\bar{n}_t/n_0$ as low as 0.005 could be reliably determined with the capacitance meter employed here.

From (14) and (16) the time-dependent capacitance for large reverse bias may be obtained as

$$g(t) = \left(\frac{C(t)}{C_0}\right)^2 - 1 = (1 - e^{-t/\tau})\left[\left(\frac{C_\infty}{C_0}\right)^2 - 1\right]. \quad (19)$$

Note that $C(t)$ is not itself an exponential function of time. Equation (19) suggests a convenient way of determining the time constant τ for measuring ϕ_T. We shall pursue this in the next section.

We consider now the average deep center density $\bar{n}_t$, defined by (13).

1) n_t Constant: When $n_t(x) = n_{t0}$, we have simply $\bar{n}_t(x) = n_{t0}$.

2) n_t/n_0 Constant: If the deep donor density is a constant multiple of the shallow doping concentration, we obtain

$$\bar{n}_t = \frac{n_t}{n_0}\frac{2}{(x_2 - \lambda(x_2))^2}\int_0^{x_2-\lambda(x_2)} x n_0(x)\, dx. \quad (20)$$

Often $n_0(x)$ may be approximated by a straight line as

$$n_0(x) = \eta_0 + \eta_1 x. \quad (21)$$

In such cases, when $\eta_1/\eta_0 \ll 1$ one finds

$$\frac{\bar{n}_t}{n_t(x_2 - \lambda(x_2))} \cong 1 - \frac{1}{3}\frac{\eta_1}{\eta_0}(x_2 - \lambda(x_2)). \quad (22)$$

Thus if the doping profile is reasonably flat, $\bar{n}_t$ approximates n_t. It may be concluded that $\bar{n}_t$ is often a meaningful approximation to the density n_t itself. Even when $\bar{n}_t \cong n_t$, the average is a useful "figure-of-merit."

V. Determination of Deep-Level Energy

The preceding analysis suggests two ways of measuring the energy of the deep level $e\phi = E_C - E_T$.

A. From C(t) Data

The measurable quantity $g(t)$ may be used to determine τ since

$$1 - \frac{g(t)}{g(\infty)} = e^{-t/\tau}. \quad (23)$$

Thus a log plot of the left-hand side yields τ. Alternatively, one may show from (23) that

$$\tau = (C_\infty^2 - C^2(t)) \Big/ \frac{dC^2}{dt}(t). \quad (24)$$

Furthermore, the time constant τ is related to the energy level by [10]

$$\frac{1}{\tau} = A \exp(-e\phi/kT), \quad (25)$$

where A is a slowly varying function of T. Measurement of τ at several temperatures thus allows ϕ to be determined. Note that τ is a very sensitive function of T since $e\phi \gg kT$.

The range of energy values ϕ accessible to measurement is governed by the temperature and the time resolution capability in obtaining $g(t)$ (and hence τ). The value of A in (25) is of the order exp (+29) at room temperature for the GaAs diodes employed in this study. Hence ϕ values as low as several hundred millivolts may be measured at room temperature if the resolution in $C(t)$ is 10 ms. Correspondingly lower limits are obtained if the temperature is decreased.

B. From λ(x₁)

Recalling (3),

$$\phi = \frac{E_c - E_f}{e} + \frac{e}{\epsilon}n_0(x_1)\frac{\lambda^2(x_1)}{2}. \quad (26)$$

Thus, knowledge of the Fermi level, n_0, and λ allows ϕ to be calculated [5]. The quantity $\lambda(x_1)$ is given by the largest value of x such that n^0 and n^∞ coincide (see (10) and Fig. 3). Implementation of this method thus requires both n^0 and n^∞ profiles, at least for small depletion widths. Since ϕ is dependent on λ^2, small errors in either n^0 or n^∞ lead to large deviations in ϕ. Thus this technique is generally less useful than the previous method.

VI. Experimental Results

A versatile capacitance–voltage measuring system was constructed for automatic acquisition of C-V data. The system was comprised of a digital C/L meter (Boonton 700C), a digital voltmeter, a data acquisition system with teletype output, and a staircase dc bias generator. The bias generator was capable of supplying voltage waveforms of the type shown in Fig. 2, with $V_1 = 0$ (return-to-zero mode) or $V_1 = V_2(k-1)$ (i.e., non-return-to-zero mode, with V_1 equal to the prior V_2). The voltage increments were programmed so that the incremental percentage change in capacitance between any two data points was constant for a uniform Schottky diode. For such a diode, we have

$$C^2 = \frac{\gamma}{V + \phi_d} \quad (27)$$

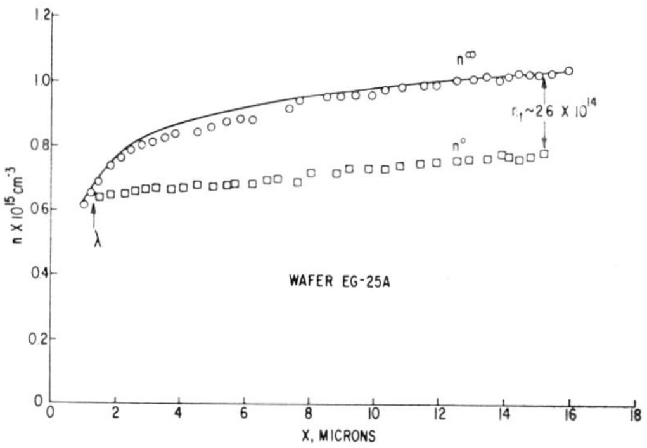

Fig. 4. Measured (data points) and calculated (solid line) profiles obtained for epitaxial n-type GaAs. The curve which is normally measured by C-V techniques n^{∞} is significantly different from the true shallow donor profile n^0.

where γ is a constant, V is the reverse bias, and ϕ_d is the diffusion potential. Then

$$\Delta V = V_{k+1} - V_k = \frac{(V_k + \phi_d)^2}{-\gamma} 2C_k \Delta C_k. \tag{28}$$

Let

$$\frac{\Delta C_k}{C_k} \equiv \frac{C_{k+1} - C_k}{C_k} = -\beta \text{ (constant)}. \tag{29}$$

Then,

$$V_{k+1} = V_k(1 + 2\beta) + 2\beta\phi_d. \tag{30}$$

If the starting voltage is $V_0 = 0$ and the final voltage is V_N, one may show

$$V_N = \phi_d[(1 + 2\beta)^N - 1]. \tag{31}$$

For a given V_N and a given β (generally made as small as practical to increase accuracy in computing dV/dC) the required number of steps N may thus be computed. With the present system it was possible to obtain up to 100 (C, V) data pairs with the voltage incremented according to (30), using $\beta = 2.5$ percent, $\phi_d = 0.8$ V.

Measurements were obtained on GaAs wafers to demonstrate the theory. The semiconducting material was in the form of a thin n layer grown by vapor phase epitaxy on an n^+ GaAs substrate. Gold contacts with known diameters up to 44 mil were vacuum deposited to form the barriers.

Fig. 4 illustrates typical measurements of $n^0(x)$ and $n^{\infty}(x)$ on the same diode. As may be seen, the deep donor density is 2.6×10^{14} cm^{-3}. The solid line was calculated from (10) using the measured $n_0(x) = n^0(x)$ and $\lambda = 1.2$ μ. The agreement is satisfactory. The deep level density calculated from capacitance data and (19) is 2.9×10^{14} cm^{-3}, which is in good agreement with the value above from $n^{\infty} - n^0$. The error was estimated from (16) and the known instrument accuracy to be $\sim \pm 5$ percent.

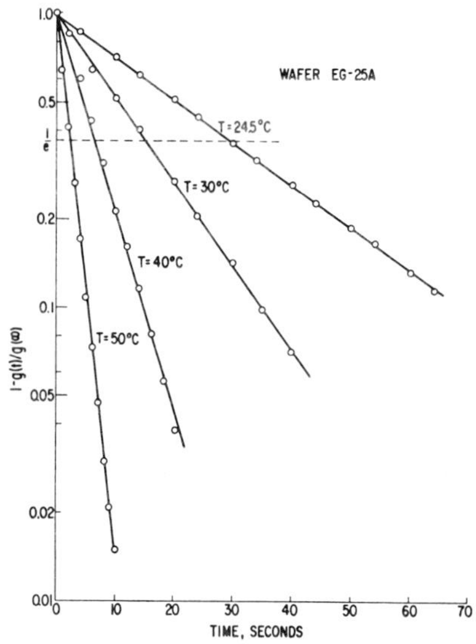

Fig. 5. Measured time variation of capacitance plotted in terms of normalized $g(t)$.

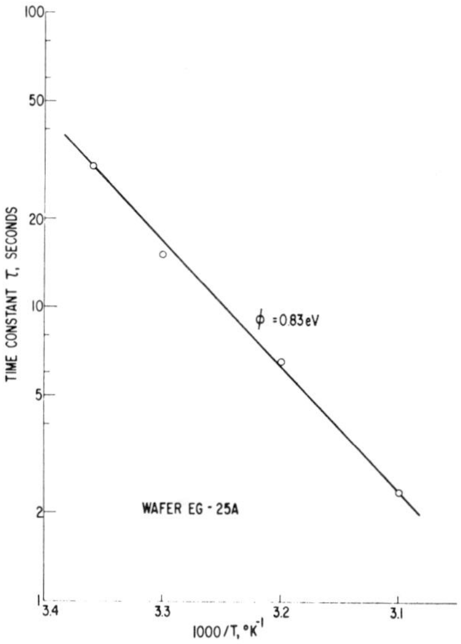

Fig. 6. Time constant τ derived from Fig. 5. A level with $\phi = 0.83$ eV is indicated.

The measured time variation of capacitance is illustrated in Fig. 5 for several temperatures. From these data the curve of Fig. 6 is obtained, yielding $\phi = 0.83 \pm .03$ eV. The error limits were estimated from the precision in measuring τ [through (23)] and temperature. This is consistent with a donor level of oxygen at

0.80 eV in bulk GaAs [11]. Similar data have been obtained for wafers with different shallow level doping concentrations grown in the same vapor-phase reactor. In one wafer, a donor level at 0.70 eV was found. Presumably, this is the same impurity observed by Williams [3] and Furukawa [2] and attributed by them to oxygen.

VII. Conclusion

A method was presented for conveniently measuring the normalized deep donor density from values of Schottky diode capacitance. The method resulted from an extension of the theory of Senechal and Basinski. The technique is implemented by measuring the transient capacitance of the Schottky diode following the application of a step increase in reverse bias. The deep level density, normalized with respect to the shallow level density, is then given in terms of the capacitance values before, immediately after, and an infinite time after the step bias increase. The energy of the impurity is obtained by making measurement as a function of temperature of the time varying capacitance at large reverse bias. A measurement on epitaxial GaAs showed a level at 0.83 eV, consistent with a donor state of oxygen in bulk GaAs. The theory also indicated a means of measuring the shallow level doping profile by C-V techniques even in the presence of deep levels. The method relies on repopulating the deep levels between measurements of C by returning the bias to some low value.

Acknowledgment

The author wishes to thank P. J. Shaver of the G. E. Solid State Physics Laboratory for supplying the Au-GaAs diodes used in the study. He is indebted to the reviewers for several comments.

References

[1] J. Hilibrand and R. D. Gold, "Determination of the impurity distribution in junction diodes from capacitance–voltage measurements," *RCA Rev.*, pp. 245–252, June 1960.

[2] Y. Furukawa, "Trap levels in gallium arsenide," *Japan. J. Appl. Phys.*, vol. 6, pp. 675–679, June 1967.

[3] R. Williams, "Determination of deep centers in conducting gallium arsenide," *J. Appl. Phys.*, vol. 37, pp. 3411–3416, Aug. 1966.

[4] R. R. Senechal and J. Basinski, "Capacitance of junctions on gold-doped silicon," *J. Appl. Phys.*, vol. 39, pp. 3723–3731, July 1968; also, "Capacitance measurements on Au–GaAs Schottky barriers," *J. Appl. Phys.*, vol. 39, pp. 4581–4589, Sept. 1968.

[5] Y. Zohta, "A new method for determination of deep level impurity centers in semiconductors," *Appl. Phys. Lett.*, vol. 17, pp. 284–286, Oct. 1970.

[6] J. A. Copeland, "Switching and low-field breakdown in n-GaAs bulk diodes," *Appl. Phys. Lett.*, vol. 9, pp. 140–142, 1966.

[7] G. Bemski, "Recombination in semiconductors," *Proc. IRE*, vol. 46, pp. 990–1004, June 1958.

[8] J. A. Copeland, "A technique for directly plotting the inverse doping profile of semiconductor wafers," *IEEE Trans. Electron Devices*, vol. ED-16, pp. 445–449, May 1969.

[9] ——, "Semiconductor impurity analysis from low-frequency noise spectra," *IEEE Trans. Electron Devices*, vol. ED-18, pp. 50–53, Jan. 1971.

[10] C. T. Sah, L. Forbes, L. L. Rosier, and A. F. Tasch, Jr., "Thermal and optical emission and capture rates and cross sections of electrons and holes at imperfection centers in semiconductors from photo and dark function current and capacitance experiments," *Solid State Electron.*, vol. 13, pp. 759–788, 1970.

[11] S. M. Sze and J. C. Irvin, "Resistivity, mobility and impurity levels in GaAs, Ge, and Si at 300°K," *Solid State Electron.*, vol. 11, p. 599, 1968.

Semiconductor Impurity Analysis From Low-Frequency Noise Spectra

J.A. Copeland
(Bell Telephone Laboratories, Inc., Murray Hill, New Jersey, US)

IEEE Transactions on Electron Devices, Vol. ED-18, No. 1 (January 1971)
Pages 50-53

key paper

Abstract—The usefulness of material for semiconductor devices can be degraded by the presence of deep impurity levels or traps which are often difficult to detect. The equations necessary for using the noise emitted by a uniform semiconductor when biased with a small dc current to find the energy, degeneracy factor, density, and time constants of such a deep level are derived. The theory was verified by an experimental study of a level in n-GaAs which was found to have an energy 0.175 eV below the conduction band and an acceptor-like degeneracy factor of 4. This technique appears to be useful for fundamental studies and as a means of monitoring material used for devices.

THE operation of n-GaAs semiconductor devices used as Gunn or LSA oscillators can be degraded by impurities with energy levels which are well within the forbidden energy band. These can be a problem in all types of semiconductor devices [1]. In bulk n-GaAs devices, which depend on a uniform fixed-charge density, they can cause a current runaway or breakdown at low voltage just above the Gunn threshold (3000 V/cm) [2] [3]. The average charge of these levels changes by impact ionization or hole capture when Gunn domains form in the device. In junction devices these levels act as generation–recombination centers which reduce minority carrier lifetime and increase leakage currents.

In n-GaAs epitaxial material used for Gunn and LSA diodes these levels are difficult to detect because they generally lie well below the Fermi level which is less than 0.3 eV below the conduction band for carrier densities greater than 10^{13} cm^{-3} and temperature less than 100°C. They do not affect resistivity versus temperature plots and if their normal charge state is neutral (deep donors) they may not affect carrier mobility at low temperature. They can be detected by infrared photoconductivity [4] and by infrared photoluminescence [5]. These latter measurements give energy levels accurately but absolute or even relative densities are hard to extract without more information than is usually available on various cross sections and transition probabilities.

The method to be described is particularly useful for n-GaAs material since it may be applied to fininshed n^+-n-n^+ devices. Other semiconductor materials could be evaluated by applying ohmic contacts to make a simple semiconductor resistor. A small current is passed through the device, the noise spectra are plotted at a number of temperatures for a frequency range of typically 20 to 600 000 Hz, and the data is evaluated as described below. Under optimum conditions, information is obtained on the energy levels, absolute densities, and time constants.

Manuscript received March 4, 1970; revised August 3, 1970

Noise Emission—Simple Model

The statistical behavior of traps and noise emission has been treated previously in great detail [1], [6]. Consider the simple case of an impurity which has two charge states (0 and +1 for a donor, −1 and 0 for an acceptor). For n-type material under thermal equilibrium conditions the primary mechanisms for changing state are the capture and emission of conduction band electrons so that

$$0 = \frac{D_f}{\tau_e} - \frac{D - D_f}{\tau_c} \tag{1}$$

where D is the density of the deep levels, D_f is the average number of filled levels, τ_e is the time constant for emission from a filled level, and τ_c is the time constant for capture by an empty level. The capture time constant τ_c is frequently expressed as $(nSv_T)^{-1}$ where n is the carrier density, S is the capture cross section, and v_T is the average electron thermal velocity.

If the density of empty deep levels is perturbed from the equilibrium density D_f by ΔD, it relaxes back to D_f with a time constant τ given by

$$\frac{d\Delta D}{dt} = \frac{-\Delta D}{\tau_e} - \frac{\Delta D}{\tau_c} = \frac{-\Delta D}{\tau} \tag{2}$$

$$\tau = \tau_e\tau_c/(\tau_e + \tau_c). \tag{3}$$

When a single deep level emits an electron, the current flowing through the uniform semiconductor device increases by I/nV where I is the dc current due to a small bias voltage and nV is the average number of free electrons in the device (V=volume). This excess current will on the average decay with the time constant τ given by (3) so that the average current as a function of time $i(t)$ created by such an event at time $t=0$ will be of the form

$$i(t) = (I/nV)\exp(-t/\tau) \qquad t > 0 \tag{4}$$

which has frequency components $i(\omega)$ given by

$$i(\omega) = \int_0^\infty i(t)\exp(j\omega t)dt = \frac{I\tau}{nV(1 + j\omega t)}. \tag{5}$$

Since the emission events occur at random, the power and current-squared spectrums are additive. The average number of emissions per second is $DV/(\tau_e+\tau_c)$ so the current-squared $\langle i^2\rangle$ emitted at frequency $f(f=\omega/2\pi)$ in a bandwidth Δf is given by

$$\langle i^2\rangle = \frac{2DV\,|\,i(\omega)\,|^2\Delta f}{\tau_e+\tau_c} \tag{6}$$

$$\langle i^2\rangle = \frac{2DI^2\tau^2\Delta f}{n^2V(\tau_e+\tau_c)[1+(2\pi f\tau)^2]}\,. \tag{7}$$

To be detectable, the deep level noise given by (7) must be larger than the thermal noise current-squared $4R^{-1}kT\Delta f$ and possible shot noise current-squared $eI\Delta f$ where R is the device resistance. During an experiment deep level noise component can be quickly identified by its dependence on the square of the dc bias current I.

Using (1), (3), and (7) it is possible to write an expression for the square of the rms noise current in a form that is convenient for interpretation of experimental data.

$$\frac{\langle i^2\rangle}{\Delta f} = \frac{2\tau I^2}{n^2V}\,\frac{D_f(D-D_f)}{D}\,\frac{1}{1+(2\pi f\tau^2)}\,. \tag{8}$$

Experiment

The noise spectra of devices being tested are plotted using the experimental setup shown in Fig. 1. The device is in a temperature controlled holder. The dc bias current I from a storage battery is controlled by a variable series resistor which is several times larger than the device resistance. A low-noise low-impedance grounded-base transistor preamplifier is used before the input of a Hewlett-Packard 3590A wave analyzer. A simple external waveform generator circuit is used to obtain an exponential frequency versus time sweep. Schematic diagrams of these are shown in Fig. 1. The HP 3590A has logarithmic outputs so that log–log plots of noise current versus frequency are drawn by an x–y plotter. The system is calibrated before each plot by injecting a known ac current across the device with the dc bias current flowing. With 10 ohm samples, noise signals $\langle i^2\rangle/\Delta f$ as low as 3×10^{-21} A^2/Hz can be plotted.

When the noise detected comes from impurities with a well-defined time constant τ as assumed in the last section, the observed noise spectrum has the shape shown in Fig. 2 which is described by

$$\frac{\langle i^2\rangle}{\Delta f} = \frac{A}{1+(f/f_0)^2}\,. \tag{9}$$

The noise has a constant value A at low frequencies, decreases by 3 dB at frequency f_0, and falls as Af_0^2/f^2 at high frequencies. The frequency f_0 is easily determined by drawing straight lines to fit the low-frequency and high-frequency parts of the log–log plot, as shown in Fig. 2, with f_0 being the frequency at the intersection. The value of τ is then determined since $\tau=1/2\pi f_0$.

To determine the density of deep levels D and the energy below the conduction band ϵ, it is necessary to

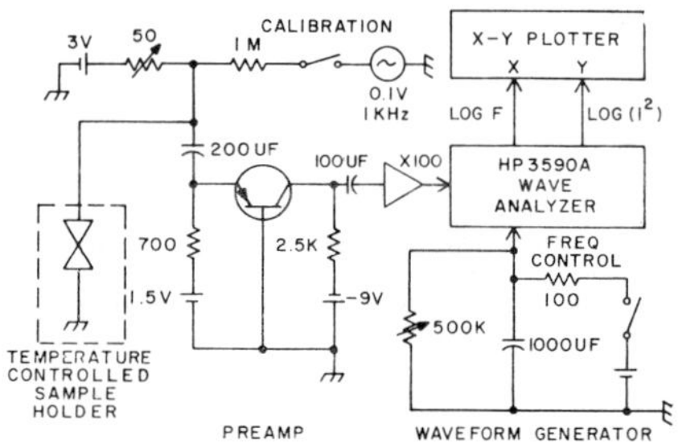

Fig. 1. Diagram of circuit used to plot noise current squared versus frequency.

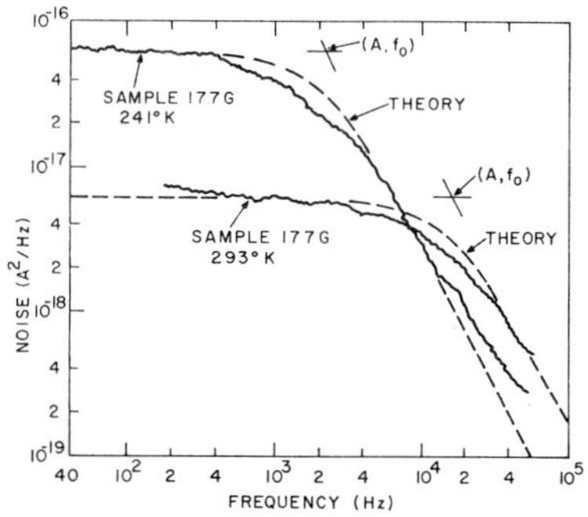

Fig. 2. Noise spectra of sample 177G at 241 and 293°K. The dashed curves are of the form $A/[1+(f/f_0)^2]$ and are fitted to the data. The corners of the experimental curves are not as sharp as those of the theoretical curves because the time constants of all the traps are not exactly the same.

determine the value of A and τ at a number of different temperatures. The volume of the device V and the carrier density n are known (n is determined from electrical resistance and magnetoresistance measurements).

This data is interpreted by using Fermi–Dirac statistics:

$$D_f = D/[1+B\exp([\epsilon-\phi]/kT)] \tag{10}$$

and assuming the semiconductor is nondegenerate so that

$$n = N_c\exp(\phi/kT) \tag{11}$$

where B is the spin degeneracy of the deep level ($B=1/2$ for a hydrogenic donor, 2 for a hydrogenic acceptor), kT is the thermal energy, ϕ is the Fermi energy with the bottom of the conduction band taken as zero, and N_c is the effective density of states in the conduction band [7]. The low-frequency ($f\ll f_0$) noise is then given by

$$A = \frac{2I^2\tau D}{n^2V}\,\frac{(BN_c/n)\exp(\epsilon/kT)}{[1+(BN_c/n)\exp(\epsilon/kT)]^2}\,. \tag{12}$$

It is convenient to combine the three temperature-

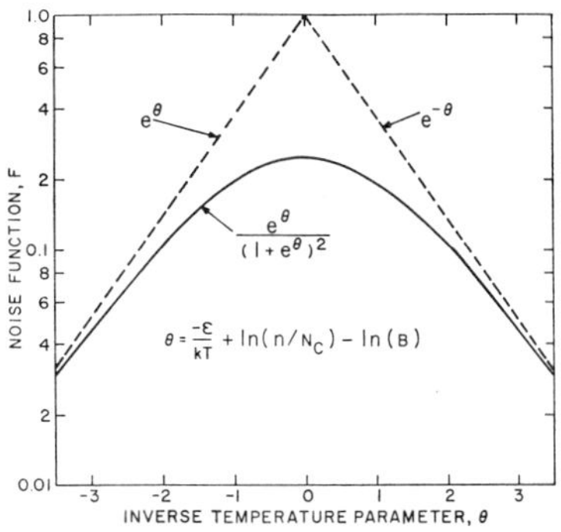

Fig. 3. A theoretical plot of how the noise function F should vary with respect to the inverse temperature parameter θ. The maximum magnitude is proportional to the density of traps and other sample parameters.

dependent experimentally determined parameters A, n, and τ into one function of temperature $F(T)$ given by

$$F(T) \equiv An^2/\tau. \quad (13)$$

From (12), F can be written as a simple function of the temperature-dependent parameter θ:

$$F(\theta) = (2I^2D/V)/[\exp(\theta) + 2 + \exp(-\theta)] \quad (14)$$

$$\theta(T) = -\epsilon/kT + \ln(n/BN_c') - 1.5\ln(T/T') \quad (15)$$

where N'_c is the value of N_c at temperature T'. A plot of $F(\theta)$ is shown in Fig. 3.

The first step in the evaluation of a set of experimental data is to plot $\ln(F)$ versus $1/T$ as shown in Fig. 4. If this plot has a maxima (such as for sample 177G), then at this temperature $\theta = 0$ and D is uniquely determined from (14) and a relationship between ϵ and B is obtained from (15) such as the example plotted in Fig. 5. To obtain separate values for ϵ and B, it is necessary to find the values which fit other points on the curve. This task is simplified if data points are available well away the maxima of F so that the curve is linear and the energy ϵ can be obtained from

$$\frac{d[\ln(F)]}{d(1/T)} = \pm\frac{d\theta}{d(1/T)} \quad (16)$$

$$= \pm(-\epsilon + 1.5kT)/k \quad (17)$$

where the initial sign is the same as the sign of the slope.

The 1.5 kT term is due to the temperature dependence of the effective density of states in the conduction band. If this effect had been neglected, the energy obtained from sample 177G would have been -0.215 eV rather than -0.175 eV.

Once B and ϵ are known, the emission time constant τ_e and capture cross section S can be found from τ by

$$\tau_e = \tau[1 + \exp(\theta)]$$

$$S = 1/nV\tau_e$$

$$= 1/nV\tau[1 + \exp(-\theta)].$$

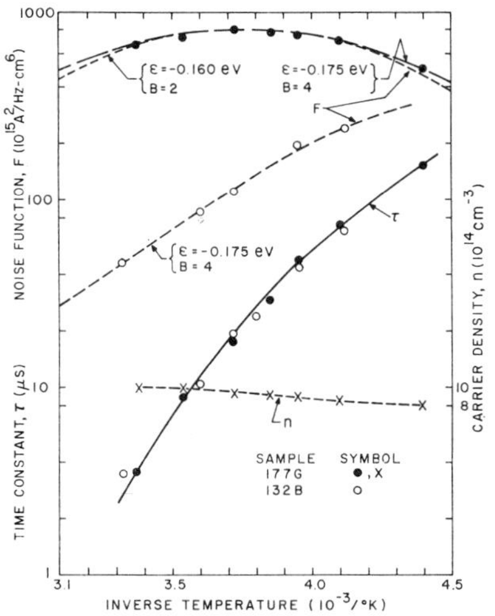

Fig. 4. Experimental and theoretical plots of the noise function F and time constant τ for two samples, 177G and 132B versus inverse temperature. The carrier density n of sample 177G is also plotted ($n = 10^{14}$ cm^{-3} for 132B).

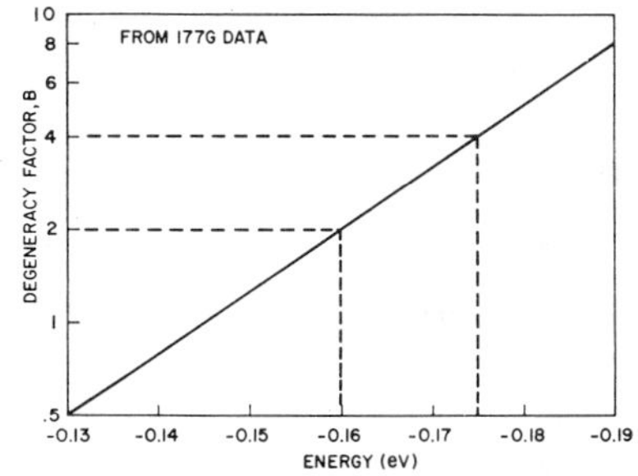

Fig. 5. The relationship between trap energy ϵ and degeneracy factor B obtained from knowing the temperature where F reached its maximum value for sample 177G (see Fig. 4). The data from 132B indicate that $\epsilon = -0.175$ eV.

The experimental results from the two samples 177G and 132B which clearly exhibited noise from a single level over the temperature range 230 to 300°K indicate the energy is 0.175 eV below the conduction band. Other results are shown in Table I. This is probably the 0.2 eV level reported previously which was produced by heat treatment and attributed to a lattice defect [8]. The degeneracy factor of 4 is the same as if the level were a simple acceptor since there are two degenerate hole bands. There is presently no satisfactory theory for such a deep level and no inference can be made about the charge. Sample 177G was grown by

TABLE I
EXPERIMENTAL RESULTS ON TWO SAMPLES

	Sample 177G	Sample 132
Energy	0.175 (0.160) eV	0.175 eV
Degeneracy factor	4 (2)	—
Deep level density	2.5×10^{14} cm^{-3}	10^{12} cm^{-3}
Carrier density	1×10^{15} cm^{-3}	1×10^{14} cm^{-3}
Volume	10^{-8} cm^3	10^{-7} cm
Mobility (cm^2/V·s)		
300°K	6850	6350
77°K	26 200	10 200
Capture cross section		
300°K	4.0×10^{-18} cm^{-2}	3.5×10^{-18} cm^{-2}
245°K	3.0×10^{-19} cm^{-2}	9.7×10^{-19} cm^{-2}
Emission time constant		
300°K	3.5×10^{-6} second	3.5×10^{-6} second
245°K	2.8×10^{-4} second	9.3×10^{-5} second
Time constant τ		
300°K	2.5×10^{-6} second	3.0×10^{-6} second
245°K	8.4×10^{-5} second	7.5×10^{-5} second

the AsH_3 process by Monsanto and 132B was grown by the water-vapor transport process at Bell Telephone Laboratories.

OTHER RESULTS ON n-GaAs

The −0.175 eV impurity level described in the last section has only been seen in a few samples. We were fortunate to have a sample such as 177G where the change in carrier concentration with temperature (Fig. 4) could be used to check the density obtained from the noise measurements. Once an impurity has been characterized (ϵ and B known), then only a single measurement is needed to obtain the density in subsequent samples when the noise from this level is identifiable.

The room temperature noise spectra from n-GaAs epitaxially grown samples obtained from a variety of sources are shown in Fig. 6. While these samples have not yet been completely studied, there are some interesting features.

The spectra from samples grown by the liquid-gallium technique have values of f_0 which are about 100 Hz or less indicating τ is about 1.5 ms or greater. The noise increases greatly when the sample is illuminated with a tungsten filament lamp (by a factor of 1000 for LC34) and τ decreases (by 1/10 for WF-135). Photoconductivity measurements reveal three levels at −1.07 eV, −0.94 eV (iron), and −0.82 eV (oxygen) with time constants of 4 ms, 4 ms, and 15 ms, respectively.

The spectra from samples grown by the $AsCl_3$ process at Bell Laboratories (E349, E409, D35, and many not shown) show noise spectra that are almost identical and seem to be independent of carrier density (10^{15} to 10^{16} cm^{-3}) and temperature (250 to 375°K) and only slightly sensitive to light. The noise decreases approximately as $f^{-1.2}$ over the observed range. The noise has been reversibly reduced by a factor of 30 by heating the sample to 100°C for 30 minutes with a bias current of 100 mA (10^3 A/cm^2) if the substrate is positive. A similar heat treatment with the substrate negative

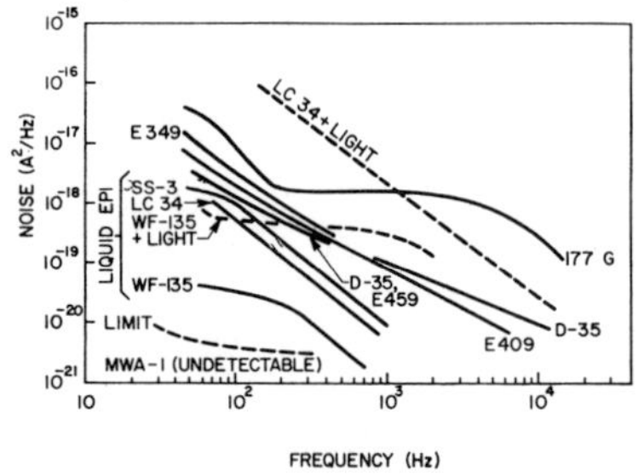

Fig. 6. Noise spectra obtained at room temperature from expitaxial n-GaAs samples obtained from many sources.

increases the noise to its original value. Premature electrical breakdown occurs at a lower voltage when these samples are in their nosier state. The high mobility of the impurity makes it likely that interstitial copper plays a role. The shape of the noise spectra implies that the value of τ for each of these impurities depends on some local parameter since they are spread over several orders of magnitude. Photoconductivity measurements indicate these impurity levels are distributed 1.0 to 1.3 eV below the conduction band. A sample of material grown by the $AsCl_3$ process at Microwave Associates (MWA-1) showed no detectable noise indicating that these impurities are not inherent in the $AsCl_3$ process.

CONCLUSION

The noise emitted by a uniform semiconductor biased with a small dc current can be used to obtain information on the energy level, degeneracy factor, and density of deep-level impurities. One such level in GaAs was investigated and found to be 0.175 eV below the conduction band with an acceptor-like degeneracy factor of 4. This technique appears to be useful for fundamental studies and for monitoring the quality of material used for devices.

REFERENCES

[1] W. Shockley, "Electrons, holes, and traps," *Proc. IRE.* vol. 46, pp. 973–990, June 1958.

[2] J. A. Copeland, "Switching and low-field breakdown in n-GaAs bulk diodes," *Appl. Phys. Lett.*, vol. 9, pp. 140–142, August 15, 1966.

[3] H. W. Thim and S. Knight. "Carrier generation and switching phenomena in n-GaAs devices," *Appl. Phys. Lett.*, vol. 11, pp. 83–85, August 1, 1967

[4] J. A. Copeland and W. C. Niehaus, "Traps and low-field breakdown in n-GaAs devices," to be published.

[5] H. J. Queisser and C. S. Fuller, "Photoluminescence of Cu-doped gallium arsenide," *J. Appl. Phys.*, vol. 37, pp. 4895–4899, December 1966; also G. E. Moore, private communication.

[6] A. van der Ziel, *Fluctuation Phenomena in Semiconductors*. London: Butterworth, 1959, pp. 22–45; K. M. Van Vliet and J. R. Fassett, in *Fluctuation Phenomena in Solids*, R. E. Burgess, Ed. New York: Academic Press, 1965, ch. VII.

[7] J. Blakemore, *Semiconductor Statistics*. New York: Pergamon, 1962, pp. 75–81, 118–119.

[8] J. Blanc, R. H. Bube, and L. R. Weisberg, "Behavior of lattice defects in GaAs," *J. Phys. Chem. Solids*, vol. 25, pp. 225–244 1964.

key paper

Electrical Characteristics Of Gallium Arsenide 'Real' Surfaces

T.M. Valahas, J.S. Sochanski*, and H.C. Gatos
(Massachusetts Institute of Technology, Cambridge, Massachusetts, US)

Surface Science, Volum 26 (1971)
Pages 41-53

*Now with Université du Québec à Trois-Rivières, Québec, Canada)

Received 12 November 1970

"Real" surfaces of gallium arsenide single crystals were investigated by pulsed-field effect techniques. A discrete energy level was found at the semiconductor–oxide interface 0.7 eV from the bottom of the conductor band (middle of the energy gap) with a large capture cross-section for electrons (10^{-14} to 10^{-12} cm^2). The parameters of the surface states were found to be insensitive to crystallographic orientation and to moderate variations in the etching treatment or in the gaseous ambient atmosphere. In the dark, the field-induced disturbance of equilibrium between the surface states and the bulk (change in surface conductance) was re-established by thermal excitation of majority carriers over the surface potential barrier. The activation energy of this process was determined from the dependence of its time constant on temperature. Under illumination, light-induced minority carriers play a dominant role in the surface conductance decay and, accordingly, the process was found to be essentially independent of temperature.

1. Introduction

Although the electrical characteristics of gallium arsenide surfaces have been extensively studied in conjunction with device applications, basic studies on these surfaces have been limited[1]). The present study is concerned with the basic electrical characteristics of gallium arsenide "real" surfaces. In spite of the uncertainties that such surfaces may present, if properly employed, they can yield results of fundamental as well as of a technological importance. This type of results, together with results on "clean" surfaces, are essential in the understanding of the overall chemical and physical surface characteristics.

2. Experimental techniques and procedures

The gallium arsenide crystals presently employed were grown in the ⟨111⟩

direction by the Czochralski method. The measured bulk electrical characteristics of the samples are summarized in table 1.

TABLE 1

Bulk electrical characteristics of gallium arsenide at 300°K

Sample	Conductivity type	Doping	Resistivity (ohm-cm)	Mobility of majority carriers (cm^2/V-sec)	Carrier concentration (cm^{-3})
S I	n	none	0.018	4200	7.0×10^{16}
S II	n	none	0.25	6400	3.8×10^{15}
S III	n	none	1.2	5800	1.2×10^{15}
S IV	n	oxygen	0.27	5460	4.1×10^{15}
S V	p	zinc	15.4	200	2.0×10^{15}

The samples were generally prepared as follows: Wafers $15 \times 2 \times 1$ mm were cut from single crystal ingots, their large planes being the surfaces of interest, i.e., {111} or {100}. Their orientation was established to within less than one degree by a back reflection X-ray Laue technique. The wafers were then lapped to the desired size – approximately $12 \times 2 \times 0.2$ mm – and subsequently polished with a 0.05 μm particle size alumina abrasive. They were cleaned ultrasonically in hot ethyl alcohol and then in dichromate solution (0.1 N K_2Cr_2O7 in 36 N H_2SO_4) to remove residual particles. Finally they were etched in 1 part 40% HF + 3 parts 70% HNO_3 + 2 parts H_2O for 15 to 30 sec until the surfaces became mirror smooth. The etchant was first "activated" with small pieces of GaAs. In this way the etching attack was very fast and rounding of the edge was minimized. The reaction was stopped by flooding one etchant with deionized water; the water was removed from the samples with filter paper and they were allowed to dry in room air; subsequently they were stored in a desiccator for varying periods of time to allow the formation of a stable oxide layer.

Tin contacts on the n-type and zinc contacts on the p-type samples were made at the ends of the samples by alloying or electroplating. Alloying was carried out in an inert atmosphere (mixture of argon and hydrogen) and hydrazine monobromide was used as a flux; for electroplating the samples were appropriately masked with a polystyrene based material. The contacts were ohmic over the whole temperature range of interest, i.e., −50°C to 180°C. Gold wire leads 0.05 mm in diameter were soldered to the contacts.

The sample-field plate condenser for the field-effect measurements was made by sandwichizing the crystal between two conducting glass electrodes. Mylar sheets 1.25×10^{-2} mm thick were used as dielectric spacers between the sample and the field electrodes. This arrangement was essentially a double capacitor allowing either the top or bottom surface of the sample to be studied under identical experimental conditions. The transparent field plates permitted the illumination of the surfaces under study while an electric field was applied. The entire crystal holder was placed in a hermetically sealed chamber which could be held at a temperature constant to better than 0.01 °C/hr.

For the experiments performed in an ultra-high vacuum (10^{-9} mm Hg) the dielectric spacers were eliminated since they could affect the slow relaxation phenomena[2]) and also prevent part of the surface from being exposed to the ambient. The field electrode was polished stainless steel which could be placed very close to the specimen by means of two alumina strips.

The electric field normal to the surface under study was applied in the form of square pulses. Surface conductance decay following the application or removal of the field was displayed on an oscilloscope or a recorder. Since the time constants of the surface conductance variation were relatively long, most measurements could be carried out with a Wheatstone bridge. In some instances, a special bridge arrangement was used[3]) to minimize the contribution of the displacement current.

3. Experimental results and discussion

3.1. Sinusoidal field-effect experiments

Large-signal sinusoidal field-effect measurements are usually carried out in the study of semiconductor surfaces. They allow the determination of the trapped surface charge, the surface potential, and the dependence of the surface charge on the surface potential, provided a minimum in surface conductance is achieved as a function of the field-induced charge[4]). No such minimum in surface conductance could be achieved in the gallium arsenide surfaces. In order to overcome the problem of the isolation of the inversion layer from the bulk[5]) and attract mobile holes to the surface region, experiments were carried out at elevated temperatures (up to 350 °K) as suggested by Frankl[6]), but again no minimum in surface conductance was observed. A minimum in surface conductance in gallium arsenide surfaces reported earlier[7]) is now believed to be due to extraneous thermal effects[8]). The reason for the absence of a surface conductance minimum in gallium arsenide surfaces is not entirely clear at this time.

3.2. Pulsed field-effect experiments

3.2.1. *Experiments in the dark*

The low-frequency pulsed field-effect experiments were carried out in the dark, under constant temperature and of course away from the surface conductance minimum, thus insuring a constant differential field effect mobility in the whole range of the conductance variation. The typical general features of the observed surface conductance decays are shown in fig. 1. It is apparent that the change of the surface conductance with time is not the same for the application and the removal of the electric field. Application of positive field (positive voltage on the field plate with the sample grounded) on the n-type samples resulted in a much smaller amplitude and faster decay

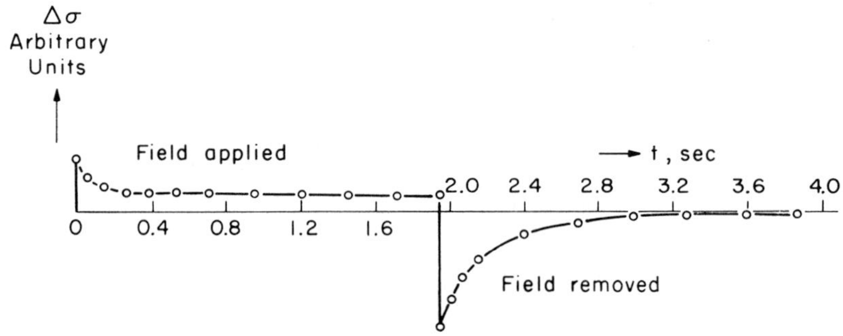

Fig. 1. Relaxation of conductance following the application and removal of dc field (in the dark); n-type sample S III (table 2); $V_F = +1600$ V.

than the removal of the field. The same behavior was observed when a negative field was applied. The decay following the application of the field was very difficult to observe and analyze in detail; the initial portion could not be seen on the oscilloscope as it took place in much less than 1 μsec. This fast decay apparently results from the accumulation of majority carriers in the surface region causing an intimate contact between the surface states and the bulk; accordingly, it is strongly dependent on the characteristics of the bulk. It will not be considered further at this time

The relatively slow decay (following the removal of the field) apparently reflects an exhaustion region at the surface; it was studied as a function of temperature; its time constant, τ, was found to decrease exponentially with increasing temperature as shown in fig. 2. In most instances this conductance decay could be expressed as follows:

$$\tau = C \exp(\Delta E/kT), \tag{1}$$

where C is a constcnt, ΔE is the activation energy of the process, k is the Boltzmann constant and T the absolute temperature.

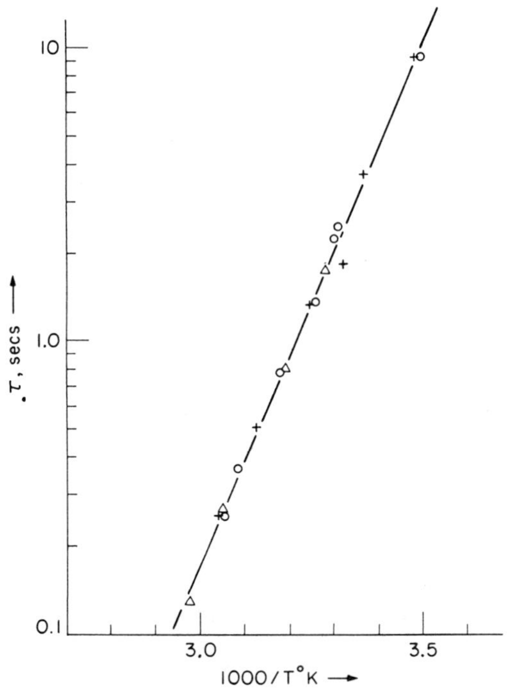

Fig. 2. Temperature dependence of the field-effect relaxation time for different ambients: (+) dry NH_3; (△) dry O_2; (○) vacuum, 10^{-19} mm Hg.

3.2.2. *Photovoltage experiments*

Photovoltage experiments were carried out to determine the direction of band bending at the surface. The surface photovoltage develops between the surface region and the bulk when electron-hole pairs are injected in the surface by light. Experimentally, it was taken to be the voltage capacitively developed on the field-effect electrode upon illumination of the grounded sample.

In all n-type samples – at room temperature – a positive photovoltage was determined showing the presence of a depletion layer (upward bending of the bands). It was not possible in these experiments to achieve the saturation value of the photovoltage which is a measure of the surface potential barrier. The reproducibility of the results above values of about $8kT$ was poor probably because of the charging of the "slow" states likely present in high densities.

3.2.3. *Energy position of surface states*

Rupprecht's[9]) analysis of field-effect decays appears most appropriate in the present case; the charge transfer between the surface states and the bulk is brought about by majority carriers which are thermally excited over the surface potential barrier. Accordingly, the coefficient C and the activation energy ΔE in eq. (1) for an n-type semiconductor are:

$$C = 1/\gamma N_c v_t \sigma_n , \tag{2}$$

$$\Delta E = E_c - E_t , \tag{3}$$

where N_c is the effective density of states in the conduction band, γ is the statistical weight factor, v_t is the thermal velocity of electrons in the conductor band, σ_n is the capture cross-section of the surface states for electrons, E_c is the energy of the bottom of the conduction band, and E_t is the energy position of the surface states. The coefficient C is not entirely independent of temperature, but it can be taken as a constant in view of the exponential temperature dependence of τ. Thus, the parameters E_t and σ_n can be determined from field-effect measurements provided the bulk parameters of the semiconductor are known. For p-type semiconductors the valence band parameters are introduced in the expressions (2) and (3).

On the basis of this model the energy positions of surface states in gallium arsenide were determined under a variety of experimental conditions (discussed below) and are summarized in table 2.

TABLE 2

Energy position and capture cross-section of the surface states of GaAs surfaces

Sample	Atmosphere	Energy position $E_c - E$ (eV)	Capture cross-section σ_n (cm^2)
S II	Vacuum, 10^{-9} mm Hg	0.70	1×10^{-14}
S II*	Vacuum, 10^{-9} mm Hg	0.71	1×10^{-14}
S III	Vacuum, 10^{-9} mm Hg	0.71	1×10^{-14}
S V	Vacuum, 10^{-9} mm Hg	0.62	5×10^{-14}
S I	Vacuum, 10^{-4} mm Hg	0.69	1×10^{-12}
**	Vacuum, 10^{-4} mm Hg	0.57	1×10^{-13}
***	Vacuum, 10^{-4} mm Hg	0.70	4×10^{-12}
S IV	Vacuum, 10^{-4} mm Hg	0.71	
	Dry oxygen	0.71; 0.45	1×10^{-12}; 1×10^{-17}
	Dry ammonia	0.70; 0.45	1×10^{-12}; 1×10^{-17}

* (100) orientation; all other samples are of the (111) orientation. For a given (111) sample the results were the same for the A and the B surface.

** Sample was etched in $HF + H_2O_2 + H_2O$; all others were treated with the standard etchant (see text).

*** Polished with 10.0 μm particle size alumina; all others were polished in the standard fashion (see text).

3.2.4. *Effect of orientation, surface treatment and ambient*

Field effect experiments on A{111}, B{111} and {100} surfaces prepared as described above were performed in ultra-high vacuum (about 10^{-9} mm Hg). No significant variation in the surface state parameters was found (table 2). Apparently, the observed surface states are not associated with the particular termination of the lattice at the surface.

The effect of surface treatments both mechanical and chemical on the surface state characteristics were also extensively studied. The following three different etchants were employed:

(a) 1 part 48% HF+3 parts 70% HNO_3+2 parts H_2O;
(b) 1 part 48% HF+1 part 30% H_2O_2+1 part H_2O;
(c) 2 parts 37% HCl+1 part 70% HNO_3+2 parts H_2O.

Etching was followed by extensive rinsing in deionized water as discussed earlier. All measurements were carried out in a vacuum of 1×10^{-4} mm Hg. No significant changes were observed in the surface state characteristics as a result of different surface treatment. When, however, the samples were rinsed in water containing small amounts of copper nitrate, after etching in solution (a), no field-effect signal could be obtained. Copper ions are known to introduce a large number of surface states interfering with the field effect as it was found, for instance, in InSb surfaces[10]).

In one instance a sample treated with etchant (b) exhibited surface states 0.57 eV below the conduction band (table 2). Surface states in this energy position were reported earlier[11]) and were attributed to lattice defects.

Abrading with 10 μm particle size alumina, rather than 0.03 μm, again had no effect on the field-effect results. No appreciable field-effect signal could be obtained with 25 μm particle size was used for abrasion. Apparently, excessive surface damage introduced a very large number of surface states obscuring the field effect.

Field-effect measurements were carried out on a number of samples treated in the standard way described above (section 2). They were placed in the temperature-controlled chamber which was then evacuated to about 10^{-5} mm Hg. Subsequently, the desired gas was introduced and circulated for about two hours. Then, circulation was stopped for about one hour before electrical measurements were performed. In general, dry oxygen and dry ammonia had no effect on the energy position and capture cross-sections of the surface states. Only in one n-type sample, doped with oxygen, a new surface state was observed with an energy position at 0.45 eV below the conduction band and a capture cross-section of 10^{-17} cm^2. This surface state apparently corresponds to the bulk oxygen donor state at 0.46 eV below the conduction band. At relatively high temperature this state is empty and, thus, capable of recapturing electrons in the space charge regions during the surface conductance decay.

3.2.5. *Analysis of the field effect results*

Of the several models proposed for the interpretation of field-effect decays that of Yunovich[12]) appears to best account for the present experimental results. This model takes into account the charge exchange between the surface states and the bulk through thermal excitation of carriers over the surface potential barrier. Assuming that the surface region is strongly depleted and that (for n-type bulk material) the equilibrium can be restored only by electron transfer from the surface states to the conduction band following the removal of the field, the Yunovich model can be expressed:

$$\mathrm{d}\Delta\Gamma_{\mathrm{n}}/\mathrm{d}t = -\,\Delta\Gamma_{\mathrm{n}}/\tau\,, \tag{4}$$

where $\Delta\Gamma_{\mathrm{n}}$ is the change in surface excess* and

$$1/\tau = c_{\mathrm{n}} n_{\mathrm{s1}}\,(1 + n_{\mathrm{t}}/\Delta\Gamma_{\mathrm{n}})\,, \tag{5}$$

$$C_{\mathrm{n}} = v_{\mathrm{t}}\sigma_{\mathrm{n}}\,, \tag{6}$$

where n_{s1} is the concentration of electrons in the conduction band when the Fermi level passes through the surface state level E_{t}; n_{t} is the density of electrons in the surface states in the steady state (no applied field); v_{t} is the thermal velocity of the electrons in the conduction band and σ_{n} is the capture cross-section of the surface states.

The present experimental results are in reasonable agreement with eq. (5), as shown in fig. 3 where $1/\tau$ is plotted against $1/\Delta\Gamma_{\mathrm{n}}$ for three temperatures. From plots such as in fig. 3, the energy position, the capture cross-section and the density of electrons in the surface states can be determined. Thus, the density of electrons, n_{t}, determined in this way as a function of temperature is shown in fig. 4. These values are in reasonable agreement with the values obtained by other methods[13]).

Exponential surface conductance decays are predicted by the Yunovich model for high applied field; for a strongly depleted space charge region:

$$|\Delta\Gamma_{\mathrm{n}}| \gg n_{\mathrm{t}}\,. \tag{7}$$

* The surface excess is defined as

$$\Gamma_{\mathrm{n}} = \int_0^{\infty} (n - n_{\mathrm{b}})\,\mathrm{d}z\,,$$

where n is the concentration of electrons in the space charge region and n_{b} the electron concentration in the bulk.

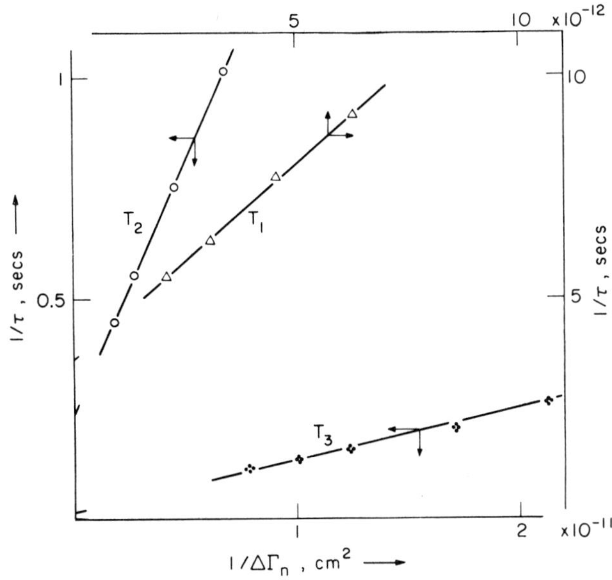

Fig. 3. Dependence of the inverse of time constant on the inverse of electron surface excess at three different temperatures. $T_1 = 320°K$, $T_2 = 299°K$; $T_3 = 280°K$.

Introducing the condition of eq. (7) in eq. (4):

$$d\Delta\Gamma_n/dt = C_n n_{s1} \Delta\Gamma_n . \qquad (8)$$

Eq. (8) gives an exponential decay of $\Delta\Gamma_n$ with a relaxation time:

$$\tau = 1/C_n n_{s1} . \qquad (9)$$

This expression corresponds to that of Rupprecht's discussed above.

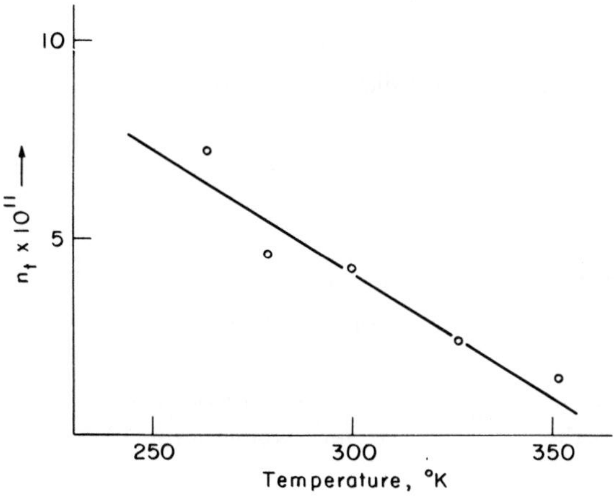

Fig. 4. The electron concentration in the trap as a function of temperature.

3.3. The effect of illumination

The field-effect surface conductance decay was further studied under white light illumination. Under constant illumination the time dependence of the change in surface conductance, $\Delta\sigma$, changed with temperature. Both linear and exponential decays were observed. Linear decays assumed an exponential time dependence at higher temperatures as shown in fig. 5, and they remained illumination dependent. The slope of the linear decays, under various experimental conditions, increases linearly with illumination as

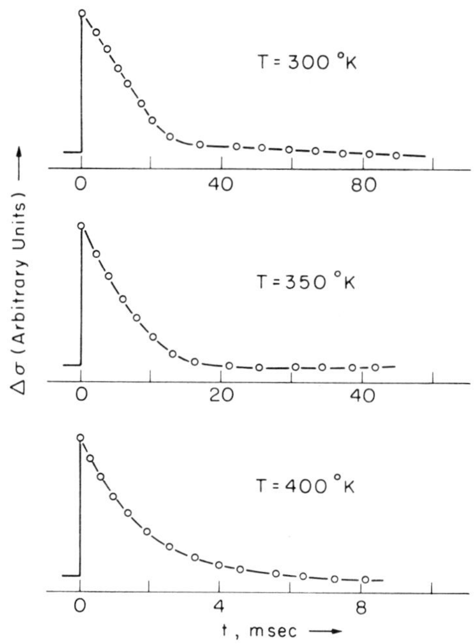

Fig. 5. Field-effect decays for three different temperatures at the same light intensity.

shown in fig. 6. In the case of the exponential decays the inverse of the relaxation time was found to vary linearly with illumination as shown in fig. 7.

The characteristics of the conductance decays under illumination can be understood by attributing a dominant role to the minority carriers in the decay process. Upon illumination the concentration of the minority carriers is significantly increased, whereas the corresponding increase in relative concentration of the majority carriers is negligible. If a negative field is applied to the surface of an n-type material, free electrons are repelled from the surface region and a depletion layer is formed. The surface states become

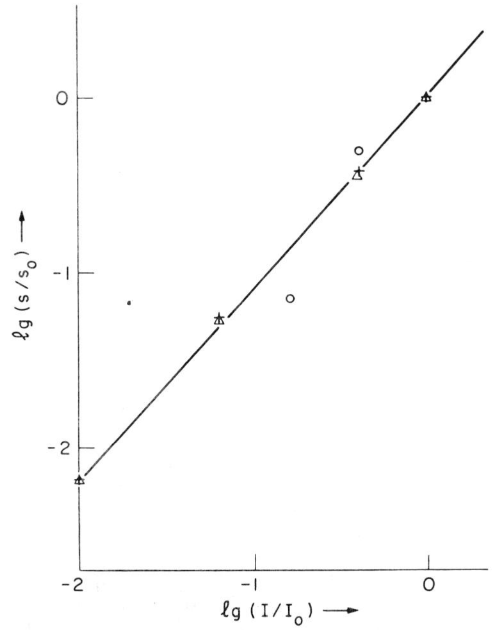

Fig. 6. The slope of the linear field-effect decay curves versus light intensity: (+) dry nitrogen (room temperature); (△) vacuum (room temperature); (○) vacuum (77°C).

over-populated because they are separated from the conduction band by the surface barrier. Whereas in the dark surface equilibrium can be brought about through thermal excitation of electrons over the barrier, under illumination

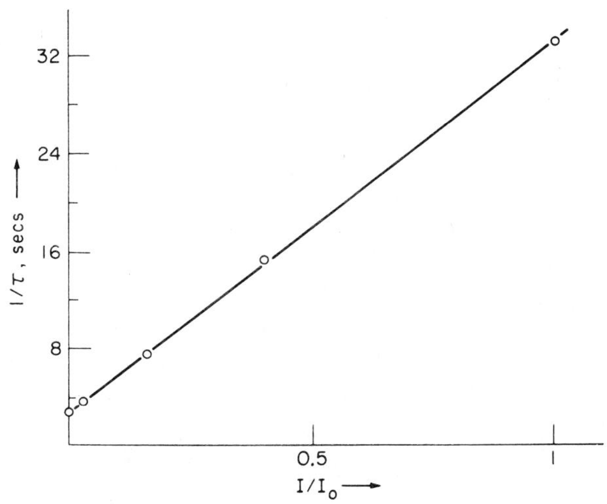

Fig. 7. The inverse of the relaxation time as a function of light intensity.

light excited holes encounter no barrier in moving toward the surface and recombining with the trapped electrons. It is expected that the capture of the holes by the surface states is very efficient.

If it is assumed that nearly all of the holes generated by light in the depletion layer recombine at the surface, then the field-induced change in conductance, $\Delta\sigma$, is proportional to the light intensity, I, and the thickness of the depletion layer, the latter being approximately proportional to the change in conductance. Thus,

$$\mathrm{d}\Delta\sigma/\mathrm{d}t = AI\,\Delta\sigma \quad \text{or} \quad \mathrm{d}\Delta\sigma/\Delta\sigma = AI\,\mathrm{d}t\,, \tag{10}$$

where A is a constant. The observed exponential conductance decays with time constants inversely proportional to the intensity of illumination as shown in fig. 7 are consistent with this relationship*.

The observed linear conductance decays can be explained by assuming that the recombining holes originate in a region of constant thickness independent of time and applied field. This condition is met when the diffusion length of the holes is greater than the space charge layer thickness whereby only the holes originating at a distance from the surface states equal to the hole diffusion length are recombining. Under these conditions eq. (10) becomes:

$$\mathrm{d}\Delta\sigma/\mathrm{d}t = AI\,.$$

This expression describes the linear conductance decays and the linear dependence of the conductance decay slopes on illumination.

It should be noted that the above model provides a means of determining the rate of carrier generation by light; thus, in conjunction with light absorption measurements it should be possible to determine the number of carriers excited per adsorbed photon. However, the model does not lead to the determination of the surface state parameters.

4. Summary

Pulsed field-effect experiments (in the dark) revealed the presence of a discrete energy level in the middle of the energy gap having a very large capture cross-section for electrons (10^{-14} to 10^{-12} cm^2). In oxygen doped gallium arsenide a surface level was found 0.45 eV below the conduction band which is believed to correspond to the bulk oxygen impurity level. Furthermore, the level of about 0.75 eV below the conduction band reported earlier[11]) was also found under certain surface preparation conditions. By means of surface photovoltage experiments the surface barrier was found

* Thermal excitation of carriers is neglected in this relationship.

to be of the order of $8kT$ and the energy bands were found to be always bent toward formation of a depletion layer.

In general, the surface states of gallium arsenide were found to be essentially insensitive to crystallographic orientation, etching treatment and gas ambient. It is believed that these surface states are associated with crystalline defects in the space charge region rather than with the particular abrupt termination of the lattice at the surface.

The mechanism of the field-effect surface conductance decays in the dark (for both n- and p-type GaAs) is believed to be based on charge transfer by majority carriers from the surface to the conduction band upon thermal excitation over the surface potential barrier. This mechanism is not applicable to the field-effect surface conductance decays under illumination. In this case the decays are independent of temperature and steady state conditions are re-established by the recombination of the light-generated minority carriers at the surface states.

Acknowledgement

The authors are indebted to the National Aeronautics and Space Administration and to the Office of Naval Research for financial support.

References

1) See for example A. Many, Y. Goldstein and N. B. Grover, *Semiconductor Surfaces* (North-Holland, Amsterdam, 1965).
Also, D. R. Frankl, *Electrical Properties of Semiconductor Surfaces* (Pergamon, New York, 1967).
2) P. Shao, Phys. Rev. Letters **2** (1959) 41;
V. E. Primachenko and O. V. Snitko, Soviet Phys.-Solid State **3** (1969) 9.
3) G. G. E. Low, Proc. Phys. Soc. (London) B **68** (1955) 1154.
4) W. L. Brown, Phys. Rev. **100** (1955) 590;
H. C. Montgomery and W. L. Brown, Phys. Rev. **103** (1956) 865.
5) I. Flinn and M. Briggs, Surface Sci. **2** (1964) 136.
6) D. R. Frankl, *Electrical Properties of Semiconductor Surfaces* (Pergamon, New York, 1967) p. 27.
7) N. L. Dmitryk, V. I. Lyashenko and T. N. Sytenko, Ukr. Fiz. Zh. **10** (1965) 745.
8) V. I. Lyashenko, private communication.
9) G. Rupprecht, Phys. Rev. **111** (1958) 75.
10) H. Huff, S. Kawaji and H. C. Gatos, Surface Sci. **5** (1966) 399.
11) S. Kawaji and H. C. Gatos, Surface Sci. **1** (1964) 407.
12) A. E. Yunovich, in: *Surface Properties of Semiconductors*, Ed. A. N. Frumkin (Consultant Bureau Enterprises, New York, 1964) p. 88.
13) Y. Furukawa, Japan. J. Appl. Phys. **6** (1967) 675.

2.4 summary

This section has represented the major methods currently employed for assessing materials and devices. These processes have contributed to the improvement in material purity and uniformity. Yet despite these advancements, the fine scale correlation of material properties to device performance is minimal, regardless of controlled experiments to identify the important factors. Most techniques discussed have already reached their practical limits in sensitivity, so that further improvements in quality are difficult to achieve.

An important factor, not sufficiently researched, is the introduction of defects, damage, or impurities during device processing. This area is likely to be a very critical one; apart from device failure analysis, there is little knowledge about small scale effects.

The future of assessment is problematic; there is a considerable lack of information concerning mobility profiles, detail on layer interfaces, the effects of small numbers of defects, impurities or vacancies, the influence of boundary conditions, and the structure of contacts. These areas present formidable problems. It is possible that the limiting factor in device performance is no longer the characteristics of the bulk material, but rather the nature of the interfaces and boundaries which form the device. If this is indeed the case, a considerable amount of ingenuity is needed for any major improvements to be made.

chapter three

device technology

This section is concerned with the conversion of the starting wafer into encapsulated and completed devices. The focus of attention is on the treatment of the wafer — the physics and technology of making electrical contacts to it, the shaping of the device, the problem of heat removal, and encapsulation of the processed device. Device manufacture is essentially a mass production process. Costs are important, often the determining factor in the choise of a particular process step.

Production facilities for compound semiconductor devices differ only in scale from those encountered in any silicon semiconductor plant. Many of the processes are different, but the general philosophy is the same. Although a wide range of devices are made from these materials, the technological problems encountered are similar. The following discussion therefore applies in general to all of the devices previously mentioned; where a process is specific to a particular device this fact is stressed.

3.1 wafer treatment

Cleaning

Cleaning is almost always the first step in wafer treatment. If wafers could be transferred directly from the epitaxy kit to device processing there would in theory be no need for further cleaning. But this is not the case, and several handling processes are involved — marking substrate side for identification, thickness measurement, etc. Thus it is necessary to clean the wafer of gross "dirt" before final cleaning. Gross contamination is removed by ultrasonic agitation in solvents, such as trich, or in a suitable proprietory cleaning agent followd by drying agent, such as methanol. The ultrasonic cleaning action is very efficient, but the contaminant remains in the solution and can redeposit on the wafer surface when drying. Several repetitive cycles with fresh solvent are needed before transferring the wafer to soxhlet-type recycling cleaning systems. Again, various cleaning agents are used, allowing sufficient time for about 10 complete cycles of each soxhlet wash chamber.

The choice of type and order of solvent cleaning is somewhat arbitrary and not easily justifiable. Many of the cleaning methods used in different industrial organisations are examples to this end. It is sufficient to say that solvent cleaning is necessary; providing that it is done carefully, methodically, and with consideration for previous and subsequent treatment. The exact permutations are not too important.

Etching

Etching is used for a number of purposes — to remove surface layers prior to contacting, to shape the device into its required structure, and

Etchants Composition	Ratio	Rate (μm/min)	Comments
(a) Etchants for GaAs			
$H_2SO_4:H_2O_2:H_2O$	3:1:1	3.1*	Standard polish etch series
	18:1:1	2.1	*Measured on 100 epitaxial surface
	8:1:1	2.8	
	9:9:2	8.7	
	1:1:0	5.0	
$NH_4OH:H_2O_2:H_2O$	1:4:20	1.8*	Polish, preferred etch for producing clean and for controlled etching of holes
$HNO_3:HF$	1:3		All of these etches polish, but main use is for delineating epitaxial layer interfaces
$HNO_3:HF:H_2O$	1:3:4		
	3:1:5		
$KOH:K_3Fe(CN)_6$	1:1		120g KOH + 500 ml H_2O 80g $K_3Fe(CN)_6$ + 500 m H_2O
$H_2O:AgNO_3:CrO_3HF$ (A-B etch)	2ml:8mg 1g:1ml	~ 4*	Etch at 65°C Also a good dislocation etch
Bromine-Methanol	Up to 10% Br	Various	Polish
(b) Etchants for InP			
$HNO_3:HCl:H_2O$	3:2:6	1	Non-preferential
	2:2:1	2	Used prior to contact evaporation
$Br:HBr:H_2O$	1:17:35	2	Polish, used for bulk shaping
Br:MeOH	Up to 10% Br	Various	Polish
$KOH:K_3Fe(CN)_6$	As Above		Delineation etches for epi layers
$HNO_3:H_2O + Fe^+$	1:1		Ferric ions can be supplied by dissolution of a nail
$HCl:HNO_3$	1:1		Saturate solution with Br Dislocation etch

Table 3-1

to form isolation regions. Each demands a different technique. Many etches have been developed for GaAs, but there are relatively few for InP. These main etches and their properties are listed in Table 3-1. Etching terminology is as follows:

Preferential — where the etch produces a surface of well defined pits whose facets are on definite crystallographic planes.

Polish — where the etchant reduces all surface features to a smooth surface.

Non-preferential — where the etchant produces a polished surface, revealing pits at dislocations.

Etching is either chemical, electrolytic, or by bombardment. Chemical etching is the easiest and most common; therefore, the table is restricted to chemical etchants.

The action mechanisms of etchants on III-V compounds are extremely complex, and not well understood. Factors influencing the rate and mode of attack are etch composition, temperature, degree of agitation, and the presence of illumination. Of added importance are the conductivity type, purity of the semiconductor, and the crystal orientation. The details in Table 3-1 are for fresh solutions of etchants at room temperature, in normal room lighting, and without any agitation (unless otherwise stated).

Mesa etching or shaping control is achieved using the sulphuric-peroxide or ammonium hydroxide etches on GaAs. Care must be taken, though, because both processes attack the surface along preferred crystal planes, and thus influence the final shape of the mesa (or hole)[52]. The bromine based etches produce very symmetrical mesa shapes on InP and GaAs, but are difficult to mask against and control.

Electrolytic etching has proven invaluable in many areas of semiconductor technology, but comparatively little work has been done on the electrolytic etching of the compounds. The basic principle is that the sample is made the anode, with an inert cathode, a suitable electrolyte, and a source of dc current. A problem arises in the choice of the electrolyte; chemical reactions in proximity to the anode cannot be predicted. The literature on the subject lists a solution of perchloric acid, with acetic acid, 1:4 for GaAs, with an operating current density of 0.3-0.4 A/cm^2 for use as a polish. For InP, a solution of $K_3Fe(CN)_6$ + KOH 20% in H_2O with 12-24 volts is listed as a polish. Both these solutions have been investigated in the RRE laboratory; findings conclude that the semiconductor surface rapidly discolours, with no apparent etching. The electrolyte used in the electrochemical characterisation of GaAs (See Sect 2.3.2) is a 10% W/V solution of KOH under illumination.

Cutting

After processing, the wafer is separated into individual devices. Every device engineer prefers wafers grown on < 100 > orientation; the < 110 > cleavage planes are then at right angles and square chips can be obtained with a diamond scribe and break. Wafers grown on < 110 > have no available suitable cleavage planes; on < 111 >, the preferred cleavage forms equilateral triangles.

The problems of cutting and subsequent handling of chips justify adhering to the < 100 > growth plane. If the material is grown on one of the other axes, chips are separated by sawing, using a thin wire (3-10 mils in diameter) with an abrasive slurry of alumina powder in cutting fluid.

Photolithography

Photolithographic processes are independent of the semiconductor material. Excellent descriptions exist of work on silicon[53] and on the various photolithographic methods[54]. Original photomask art work is produced using a computer aided design facility from which commercial step-and-repeat reductions are made to produce either emulsion or chrome masks. These are in-contact exposure plates for use in conjunction with a conventional mask alignment system.

Metallic contacts to the compound semiconductors are thicker than with silicon; they require sintering schedules at high temperature; and they have a poorer "sticking coefficient." Positive working photoresist helps to alleviate some of these problems by giving more desirable mask configurations for relatively uncomplicated device structures. Exposed positive working resist dissolves on development – dark areas on the mask remain on the wafer. A great variety of commercial photo-chemicals exist, allowing choice among many types, viscosities, resolutions, and prices.

The resist can be dispensed from a hypodermic syringe, fitted with a filter to ensure clean deposition. A coating is achieved by spinning under conditions determined by the required thickness and uniformity; the emulsion is then dried before exposure and final development. To achieve a more resistant film where prolonged etching is needed, the film may be post-baked after development.

Lack of process control and deterioration of the photoresist materials (through extended storage) are the primary causes of problems. Resist viscosity, resolution, and final thickness all need periodic checking. These aspects are critical to achieving high device yields in the "float-off" technique described later. Exposure and development of the images must be optimised in accordance with schedules described in relevant commercial literature. Development is achieved either by dip, spray, or ultrasonic procedures. For high resolution and ensurance of complete removal of the exposed pattern, the ultrasonic method is generally favored.

3.2 metal semiconductor contacts

The majority of the techniques mentioned thus far apply to all semiconductor materials. They largely have developed from the investment in silicon growth, assessment, and technology. But, with semiconductor contacts this is not the case – most silicon devices are bipolar or MOS and the metal semiconductor contact is to a very heavily doped diffused region. The contact problems are thus transferred from the metal-semiconductor boundary to the junction itself. But with GaAs and InP TED's, FET's, and IMPATT's, the metal contact is made to the semiconductor in the doping range 10^{14}-10^{18} cm^{-3}. The control of the contact properties becomes infinitely more difficult – particularly in the lightly doped part of the range. Two types of contact are of interest – the Schottky barrier (with step junction properties for avalanche devices, mixers, and varactors) and the "ohmic" contact (for transferred electron devices and for the more heavily doped junction structures). The barrier contact is predominantly to n-type material, whereas ohmic contacts are required to lightly doped n-type and to p^+ and n^+ heavily doped material.

But the classification is not quite this simple. Metallic contact properties can vary progressively over the whole range from truly "ohmic" to the classical Schottky barrier. Recently, this fact has become important in terms of controlling the cathode field in TED's[55].

The metallurgy of metal contacts to binary semiconductors is not understood, and is complicated by the volatility of the group V element, the different eutectics that can exist, vacancy disorder, introduced impurities, and the wide range of process variables.

Similarly, the theory of contact behaviour only has been established for a very limited set of conditions; where these conditions do not apply, the theory is at best qualitative, and often is not supported by experimental observation.

It is clear that the understanding of metal semiconductor contacts leaves much to be desired. This section attempts to summarise the present state of contact theory, metallurgy, technology, and characterisation of "ohmic" and rectifying contacts.

3.2.1 contact theory

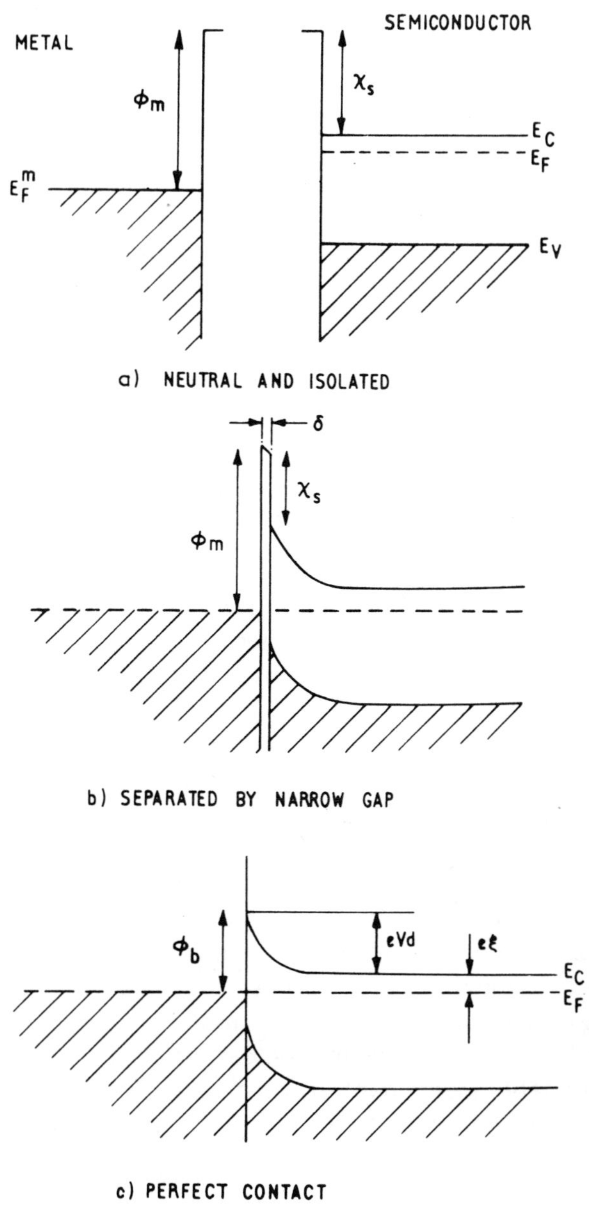

Figure 3-1/Formation of Schottky Barrier

When a metal is deposited on a semiconductor, an electrostatic barrier is established. If this barrier is of both sufficient energy and width, the current is dominated by thermionic emission and rectification results. But, if the barrier is low and narrow, electrons of low thermal energy can tunnel through the barrier to such an extent that the contact appears "ohmic" (depending on the magnitude of the tunnelling component). This section starts with the pure thermionic emission case; the theory is developed through to the field emission regime.

If two metal plates are placed in a vacuum and a voltage is applied between them, very little current flows. This minimal reaction is not due to there being no electrons in the metal, nor is it a result of free electrons that are unable to flow between the plates. The reason is that electrons have difficulty leaving the metal because of an energy barrier between the highest energy electron in the metal and the lowest energy electron that can exist in the vacuum. The energy difference is termed the "work function" of the metal. In a metal, the electrons are spread in energy — the highest is of the Fermi level. But in a semiconductor, the electrons are concentrated in a narrow energy range at the edge of the conduction band. Again, there is an energy barrier to the vacuum, known as the "electron affinity" of the semiconductor.

When the metal and semiconductor are brought into contact, electrons flow from the semiconductor into the metal; the Fermi levels are forced into coincidence. There then must be a negative charge in the metal which must be balanced by a positive charge in the semiconductor. For an n-type semiconductor this positive charge is provided by conduction electrons receding from the surface and leaving a depletion region of uncompensated donors. (The case of the p-type semiconductor is analogous.) Formation of the barrier is outlined in Figure 3-1. In the ideal circumstance the barrier height is given by the energy difference between the work function of the metal and the electron affinity of the semiconductor. That is:

$$\phi_b = \phi_m - \chi_s$$

Under this condition, the barrier energy on any semiconductor is related directly to the metal work function. Data substantiates this finding for the more ionic semiconductors such as CdS, ZnS, and ZnO. But in cases of the covalent semiconductors (Si, Ge, GaAs, InP, etc.), the barrier energy is virtually independent of the metal; it is more likely a property of the semiconductor itself. The key paper by Mead investigates this situation. It is interesting to note that for the majority of materials:

$$\phi_b = \frac{2}{3} \cdot E_g,$$

where E_g = energy gap. (This equation applies for GaAs, but not for InP.)

The departure from the ideal dependence of barrier height on work function of the metal is caused by the influence of surface states in the semiconductor. Assume that these surface states are continuously distributed in energy within the forbidden gap, with a neutral level (ϕ_o) equivalent to the Fermi level of the surface states. If, as in Figure 3-2, $\phi_o > E_F$ at the surface, there is a net positive charge in the surface states which reduces the barrier height and narrows the space charge region. In the same way, if $\phi_o < E_F$, the barrier height is increased. The total effect is to reduce the difference between ϕ_o and E_F; if the surface state density is very large, ϕ_o is indeed locked to E_F, making the barrier height completely independent of the metal. The approximate equation for the barrier height is now:

$$\phi_b = \gamma\,(\phi_m - \chi_s) + (1 - \gamma)\,(E_g - \phi_o)$$

where $\gamma = \epsilon/(\epsilon + q\delta D_s)$; and ϵ is the permitivity of the oxide film, δ is its thickness, and D_s is the density of surface states per eV unit area.

The complete equation is more complex[56], but the above approximation describes the general features — if $D_s = 0$, $\phi_b = \phi_m - \chi_s$; if D_s is large, $\phi_b = E_g - \phi_o$. Presently, it is not known how to calculate the barrier energy of any metal semiconductor system from first principles, but a considerable amount of experimental information exists.

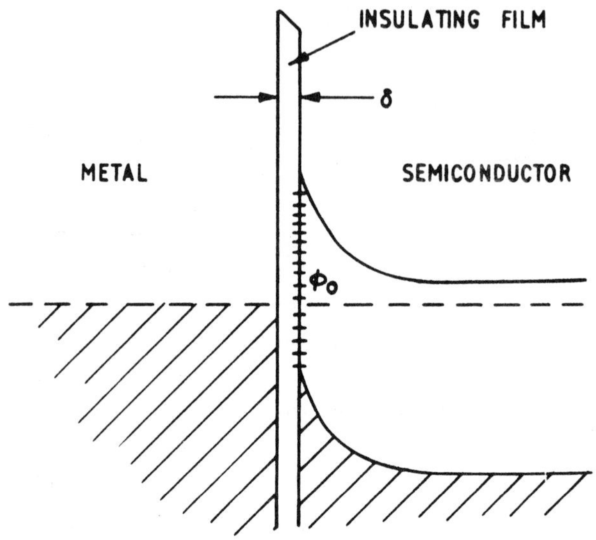

Figure 3-2/Metal Semiconductor Contact with Surface States

The current voltage characteristic as given by the diode theory is:

$$I = AST^2 \exp(-q\phi_b/kT)\,[\exp(qV/kT) - 1]$$

where A is the Richardson constant, S is the diode area, and T is the absolute temperature.

The equation indicates that the reverse current should be independent of bias if it is greater than a few (kT/q). This would be the case if the barrier height were independent of bias; however, effects such as image forces, tunnelling, and surface state penetration change the barrier height. The result is the lack of current saturation in the reverse I-V characteristic. The change in barrier height resulting from image forces is:

$$\Delta\phi_b = \left(q^3 N\,[Vd - V]\,/8\pi^2\epsilon_o^3\epsilon_s^3\right)^{1/4}$$

and that due to tunnel penetration of the top of the barrier:

$$\Delta\phi_b = X_c\,[2qN(\phi_b - V - \xi_1)/\epsilon]^{1/2}$$

where X_c is a critical tunnelling length of the order of a few angstroms.

The barrier height lowering is therefore not only a function of applied bias (height decreasing with increasing reverse bias), but also of carrier concentration (height decreasing with increasing N). These changes in barrier height with bias are reflected in the forward characteristic of the junction. The slope of the log I-V characteristic, instead of being (q/kT), has an additional dimensionless parameter n where the slope is (q/n kT) and $n = 1 + (\partial\phi_b/\partial V)$. The parameter n is found empirically to be about 1.06 at room temperature. But a further complication exists in that n is strongly temperature dependent. The work of Padovani[57] showed that n could be represented by:

$$n = 1 + \frac{T_o}{T},$$

where T_o is a constant in temperature and $T_o = 45 \pm 8°K$ for GaAs-Au barriers. The full I-V characteristic was developed to be:

$$I = A^* ST^2 \exp\left[-q\phi_b/k(T + T_o)\right] \left(\exp\left[qV/k(T + T_o)\right] - 1\right)$$

A* is the modified Richardson constant.

This expression is purely empirical; it fits the observed data though it cannot be explained theoretically because:

1./At any bias and temperature the forward current with bias is smaller than predicted by diode theory.
2./The rate of increase in current with bias is smaller than predicted by diode theory.
3./The difference between theory and experiment is greater the lower both the temperature and applied bias.
4./The experimental characteristic becomes independent of temperature at low temperatures.

The last point suggests that the observed characteristic is due to tunnelling of electrons through the barrier. At room temperature, though, the I-V characteristic can be simplified to:

$$I = I_s \left[\exp (qV/n\,kT) - 1\right]$$

The theory discussed so far applies specifically to low doped semiconductors ($< 10^{16}$). The barrier height is lowered by increasing N; specifically through the tunnelling component. Tunnelling results from the fact that the electron probability distribution in the forbidden region is damped exponentially with distance; it can penetrate a barrier when that barrier is sufficiently thin. The damping of this distribution is characterised by a constant q — a function of electron energy. Near the conduction and valence band edges, q is small; the electron penetrates deeply into the semiconductor. But near the gap centre, q is large and the electron distribution damps rapidly. The relationship is of the form:

$$E^2 \approx \left(\frac{E_g}{2}\right)^2 + \frac{n^2 q^2 E_g}{2m^*}$$

where E is the electron energy measured from the centre of the forbidden gap and m* is the effective mass of the majority carriers[57].

The key paper by Crowell et al. discusses the thermionic field emission regime.
Figure 3-3 shows the log I-V forward characteristic as a function of the carrier concentration of the material. At low N_D, the contact functions as a Schottky diode with a slope characteristic proportional to q/kT; as N_D increases the logarithm of the current becomes proportional (to a first approximation) to $N^{-1/2}$ V. That is, the current decreases exponentially with an increase in the width of the depletion layer through which the electron must tunnel. We would therefore expect the ohmic contact resistance to depend exponentially on the zero bias depletion width, and, hence, as the square root of carrier density. Figure 3-4 illustrates experimental results for gold on GaAs[58] which demonstrates this dependence.

Thus, at low carrier concentrations, the current transport can be calculated by using the empirical relations given above. These equations apply only in the case of the metal contact to the *clean* semiconductor surface, as any surface contamin-

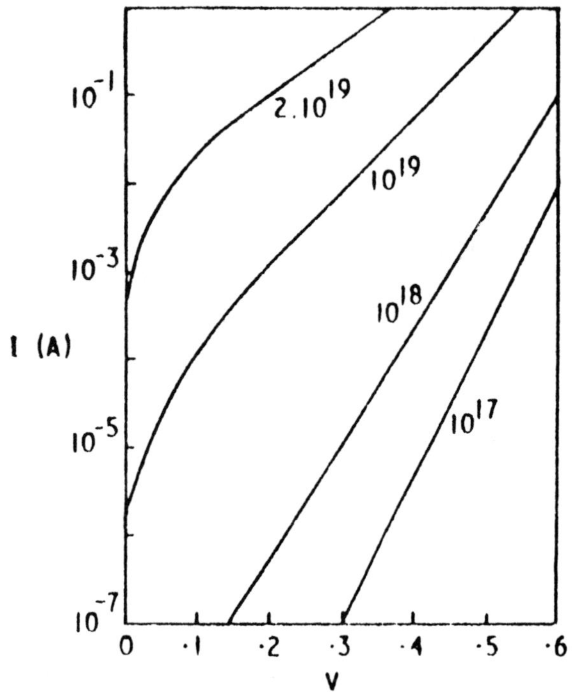

Figure 3-3/Plot of Log I-V as a Function of Semiconductor Doping (Schematic)

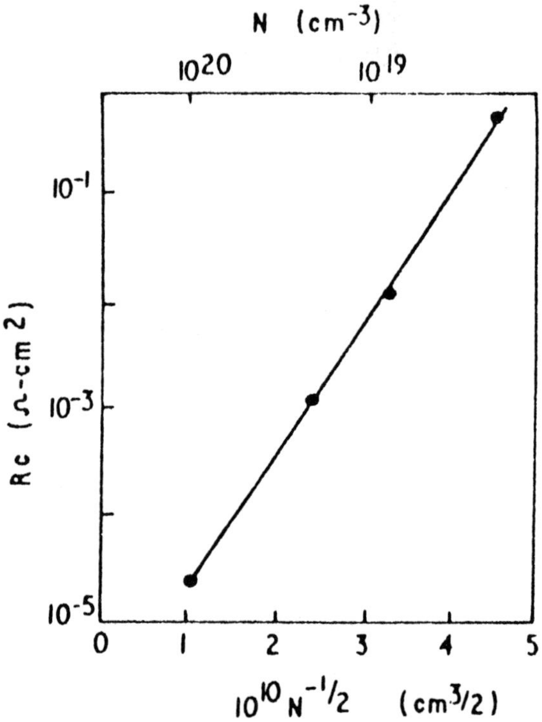

Figure 3-4/Contact Resistance of Gold on GaAs[58]

ation increases the value of n. At high carrier concentrations, the thermally assisted tunnelling regime is introduced, and analysis becomes more complex; but the contact becomes "ohmic" in the sense that the current-voltage is symmetrical through the origin up to high currents. The device engineer attempts to minimise the resistance associated with the contact.

The most interesting feature of contact theory is the gradual transition from the ideal Schottky barrier to the ohmic contact. The theoretical analysis and the interpretation of physical contact behaviour are difficult because of this occurrence. In the last part of this section the methods available for characterisation of contacts are discussed.

The last question is how to make ohmic contact to low doped material. One obvious way is to grow a heavily doped semiconductor contact to form an n^+-n structure, and complete the device by forming a metallic contact to the heavily doped material. It is often desirable to contact directly to the low doped region. If possible, the reduction of the barrier height to a low value ($<$ 0.25 eV) would result in essentially ohmic behaviour; but, on GaAs and InP, the barrier height is independent of the metal work function. The only solution then is to look for a metallic system that reacts with the semiconductor and produces a region of high doping adjacent to the metal.

key paper

Metal-Semiconductor Surface Barriers*

C.A. Mead
(California Institute of Technology, Pasadena, California, US)

Solid State Electronics, Volume 9 (1966)
Pages 1023-1033

*Supported in part by the Office of Naval Research under Contract No. Nonr-220(42) and the International Telephone and Telegraph Company

Abstract—The physical principles underlying the metal–semiconductor barrier are discussed in the light of recent experimental results. A semi-empirical approach for predicting the type of contact to be expected at an arbitrary metal–semiconductor interface is presented.

Résumé—Les principes de physique sur lesquels se base la barrière de potential d'un contact métal–semiconducteur sont discutés en relation avec de récents résultats expérimentaux. Un procédé semi-empirique est présente qui permet de prédire le type de contact qu'un joint métal–semi-conducteur arbitraire est sensé avoir.

Zusammenfassung—Die physikalischen Prinzipien, die dem Potentialwall des Metall–Halbleiterkontaktes zu Grunde liegen, werden im Lichte neuerer experimenteller Resultate diskutiert. Ein halb-empirisches Verfahren wird beschrieben, nach welchem für einen beliebigen Metall–Halbleiterübergang der zu erwartende Typ des Kontaktes vorausgesagt werden kann.

INTRODUCTION

IN THE study of metal–semiconductor contacts, one striking fact is immediately apparent. Given an arbitrary metal deposited upon an arbitrary surface of an arbitrary semiconductor, the result has a probability of well over 90 per cent of being a rectifier. This fact has been realized in a vague way for many years. Since the advent of our modern semiconductor technology it has represented to many a device development engineer one of the prime examples of the innate perversity of nature. In the past few years several new device ideas have been proposed which depend in a very direct way upon metal–semiconductor or metal–insulator contacts. Largely because of such device considerations, a great deal has been learned about the behaviour of contacts from a fundamental point of view. It is the purpose of this paper to review our present understanding of surface barrier contacts, with emphasis on the barrier energies, in as concise and systematic way as possible. Although an extensive bibliography is appended, no attempt has been made to be exhaustive. In particular the paper does not purport to discuss many of the alternative points of view and peripheral issues, but rather to set forward as clearly as possible the main features. For those interested in more detail or in the evolution of ideas in the field, a careful reading of the references in approximately the order given will prove worthwhile. In addition the paper will be confined to the "clean" metal–semiconductor contact where the materials are in intimate contact. Hopeless complications present themselves when an interfacial layer is present, as attested to by a vast literature, none of which will be considered here.

BEHAVIOUR AT THE INTERFACE

When a metal is brought into intimate contact with a semiconductor, the valence and conduction bands of the semiconductor are brought into a definite energy relationship with the fermi level in the metal. This relationship is essentially independent of the doping of the semiconductor. Once this relationship is known, it serves as a boundary condition on the solution of Poisson's equation in the semiconductor, which proceeds in exactly the same way as in the *p–n* junction. The energy band configurations for typical contacts on *n*- and *p*-type material are shown in Fig. 1. When the device is in the forward direction, the diffusion

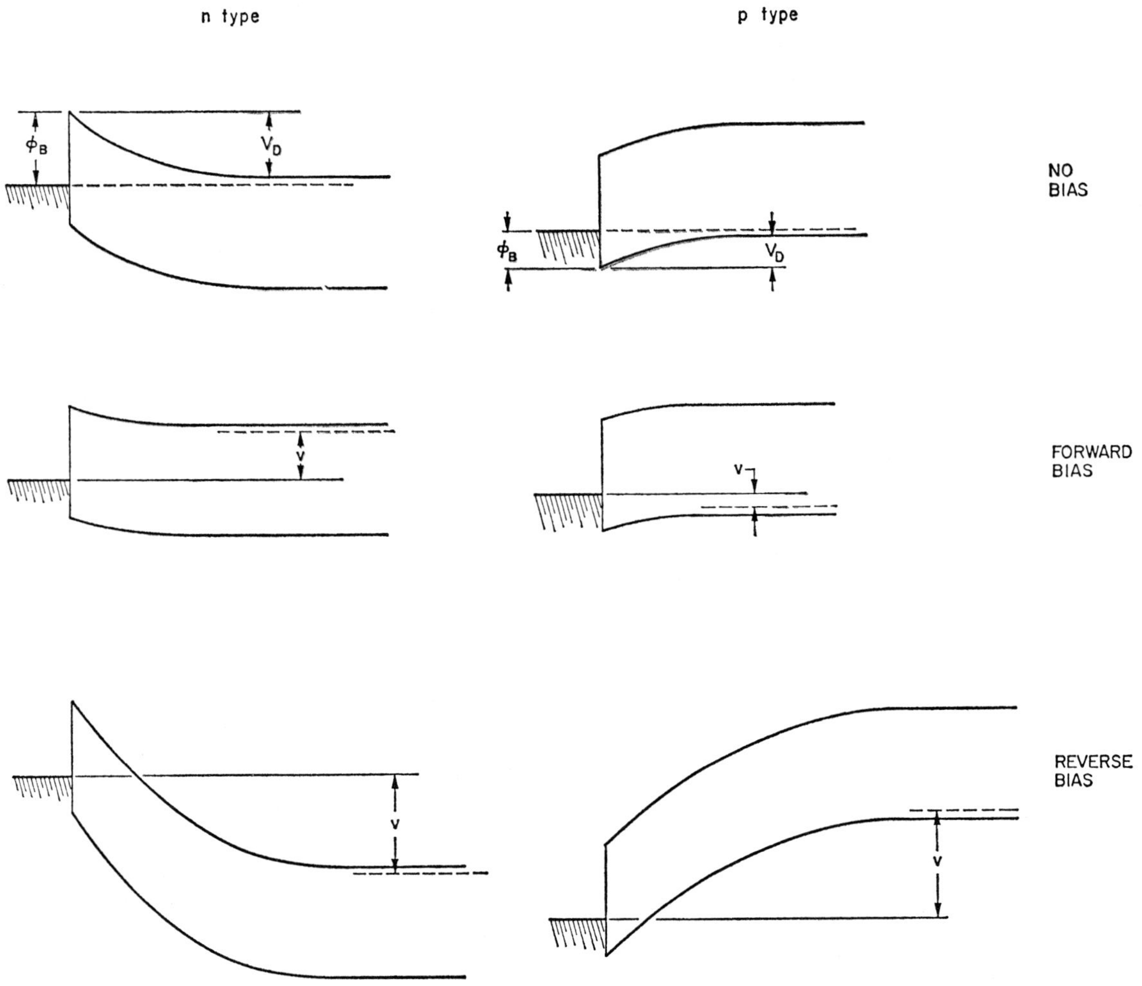

FIG. 1. Energy band diagrams for metal contacts on *p*- and *n*-type semiconductor for various bias conditions.

potential in the semiconductor is decreased and majority carriers are injected into the metal. In the reverse bias case, the diffusion potential is increased. A saturation current then flows which consists of two components:

(*a*) the minority carriers generated in the semiconductor in or within a diffusion length of the space charge layer

(*b*) electrons or holes in the metal incident with enough energy to surmount the barrier.

The entire behaviour of the device is quite analogous to that of an abrupt *p–n* junction. The details of the electrical behaviour are discussed elsewhere,[1,2] however a few approximate but important results will be given here.

(*a*) *I–V characteristic*

The current density is of the form

$$J = J_0\, e^{-\phi_B/kT}(e^{qv/kT} - 1) \qquad (1)$$

where $J_0 \approx 120\, T^2\, (\text{A/cm}^2)$ for semiconductors with effective masses not too different from the free electron mass.[34] At room temperature the current density is 1 A/cm^2 for $\phi_B - qv$ approximately 0·3 eV. Thus it is possible to estimate ϕ_B directly from the I–V curve (Fig. 2). Alternatively it is clear from equation (1) that the thermal activation energy is $\phi_B - qv$, which can be determined from the slope of the log J vs. $1/T$ curve as shown in Fig. 3.

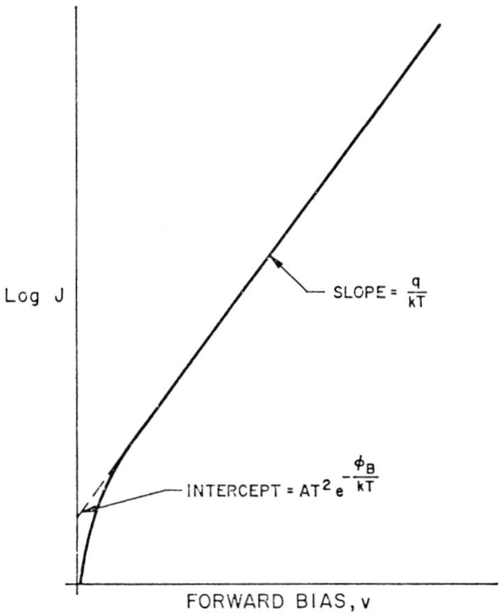

FIG. 2. I–V characteristic of forward biased metal–semiconductor contact.

(b) Capitance–voltage

When a small a.c. voltage is superimposed upon a d.c. reverse bias, charge of one sign is induced in the metal surface and of the opposite sign in the semiconductor by a slight widening of the depletion layer. Thus the space charge layer may be represented by an effective capacitance $C = \epsilon/d$ per unit area, where ϵ is the permittivity of the semiconductor and d the depletion layer width. As larger reverse bias voltage is applied, the depletion layer expands as

$$d = \frac{2\epsilon(V_d - v)^{1/2}}{qN_0} \tag{2}$$

and consequently the capacitance decreases

$$C = A\frac{qN_0\epsilon^{1/2}}{2(V_d - v)} \tag{3}$$

where A is the sample area. It can be seen for equation (3) that if $1/C^2$ is plotted vs. the reverse bias voltage $-v$, a straight line is obtained whose intercept is V_d, as shown in Fig. 4.

It should be noted that this procedure does not give the barrier energy directly but only the diffusion potential V_d, which differs by the fermi energy. The fermi energy can be computed if the net doping density N_0 and the effective mass are known. The carrier density itself can also be conveniently determined from the slope of the $1/C^2$ vs. v plot.

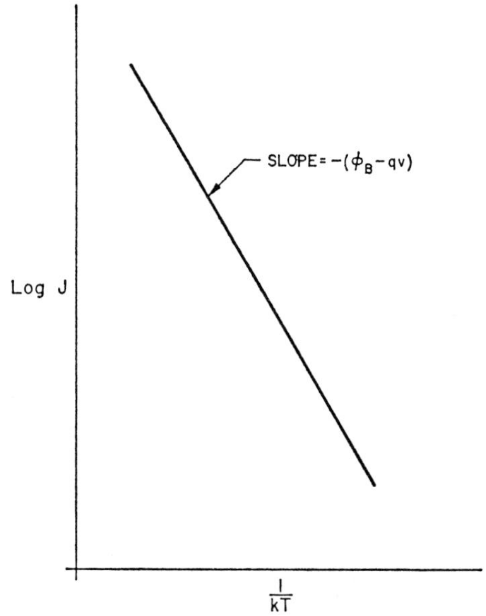

FIG. 3. Activation energy plot of forward biased metal–semiconductor contact.

$$\frac{d(1/C^2)}{dv} = \frac{2}{A^2 q\epsilon N_0} \tag{4}$$

In many cases the easiest way of determining the carrier concentration and sign of carriers is by use of the capacitance and I–V characteristics of a small surface barrier contact.

(c) Photoresponse

A very powerful method for investigating the surface is the use of monochromatic light to induce a photocurrent in the device. Light may be incident upon the metal surface in which case the metal should be thin enough to transmit as much light

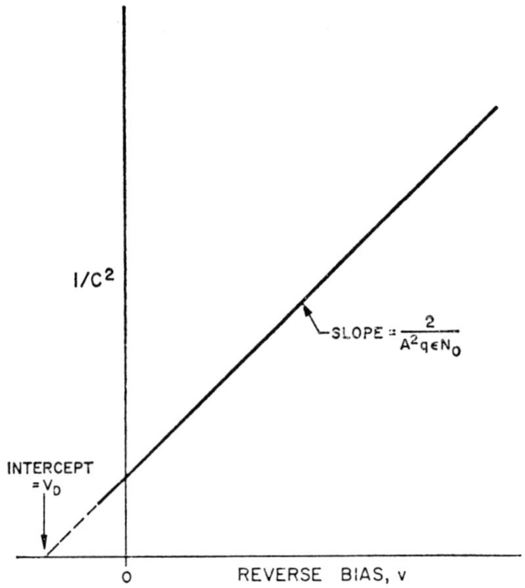

FIG. 4. Capacitance–voltage characteristic of reverse biased metal–semiconductor contact.

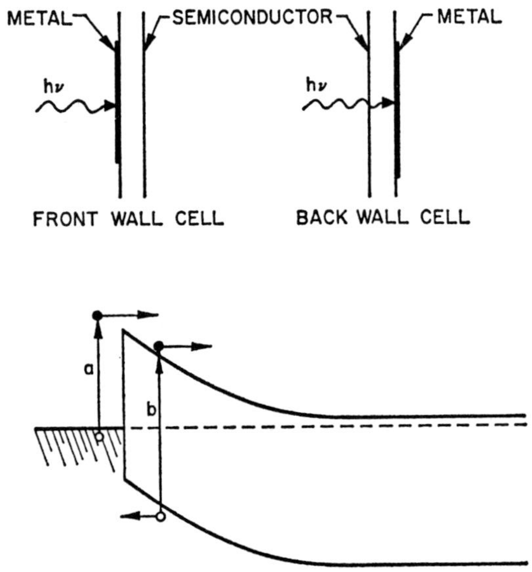

FIG. 5. *Top*: Geometry of samples for photoresponse measurements.
Bottom: Photo-excitation processes resulting in zero bias photocurrent.
(*a*) Photoemission from metal.
(*b*) Band to band excitation in semiconductor.

as possible. Alternatively the light may pass through the semiconductor to the interface. In either case two distinct photoexcitation processes can occur, as shown in Fig. 5.

(*a*) photoemission of excited electrons in the metal over the barrier

(*b*) band to band excitation of hole–electron pairs in or within a diffusion length of the semiconductor depletion layer.

These two processes can be seen graphically in the photoresponse of a typical metal semiconductor barrier in Fig. 6. Process (*a*) corresponds closely to photoemission from a metal surface into a vacuum. For this case the dependence of the photocurrent on photon energy for photon energies more than about $3kT$ larger than ϕ_B reduces to

$$J \propto (h\nu - \phi_B)^2. \tag{5}$$

Hence if the square root of the photocurrent (normalized to the incident photon flux) is plotted vs. the photon energy, a straight line should result whose intercept is ϕ_B. An example of this procedure is shown in Fig. 7. This is under most circumstances the most accurate and reliable method of determining ϕ_B. However, if possible at least two methods should be used, as any one is subject to difficulties and can give misleading results.

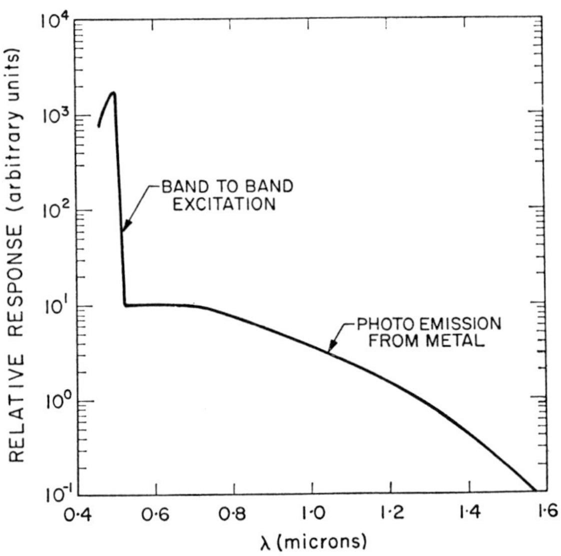

FIG. 6. Spectral response of metal–semiconductor contact showing regions of different excitation processes.

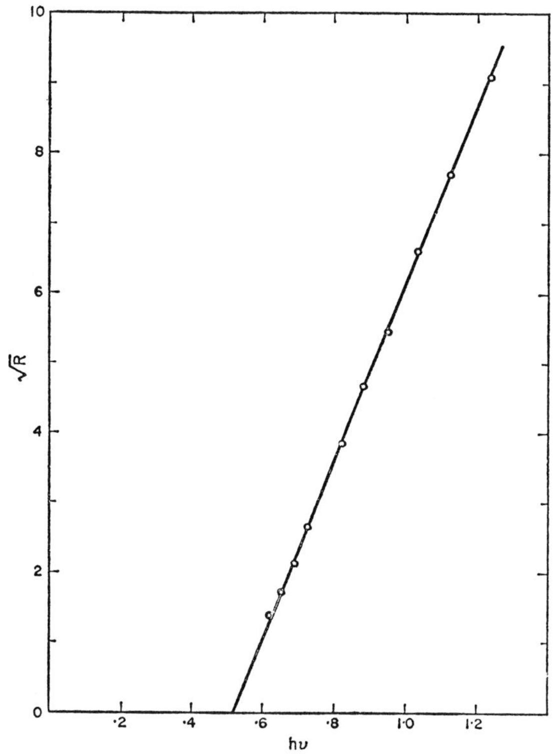

FIG. 7. Square root plot of long wavelength photoemission data showing extrapolation to obtain barrier energy.

THE PHYSICS OF THE BARRIER

In order to more fully understand the results of the measurements mentioned above, we will briefly discuss how one would expect, *a priori*, the barrier energies to be related to the characteristics of the metal and semiconductor.

If one performs the imaginary experiment of bringing a clean metal plate closer and closer to a clean semiconductor surface, one can see immediately that there will be an electric field between the surfaces given by the difference between the work functions of the materials divided by their separation. This electric field requires an electric change in the two surfaces, which increases as the separation decreases. In the semiconductor this charge can arise in two ways.

(*a*) Ionized impurity atoms, causing a space charge layer very similar to that in a *p–n* junction

(*b*) Surface charge induced in surface states which arise because of the termination of the crystal lattice.

In general some charge will arise from both sources. However, it is possible to distinguish two broad classes of barriers, one in which the surface charge is dominant, and one in which the surface charge is negligible.

In the latter case, as the metal is brought closer and closer to the semiconductor, the induced charge in the semiconductor will increase, the space charge layer will widen, and an ever increasing fraction of the contact potential difference will appear across the space charge layer within the semiconductor, and less will appear across the gap. In the limit the entire potential drop is within the semiconductor. In this case one would expect the barrier energy to depend directly upon the work function of the metal. In many semiconductors such a dependence is indeed a reasonably good approximation. However one additional complication is present at this point. The work function may be thought of as composed of two parts. One is related to the chemical electronegativity of the metal and the other is a surface dipole layer caused by a distortion of the electron cloud at the surface. The surface dipole layer is certainly not present in anything like the same form when the metal is in intimate contact with a semiconductor as it is when the metal faces a vacuum. Hence we will use the electronegativity of the metal as its reference property rather than the work function. However for the metals discussed here, differences are small except for an arbitrary shift in the zero.

In Fig. 8, the barrier energies of various metals on ZnS are plotted as a function of the electronegativity of the metal. The result is a straight line with unity slope as expected. Also shown are the barriers obtained on GaAs, a material in which the surface charge is dominant. Here it can be seen that the effect of the metal electronegativity on the barrier energy has been reduced an order of magnitude by the presence of surface states.

In order to see how the surface states affect the barrier energies, we can argue as follows. Suppose the surface states are distributed in energy somewhere in the forbidden gap of the semiconductor, and when the free surface is charge neutral they are filled to some energy E_0, which will be the location of the fermi level at the surface. Now let us apply an electric field to the surface. The field will induce charge in the semiconductor, primarily in the surface states if their density is large, and the fermi level at the surface will shift from E_0, but by only a small amount compared with the movement in the absence of the surface states. The larger the number of surface states per unit energy interval, the smaller movement of the fermi level for a given induced charge, and hence the less effect an electric field will have on the interior of the semiconductor. The same argument

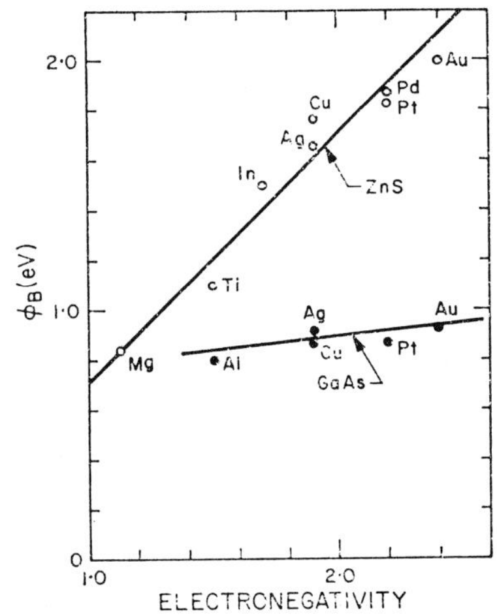

FIG. 8. Barrier energies of various metals on ZnS (electronegatively controlled) and GaAs (surface state controlled).

may be applied to the energies of metal–semiconductor barriers. The larger the density of surface states, the less the barrier energy will depend upon the electronegativity of the metal. Hence in the limit of very high surface state density, the fermi level at the surface is fixed at E_0, and the barrier energy is just the difference between the band edge and E_0, independent of the contact potential difference between metal and semiconductor.

For practical purposes, materials like GaAs (Fig. 8) in which the barrier energies change only slightly with different metals will be considered to fall in the surface-state determined category although there is obviously some dependence upon the metal.

We are now in a position to characterize systematically any semiconductor with regard to its barrier properties.

(*a*) If the material has a large density of surface states, the value of E_0 can be used to approximate the barrier energy.

(*b*) If the material behaves as if it had no surface states, the location of the intercept on the electronegativity axis of a plot similar to Fig. 8, together with the electronegativity of the metal serves to specify the barrier energy. (This intercept would be called χ, the electron affinity of the semiconductor, if a work function plot had been used.)

Thus in either case one number suffices to characterize the semiconductor and can determine quite closely the position of the fermi level at the interface relative to the band edges of the semiconductor, for any metal contact. Hence the expected barrier energy on *n*- or *p*-type material can be predicted.

MATERIALS WITH SURFACE STATES

Experimentally, more work has been done on materials with a high density of surface states. In general the E_0 value is taken as that obtained with a Au electrode. For other metals, the fermi level at the interface will in general be slightly closer to the conduction band edge, E_c. For the Group IV and III–V semiconductors, the E_0 values relative to the conduction band edge are shown as a function of the energy gap E_g in Fig. 9. The straight line is $E_c - E_0 = 2/3E_g$ and is an excellent approximation for most of the materials.(16,17) To the author's knowledge only one detailed theoretical calculation has been attempted; that of PUGH(5) for the (111) surface of diamond. The results of that computation have been converted to a rough density of states plot in Fig. 10. It can be seen that the surface states effectively form a narrow band centred slightly below the centre of the forbidden gap. These states would be half full for a charge neutral surface and hence E_0 should be slightly below the peak in the curve, or about one third of the gap up from the valence band. From the experimental results, one can infer that a similar situation exists for the other materials in this same general category. Although the "one third rule" only applies to certain semiconductors, many others are dominated by surface states, with different energies relative to the band edges. Such materials are InAs, InP, GaSb, CdTe and CdSe. Detailed data on these materials as well as those in Fig. 9 are given in the Appendix. A discussion of how the metal layer is expected to modify the surface states themselves is given by HEINE.(6)

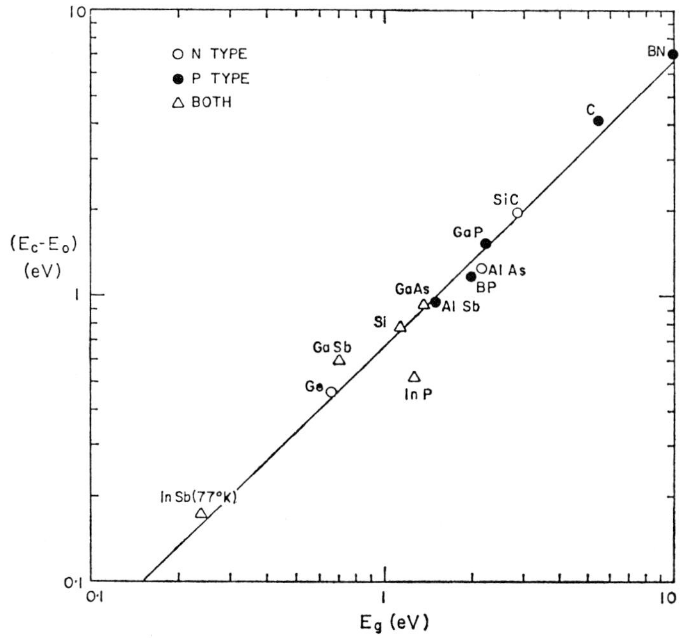

FIG. 9. Location of interface fermi level relative to conduction band edge for gold contacts on various surface state controlled materials. The line is $E_c - E_0 = 2/3 E_g$.

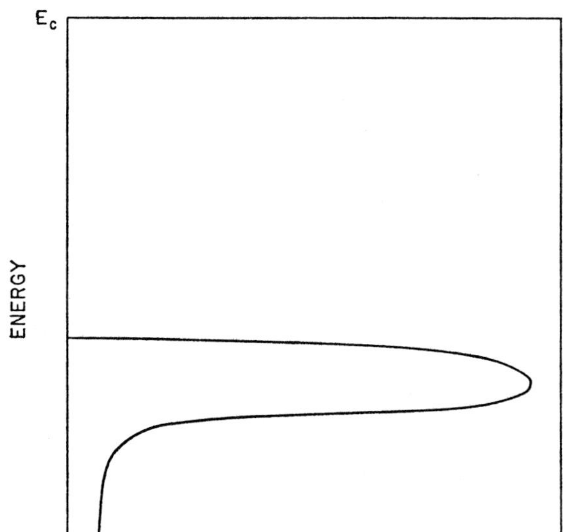

FIG. 10. Calculated approximate density of surface states vs. energy for (111) surface of diamond (after PUGH[5]).

MATERIALS WITHOUT SURFACE STATES

Although no detailed computations of the type mentioned above have been done on specific materials other than diamond, there has been a great deal of general theoretical work on the properties of surface states (see references). One general result of all of this work is that *surface states are an unavoidable result of the termination of the lattice*. Hence one should observe surface states on all semiconductors, not just certain types. It is the materials like ZnS which behave as if they had no surface states which are anomalous, not those which behave as if they did. A solution to this curious dilemma can be found in a classic paper on surface states by SHOCKLEY.[4] A figure from this paper is reproduced in Fig. 11.

This plot shows the traditional splitting of atomic energy levels as the wave functions of the atoms overlap. In addition it shows the energy of the surface states as a function of lattice spacing. One remarkable fact is apparent, there are two distinct regions of surface state behaviour. At large separations, the wave functions overlap only slightly and the surface states lie at the band edges.

At small separations the wave functions overlap strongly and the surface states are near the centre of the forbidden gap. This provides a valuable clue as to the behaviour of different types of semiconductors with respect to surface states. In essentially ionic materials, the wave functions are atomic in nature and the overlap is small. To a first approximation the valence band can be identified with the anions while the conduction band can be identified with the cations. Surface states are formed by the two types of atoms at

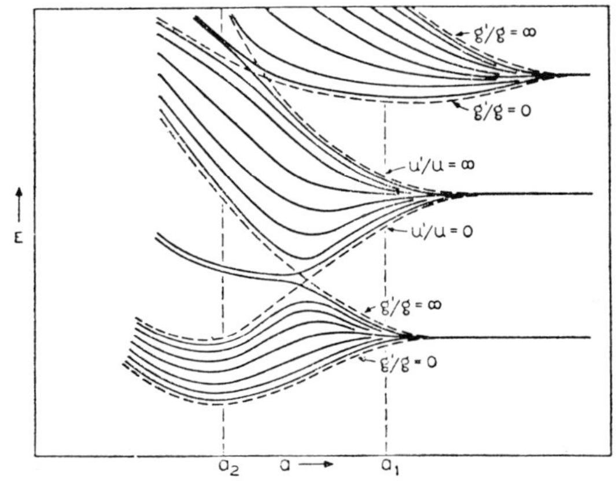

FIG. 11. Energy bands as a function of interatomic spacing for a simple linear chain (from SHOCKLEY[4]).

the surface of the crystal and their energies lie close to the valence and conduction band edges. The upper surface states are unoccupied and the lower surface states are fully occupied. The fermi level at the surface may move from near the conduction band edge to near the valence band edge without creating any surface charge. This behaviour is quite consistent with our oversimplified picture of ionic crystals. However it is surprising that it works well for materials like CdS which are not normally thought of as totally ionic. No detailed calculations have been made on such materials, and a complete answer will not be available until they are. In the meantime, one can argue by analogy with Shockley, that until the wave function overlap reaches a critical (quite large) value, the surface states lie at the band edges. A small increase in overlap above this critical value causes the surface states to move rapidly to a region near the centre of the forbidden gap. Thus there should be two well defined types of semiconductor, those whose overlap is larger than the critical value and those whose overlap is smaller. These ideas are quite in accord with experimental results and provide a substantial basis for making predictions concerning the behaviour of any given material.

In order to check out the preceding picture, one would very much like to be able to vary the overlap integral in one material (by e.g. varying the lattice spacing) and study the behaviour of the surface states. Such an experiment may actually be possible, but another equally elegant study has been used to accomplish essentially the same result.[32] Suppose we start with an ionic-like material whose overlap is nearly, but not quite, the critical value. Then let us substitute for some fraction of the cations of this crystal other cations which greatly increase the wave function overlap. The "average overlap" of the crystal may be controlled by the fraction of new cations substituted. Such an experiment was done using CdS and substituting Se ions for the S. Barrier heights obtained for several metals are shown as a function of semiconductor composition in Fig. 12. The barrier heights on the CdS vary widely with the metal used, approximately as the metal electronegativity. In this case the surface states are close to the band edges, and the fermi level at the interface can move up or down in energy over a relatively large range without requiring any surface charge to fill or empty these states. On the CdSe end, however, the surface states lie in a narrow band in the forbidden gap, and the fermi level at the interface cannot move appreciably without filling or emptying a large number of surface states and hence is "pinned" at a given energy. As the crystal composition is made richer in selenium, the overlap between adjacent atoms is increased, the surface states move towards the centre of the gap, the fermi level at the surface is "squeezed" between them, and a continuous transition is observed between the electronegativity-controlled case and the surface-state-controlled case.

CONCLUSION

We have shown how, by using simple ideas of covalent and ionic crystals, one can make meaningful qualitative predictions of the behaviour to be expected from an arbitrary metal on an arbitrary

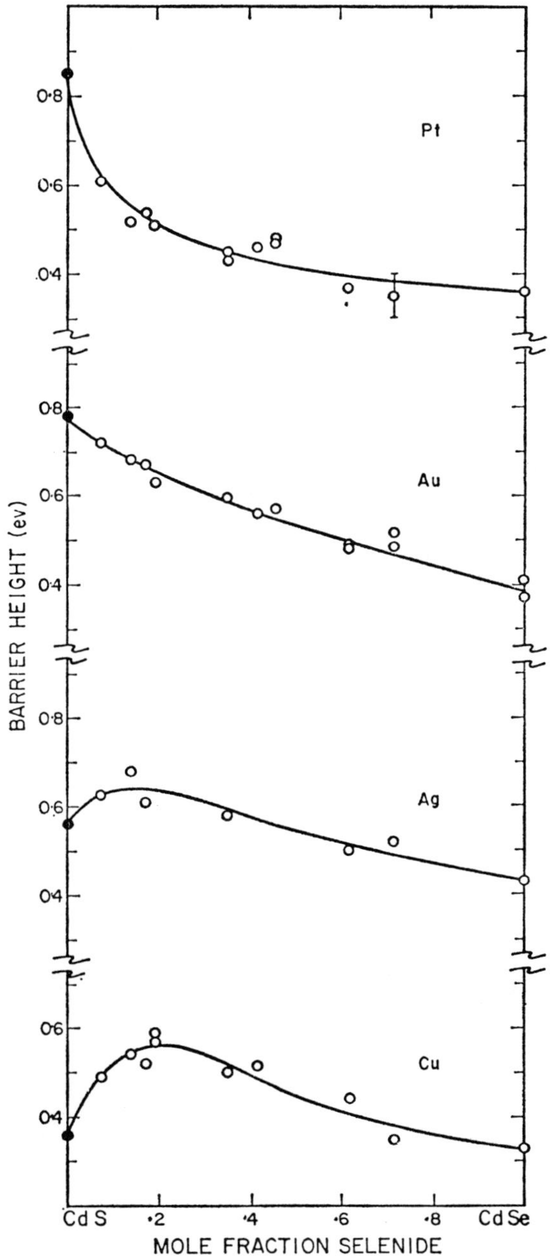

FIG. 12. Barrier energies as a function of composition on Cd (S–Se) mixed crystals. The Se rich end represents larger wave function overlap hence surface state dominated behaviour.[32]

semiconductor. Further detailed understanding must await a more refined theoretical analysis. However at present our ideas are sufficiently clear to provide a framework within which to interpret new experimental findings and also to serve as a guide for engineering applications.

REFERENCES

General

1. H. K. HENISCH, *Rectifying Semiconductor Contacts*, Ch. 7, Oxford University Press (1957).
2. E. SPENKE, *Electronic Semiconductors*, Ch. 4, McGraw-Hill, New York (1958).

Surface states

3. E. T. GOODWIN, *Proc. Comb. Phil. Soc.* **35,** 205, 221, 232 (1939).
4. W. SHOCKLEY, *Phys. Rev.* **56,** 317 (1939).
5. D. PUGH, *Phys. Rev. Lett.* **12,** 390 (1964).
6. V. HEINE, *Phys. Rev.* **138,** A1689 (1965).
7. J. VAN LAAR and J. J. SCHEER, *Surface Sci.* **3,** 189 (1965).

Metal–semiconductor contacts

8. W. E. MEYERHOFF, *Phys. Rev.* **71,** 727 (1947).
9. J. BARDEEN, *Phys. Rev.* **71,** 717 (1947).
10. R. J. ARCHER and M. M. ATALLA, *Ann. N.Y. Acad. Sci.* **101,** 697 (1963).
11. R. WILLIAMS and R. BUBE, *J. appl. Phys.* **31,** 968 (1960).
12. E. FABRICIUS, *J. appl. Phys.* **33,** 1597 (1962).
13. H. GRIMMEISS and R. MEMMING, *J. appl. Phys.* **33,** 2217 (1962).
14. C. A. MEAD and W. G. SPITZER, *appl. Phys. Lett.* **2,** 74 (1963).
15. W. G. SPITZER and C. A. MEAD, *J. appl. Phys.* **34,** 3061 (1963).
16. C. A. MEAD and W. G. SPITZER, *Phys. Rev. Lett.* **10,** 471 (1963).
17. C. A. MEAD and W. G. SPITZER, *Phys. Rev.* **134,** A713 (1964).
18. C. R. CROWELL, S. M. SZE and W. G. SPITZER, *Appl. Phys. Lett.* **4,** 91 (1964).
19. C. R. CROWELL, J. C. SARACE and S. M. SZE, *Trans. metall. Soc. AIME* **233,** 478 (1965).
20. D. KAHNG, *Solid-St. Electron.* **6,** 281 (1963).
21. S. M. SZE, C. R. CROWELL and D. KAHNG, *J. appl. Phys.* **35,** 2534 (1964).
22. A. M. GOODMAN, *J. appl. Phys.* **34,** 329 (1963).
23. C. R. CROWELL, W. G. SPITZER and H. G. WHITE, *Appl. Phys. Lett.* **1,** 3 (1962).
24. H. G. WHITE and R. A. LOGAN, *J. appl. Phys.* **34,** 1990 (1963).
25. M. COWLEY and H. HEFFNER, *J. appl. Phys.* **35,** 255 (1964).
26. A. M. GOODMAN, *J. appl. Phys.* **35,** 573 (1964).
27. A. M. GOODMAN, *J. appl. Phys.* **36,** 1411 (1965).
28. A. M. COWLEY and S. M. SZE, *J. appl. Phys.* **36,** 3212 (1965).
29. A. M. GOODMAN, *Surface Sci.* **1,** 54 (1964).

30. C. A. Mead, *Phys. Lett.* **18,** 218 (1965).
31. M. Aven and C. A. Mead, *Appl. Phys. Lett.* **7,** 8 (1965).
32. C. A. Mead, *Appl. Phys. Lett.* **6,** 103 (1965).
33. R. J. Archer *et al. Phys. Rev. Lett.* **12,** 538 (1964).
34. C. R. Crowell, *Solid-St. Electron.* **8,** 395 (1965).
35. C. R. Crowell, *et al. J. appl. Phys.* **36,** 3843 (1965).
36. J. W. Surhigh and C. A. Mead, *Phys. Lett.* **20,** 367 (1966).
37. J. van den Broek, *Phillips Res. Rpts.* **20,** 624 (1965).

APPENDIX

The barrier energies in (eV) of metals on semiconductors are shown opposite measured by I–V characteristic, capacitance intercept (cap) or photoresponse (photo). Sample temperatures were room temperatures except where noted otherwise. Details of measurements should be checked in the references.

Ref.	Semiconductor	Metal	I–V	Cap	Photo
17	*n* AlAs (Vac Cleave)	Au			1·2
		Pt			1·0
17	*p* AlSb (Vac Cleave)	Au		0·53	0·55
		Au (77°)		0·59	
17	*p* BN (Chem)	Au		3·1	
33	*p* BP	Au			0·87
17	*n* CdS (Vac Cleave)	Au		0·80	0·78
		Cu		0·35	0·36
		Ni			0·45
		Al (77°)			
		Ag		0·58	0·56
		Pt		0·86	0·85
26	*n* CdS (Chem)	Pt	1·2	1·2	1·1
		Au	0·68	0·66	0·68
		Pd	0·61	0·59	0·62
		Cu	0·47	0·41	0·50
		Ag		≃0·35	
		Al		None detected	
32	*n* CdSe (Vac Cleave)	Pt			0·37
		Au			0·49
		Ag			0·43
		Cu			0·33
17	*n* CdTe (Vac Cleave)	Au		0·71	0·60
		Pt		0·76	0·58
		Ag		0·81	0·66
		Al			0·76
17	*p* Diamond	Au		1·35	
17	*n* GaAs (Vac Cleave)	Au		0·95	0·90
		Pt		0·94	0·86
		Be		0·82	0·81
		Ag		0·93	0·88
		Cu		0·87	0·82
		Al		0·80	0·80
		Al (77°)		0·88	
17	*p* GaAs (Vac Cleave)	Au		0·48	0·42
		Au (77°)		0·46	
		Pt (77°)		0·48	
		Ag (77°)		0·44	
		Cu (77°)		0·52	
		Al		0·63	0·50
		Al (77°)		0·61	
19	*n* GaAs	W	0·71	0·77	0·80
23	*p* GaP (Chem)	Au			0·715
24	*p* GaP	Au	0·68	0·75	0·72
25	*n* GaP	Au	1·1	1·3	1·3
28	*n* GaP (Chem)	Cu		1·34	1·20
		Al		1·14	1·05
		Au		1·34	1·28
		Pt		1·52	1·45
		Mg		1·09	1·04
		Ag			1·20
17	*n* GaSb (Vac Cleave)	Au		0·61	0·60
		Au (77°)		0·75	
	p GaSb (Vac Cleave)	Au (77°)	ohmic		
17	*n* Ge (Vac Cleave)	Au		0·45	
		Au (77°)		0·50	
		Al		0·48	
19	*n* Ge	W	0·48		
17	*n* InAs (Vac Cleave and Chem)	Au, Ag, Al (77°)	ohmic		
	p InAs (Vac Cleave and Chem)	Au (77°)		0·47	
17	*n* InP (Vac Cleave)	Au		0·49	0·52
		Au (77°)		0·56	
		Cu (77°)		0·5	
		Ag		0·54	0·57
		Ag (77°)		0·50	
	p InP (Vac Cleave)	Au		0·76	
		Au (77°)		0·78	
17	*n* InSb (Vac Cleave and Chem)	Au (77°)		0·17	
		Ag (77°)		0·18	
	p InSb (Vac Cleave and Chem)	Au (77°)	ohmic		
37	*n* PbO	Ag			0·95
		Bi			0·94
		Ni			0·96
		Pb			0·95
		In			0·93
18	*n* Si (Chem)	Au			0·78
20	*n* Si (Chem)	Au	0·79		
20	*n* Si (Chem) (200)	Au			0·82
19	*n* Si	W	0·67	0·65	0·65
17	*n* SiC (hexag) (Chem)	Au		1·95	
		Al		2·0	
36	*n* SnO_2	Au			0·98
		Ag			0·65
		Cu			0·47
30	*n* ZnO (Vac Cleave)	Au			0·65
		Pt			0·75
		Pd			0·68
		Ag			0·68
		Cu	0·45		
		In	<0·3		
		Al	0		
		Ti	<0·3		
31	*n* ZnS (Vac Cleave)	Au			2·0
		Pd			1·87
		Pt			1·84
		Cu			1·75
		Ag			1·65
		In			1·50
		Al			0·8
		Ti			1·1
		Mg			0·82
30	*n* ZnSe (Vac Cleave)	Au			1·36
		Pt			1·40
		Cu			1·10
		Mg	<0·4		0·70

Normalized Thermionic-Field (T-F) Emission Metal-Semiconductor (Schottky) Barriers*

C.R. Crowell and V.L. Rideout**
(University of Southern California, Los Angeles, California, UK)

Solid-State Electronics, Volume 12 (1969)
Pages 89-105

key paper

*Supported in part by the Technical Committee of the Joint Services Electronics Program under Grant No. AF-AFOSR-496-68.
**National Aeronautics and Space Administration trainee

Abstract—Thermionic field (T-F) emission in uniformly doped metal-semiconductor (Schottky) barriers is analyzed to yield a normalized solution in closed form for the forward and reverse current (I)–voltage (V) relationship. A quasi one-dimensional approach and Maxwell–Boltzmann statistics are used. The formulation is expressed in terms of the 'flat-band' current density I_m, the band bending E_b in the semiconductor depletion region, the materials constant E_{00} ($\ln[I/I_m] = -E_b/E_{00}$ at 0°K in the WKB approximation), and kT. The kinetic energy in units of E_b at which the maximum injection of carriers occurs in the semiconductor is shown to be $\cosh^{-2}(kT/E_{00})$. Current flow in the temperature range between pure thermionic emission ($kT/E_{00} \gg 1$) and pure field emission ($kT/E_{00} \ll 1$) is analyzed and criteria for the transition of T-F emission to thermionic and to field emission are given. Computer solutions for the energy distribution of the injected carriers and for the normalized I–V characteristic are presented in graphical form. The results permit a straightforward calculation of the barrier height and the impurity concentration in the semiconductor from measurements of current density and differential resistance at a single applied bias. Application of these results explains a reported discrepancy between barrier heights deduced from photothreshold, C–V and I–V characteristics of W–GaAs and Au–GaAs Schottky barriers. A relatively constant excess temperature T_0 (i.e., $\ln I \propto (T+T_0)$ when $V \gg kT/q$) is predicted in the case of large E_b/E_{00} in the higher kT/E_{00} range where thermionic emission is nearly predominant. $I \propto [\exp(qV/kT)-1]$ is shown to be a general expectation for all Schottky barriers near zero bias when the I–V characteristic is dominated by either thermionic or thermionic-field emission. The assumption of a Gaussian energy distribution of carriers leads to values for the slope of ln I vs. V in reasonable agreement with the results of the computer analysis, but the prediction of the absolute value of the current density deviates rapidly from the computed value when kT/E_{00} departs appreciably from unity. The Gaussian distribution also does not provide the smooth transition from T-F to thermionic emission characteristic of the computer solution.

Résumé—L'émission à champ thermionique (T-F) dans les barrières métal-semiconducteur dopées d'une manière uniforme est analysée pour produire une solution, normalisée en forme fermée, de la relation tension (V)-courant direct et inverse (I). Une approche quasi-unidimensionnelle et les statistiques Maxwell–Boltzmann sont employées. Les formules sont exprimées en termes d'une densité de courant à bande plate I_m, la courbure de bande E_b dans la région d'épuisement du semiconducteur, la constante de matériau E_{00} ($\ln[I/I_m] = -E_b/E_{00}$ à 0°K dans l'approximation WKB) et kT. L'énergie cinétique en unités E_b dans laquelle se produit l'injection maximum de porteurs dans le semiconducteur est démontrée comme étant $\cosh^{-2}(kT/E_{00})$. L'écoulement du courant dans la gamme de température entre l'émission thermionique pure ($kT/E_{00} \gg 1$) et l'émission de champ pure ($kT/E_{00} \ll 1$) est analysé et les critères pour la transmission d'émission T-F aux émissions thermionique et de champ sont donnés. Des solutions d'ordinateur pour la distribution d'énergie des porteurs injectés et pour la caractéristique C–T normalisée sont présentées en forme graphique. Les résultats permettent le calcul direct de la hauteur de barrière et de la concentration d'impuretés dans le semiconducteur par les mesures de la densité de courant et de la résistance différentielle á une polarisation simple appliquée. L'application de ces résultats explique une différence reportée entre les hauteurs de barrière déduites des photolimites des caractéristiques tension-capacité, courant-tension des barrières Schottky en W–AsGa et Au–AsGa. Une température excédente relativement constante T_0 (c'est à dire $\ln I \propto (T + T_0)$ quand $V \gg kT/q$) est prédite dans le case des rapports élevés E_b/E_{00} dans la gamme élevée kT/E_{00} où l'émission thermionique est presque prédominante. On démontre que $I \propto [\exp(qV/kT) - 1]$ est souvent la règle générale pour toutes les barrières Schottky à polarisation d'environ zéro quand la caractéristique courant-tension est dominée par l'émission thermionique ou thermionique-champ. L'hypothèse d'une distribution d'énergie gaussienne des porteurs mène à des valeurs pour l'inclinaison de $\ln I/V$ en bon accord avec les résultats obtenus par l'analyse d'ordinateur, mais la prédiction de la valeur absolue de la densité de courant dévie rapidement de la valeur calculée quand kT/E_{00} s'éloigne de l'unité. La distribution gaussienne ne produit pas non plus une transition aplanie de T-F à la caractéristique d'émission thermionique de la solution d'ordinateur.

Zusammenfassung—Thermionische Feld-(TF-)Emission in Metall-Halbleiter-(Schottky-)Übergängen mit gleichförming dotiertem Halbleitermaterial wird mit dem Ziel analysiert, einen geschlossenen Ausdruck für die normalisierte Kennlinie in Durchlass- und Sperr-Richtung zu erhalten. Ein quasi-eindimensionales Modell und Boltzmannstatistik werden verwendet. In die Formulierung gehenein die "Flachbandstromdichte" I_m, die Bandverschiebung E_b in der Verarmungszone des Halbleiters, die Materialkonstante E_{00} ($\ln[I/I_m] = -E_b/E_{00}$ bei 0°K in der WKB-Näherung) und kT. Die kinetische Energie in Einheiten von E_b, bei welcher die maximale Trägerinjektion in den Halbeiter stattfindet, ist $\cosh^{-2}(kT/E_{00})$. Der Strom in Temperaturbereich zwischen rein thermionischer Emission ($kT/E_{00} \gg 1$) und reiner Feldemission ($kT/E_{00} \ll 1$) wird analysiert. Kriterien für den Übergang von TF-Emission zu thermionischer und zu Feldemission werden angegeben. Die Computer-Lösungen für die Energieverteilung der injizierten Träger und für die normalisierten IV-Kennlinien werden in graphischer Form dargestellt. Die Ergebnisse ermöglichen eine Berechnung des Potentialunterschieds am Übergang und die Störstellenkonzentration im Halbleiter aus Messungen der Stromdichte und des differentiellen Widerstands bei einer einzigen Spannung. Die Anwendung dieser Ergebnisse erklärt eine früher bereichtete Diskrepanz zwischen Potentialunterschieden am Übergang, welche aus der Photoschwelle, Kapazitäts-Spannungs-Messungen und Strom-Spannungs-Kennlinien an W–GaAs-und Au–GaAs–Schottky-Übergängen abgeleitet wurden. Eine relativ konstante Zusatztemperatur (d.h. $\ln I \sim (T + T_0)$ für $V \gg kT/q$) wird im Fall grösserer Werte E_b/E_{00} im höheren kT/E_{00}-Bereich vorausgesagt, wo thermionische Emission überwiegt. $I \sim [\exp(qV/kT) - 1]$ ist allgemein für alle Schottky-Übergänge in der Nähe verschwindender äusserer Spannung zu erwarten, wenn thermionische oder thermionische Feldemission dominiert. Die Annahme einer Gaußschen Energieverteilung der Träger führt zu Werten für die Steigung von $\ln I$ gegen V, die sich in vernüftiger Übereinstimmung mit den Werten der Computerrechnung befinden, aber der daraus gewonnene Absolutwert für die Stromdichte weicht schnell von dem entsprechenden Computerwert ab, sobald sich kT/E_{00} merklich von 1 unterscheidet. Die Gaußsche Energieverteilung führt ausserdem nicht zu dem allmählichen Übergang zwischen den Kennlinienformen der TF- und der thermionischen Emission, den die Computerlösung aufzeigt.

1. INTRODUCTION

MANY experimental and theoretical studies of current flow in metal-semiconductor (Schottky) barriers have been reported.[1-11] CROWELL and SZE[7] coupled Schottky's diffusion theory, Bethe's thermionic emission theory, effects of electron–phonon scattering and quantum-mechanical transmission into a single thermionic-diffusion theory for majority carrier current flow. This work, which is applicable at higher temperatures, explicitly included the effects of image force rounding of the barrier and the potential transition from the image force to the conduction band in the metal. PADOVANI and STRATTON[8] analyzed tunneling currents in Au–GaAs barriers from the standpoint of field and thermionic-field (T-F) emission using a one-dimensional WKB transmission of the barrier. In their analysis image force effects were neglected and a simple parabolic barrier shape was employed. In a zero temperature approximation, CONLEY *et al.*[10] neglected image force effects in obtaining a solution of the one-dimensional Schrödinger equation for a parabolic potential. Elaborate computer programs were required to obtain expressions for the tunneling current. A maximum in the incremental resistance (dV/dI) was predicted at a bias potential corresponding to the Fermi degeneracy of the semiconductor. CONLEY and MAHAN[11] subsequently interpreted low temperature observations of current flow in

Au–GaAs barriers in terms of a theory using the WKB tunneling approximation in which both a parabolic and an exponential-parabolic potential (which allowed an analytic solution of the WKB integral) were used. Effects of image forces were neglected.

In this paper, current flow in the temperature range between nearly pure thermionic emission[7] and pure field emission[10] is analyzed in a normalized formulation using one of the simpler models for T-F emission.[8]. The normalizing parameters all have simple physical interpretations. Results of computer calculations are expressed in graphical form. While there are several artificial features of the model, we feel it is desirable to document the full implications of the model because the ease with which the results can be used provides valuable physical insight into current transport processes and barrier heights in metal–semiconductor systems. The present analysis bears closer resemblance to the work of PADOVANI and STRATTON[8] than to any of the other works mentioned above. Accordingly, when appropriate, the results of the present analysis are compared with improved versions of the approximations deduced by Padovani and Stratton. These expressions are treated in the Appendix.

2. SCHOTTKY BARRIER MODEL

(a) Assumptions

A Maxwellian distribution of charge carriers is assumed to be incident on a space charge depletion region of width w in a semiconductor of uniform doping concentration N and static permittivity ϵ. The surface of a metal film is located at $x = w$ (cf. Fig. 1). The electrostatic potential in the depletion region is assumed to be independent of the mobile carriers. Thus the charge carrier potential energy in the depletion region, $qV(x)$, relative to the charge carrier potential energy in the bulk semiconductor, $q\psi_c$, is given as a function of distance by

$$qV(x) = q^2Nx^2/2\epsilon, \tag{1}$$

for $0 \le x \le w$. Image force corrections[4] to the barrier shape are not considered. For a flow of carriers with an effective mass component m^* in the direction of the flow and with an energy E associated with the component of momentum

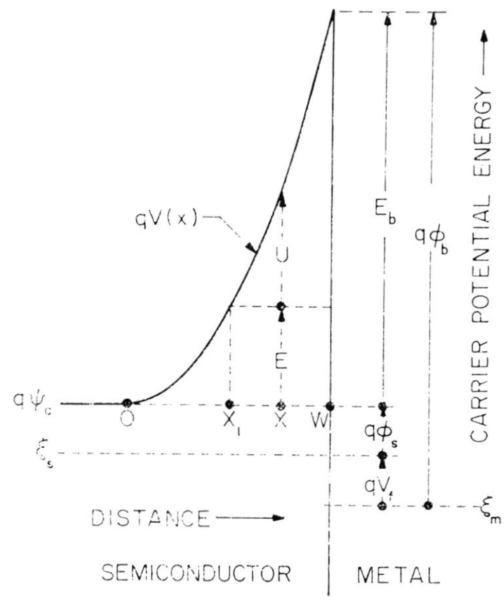

FIG. 1. Potential energy vs. distance for carriers in a forward biased metal-semiconductor (Schottky) barrier diode.

normal to the metal-semiconductor interface, the one-dimensional time-independent WKB transmission probability for the barrier is[12]

$$\tau(E) = \exp \frac{-4\pi}{h} \int_{x_1}^{w} \{2m^*[qV(x) - E]\}^{1/2} \mathrm{d}x, \tag{2}$$

for values of E less than the potential energy barrier E_b (i.e., for T-F (thermally excited tunnel) emission). For thermionic emission (i.e., $E > E_b$) $\tau(E)$ is assumed to be unity. Tunneling occurs in the interval from x_1 to w where $qV(x_1) = E$ and $qV(w) = E_b$ respectively (cf. Fig. 1). In $\tau(E)$ we do not include any term in the effective mass in the metal because the tunneling path in the true barrier, that with the image force, terminates within the body of the semiconductor. In practice both the image force lowering and a more precise calculation of $\tau(E)$ are required when $E \approx E_b$. The lack of these factors, however, is somewhat compensating (cf. CROWELL and SZE[7]). Periodic deviations such as have been observed in thermionic emission into vacuum are very unlikely in Schottky barriers where the charge carrier effective mass is usually less than in vacuum and where the dielectric screening reduces the range of the image force.[13]

The net current density I due to the Maxwell–Boltzmann tails of thermally excited distributions of carriers in the semiconductor and in the metal (after integration over momentum transverse to the metal-semiconductor interface) is given by

$$I = I_f - I_r = \frac{A^*T}{k}\int_0^\infty f_s(E)\tau(E)\,dE - \frac{A^*T}{k}\int_0^\infty f_m(E)\tau(E)\,dE, \quad (3)$$

where A^* is the Richardson constant,[3,14] $f_s(E)$ is the occupation probability for states of zero transverse energy in the semiconductor, and $f_m(E)$ is the occupation probability for states in the metal which will accept carriers of zero transverse energy in the semiconductor. I_f and I_r are components which approximate the flux of carriers in the semiconductor depletion region flowing respectively towards and away from the metal surface. Then

$$f_s(E) = \exp[-(q\varphi_s+E)/kT] = f_m(E)\exp(qV_f/kT), \quad (4)$$

where V_f is the applied forward bias and $q\varphi_s$ is the potential energy of the charge carriers (for $x < 0$) relative to ξ_s, the Fermi energy in the semiconductor (cf. Fig. 1). Hence

$$I_f = I_r\exp(qV_f/kT) = \frac{A^*T}{k}\int_0^\infty \tau(E) \times \exp-\left[\frac{E+q\varphi_s}{kT}\right] dE. \quad (5)$$

The Richardson constant of the semiconductor for thermionic and thermionic-field emission in a semiconductor with ellipsoidal constant energy surfaces is

$$A^* = \frac{4\pi qk^2}{h^3}[l^2m_ym_z + m^2m_zm_x + n^2m_xm_y]^{1/2}, \quad (6)$$

where l, m and n are the direction cosines of the current flux density relative to the principle axes of the constant energy ellipsoid and m_x, m_y and m_z are the corresponding components of the effective mass tensor.[3,14]

Note that if a more precise expression for the transmission coefficient than equation (2) is used, the value of A^* should reflect the band structure and the potential energy in both the metal and the semiconductor.[7,14] Since the tunneling coefficient for transitions which occur between states in the semiconductor and in the metal and which conserve energy and transverse momentum must be equal, and since the principle of detailed balance applies at zero bias, the value of A^* must be identical for both the forward and reverse fluxes. For example, in the simple case of isotropic effective masses and in the absence of a potential energy discontinuity, the smaller effective mass dominates. This occurs because transmission across the boundary is assured for any angle of incidence in the light mass region, but the carrier distribution appears within a solid angle of less than 2π after refraction into the heavy mass region. Carriers in the heavy mass region with the same energy but with angles of incidence outside this solid angle suffer internal reflection. In general the semiconductor effective mass dominates the current flow because almost any electron which reaches the semiconductor surface can cross into the metal and still conserve energy and transverse momentum. Crowell[14] has considered this point in greater detail.

Throughout the present treatment we have assumed that the tunneling effective mass is not a function of energy. This assumption in general becomes invalid if an appreciable portion of the charge carriers approach the middle of the semiconductor bandgap during the process of tunneling. When this occurs the WKB tunneling approximation is no longer independent of transverse momentum in the semiconductor and a two band approach must be employed.[9,11]

(b) Parameters

Three convenient parameters for normalizing the I–V characteristic are considered below.

(1) I_m is the 'flat-band' current density associated with the Maxwellian distribution of carriers

incident on the depletion layer from the body of the semiconductor:

$$I_m \equiv A^* T^2 \exp(-q\varphi_s/kT). \quad (7)$$

Thus I_f/I_m is the transmission coefficient of the barrier averaged over the carrier distribution [cf. equations (6) and (7)].

(2) E_{00} is a constant of the material and is associated with the WKB expression for the transmission of the barrier for carriers of energy $E = 0$. Note that according to HARRISON[15] the actual transmission coefficient for $E = 0$ is zero. From equations (1) and (2),

$$\tau(E=0) = \exp(-E_b/E_{00}) \quad (8)$$

and

$$E_{00} = \frac{qh}{4\pi}\left[\frac{N}{m^*\epsilon}\right]^{1/2} = 18.5\times 10^{-12}\left[\frac{N}{m_r\epsilon_r}\right]^{1/2} \text{eV} \quad (9)$$

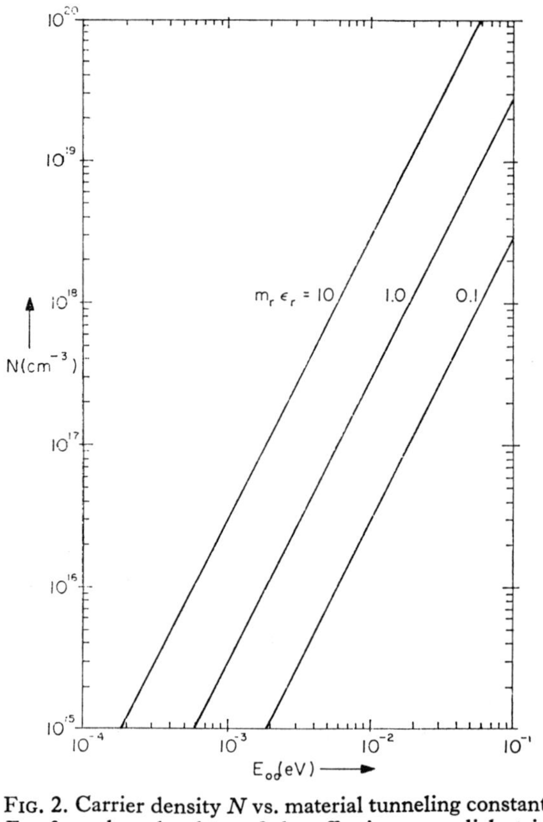

FIG. 2. Carrier density N vs. material tunneling constant E_{00} for selected values of the effective mass-dielectric constant product $m_r\epsilon_r$.

where m_r is the tunneling effective mass measured in units of the free electron mass, N the impurity concentration in units of cm^{-3}, and ϵ_r the static dielectric constant. The N vs E_{00} relationship is plotted in Fig. 2 with $m_r\epsilon_r$ as a parameter. The term E_{00} is identical to the term E_{00} used by PADOVANI and STRATTON[8] and to the MKS equivalent of the term E_0 used by CONLEY and MAHAN.[11]

(3) E_b, the band bending in the semiconductor depletion region, is related to the metal–semiconductor barrier height, φ_b, and to φ_s and V_f by the relationship

$$E_b = q(\varphi_b - \varphi_s - V_f). \quad (10)$$

In the high temperature range, where thermionic emission is important, it is convenient to measure E_b in units of kT. At low temperatures, where pure field emission is important, E_b can more appropriately be measured in units of E_{00}.

(c) Normalized formulation

Using equations (1) and (9) we can normalize and transform the integration over distance in equation (2) into one over energy. The result is

$$\tau(E) = \exp\left\{-E_{00}^{-1}\int_0^{E_b-E}\left[\frac{U}{E+U}\right]^{1/2}\mathrm{d}U\right\}$$

$$= \exp[-E_b y(\alpha)/E_{00}], \quad (11)$$

where $U \equiv qV(x) - E$, $\alpha \equiv E/E_b$, and

$$y(\alpha) \equiv (1-\alpha)^{1/2} - \alpha \ln\left[\frac{1+(1-\alpha)^{1/2}}{\alpha^{1/2}}\right]$$

$$= \tanh\theta - \theta\cosh^{-2}\theta, \quad (12)$$

where $\alpha \equiv \cosh^{-2}\theta$. Then, from equations (5), (7), (11) and (12),

$$\frac{I_f}{I_m} = \frac{E_b}{kT}\int_0^1 \exp\left\{-\frac{E_b}{kT}[\alpha + kTy(\alpha)/E_{00}]\right\}$$

$$\times\,\mathrm{d}\alpha + \exp(-E_b/kT). \quad (13)$$

Thus I_f/I_m is solely a function of the dimensionless parameters E_b/kT and kT/E_{00}. Note that $E_b/E_{00}[= (E_b/kT)\times(kT/E_{00})]$ and kT/E_{00} also provide a satisfactory pair of dimensionless variables.

3. THEORETICAL RESULTS

(a) *Energy distribution of carriers*

For the interpretation of measurements of current transport in metal-base transistors,[16–18] and in general for the determination of the predominance of T-F emission, it is of interest to know the energy distribution of the carriers injected into the metal from the semiconductor. In the present model the energy $E = E_M$ at which the maximum contribution to the current occurs is obtained from the maximum of the integrand in equation (13) (i.e., the maximum can never occur within the classically allowed region). The value of α at which this maximum occurs, α_M, is dependent only on kT/E_{00} (cf. Fig. 3);

$$\alpha_M \equiv E_M/E_b = [\cosh(E_{00}/kT)]^{-2}. \quad (14)$$

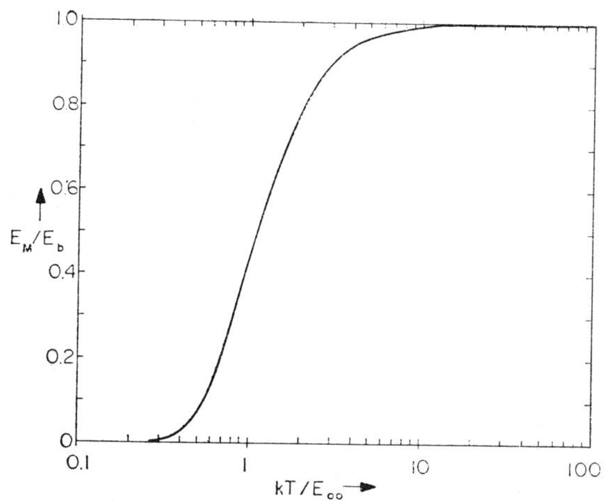

FIG. 3. Potential energy E_M (relative to the band bending E_b) at which the maximum number of carriers traverse the barrier as a function of the dimensionless parameter kT/E_{00}.

From equations (10) and (14), carriers of energy E_M relative to $q\psi_c$ (cf. Fig. 1) have an energy $E_{M\text{-metal}}$ relative to the Fermi level, ξ_m, in the metal given by

$$E_{M\text{-metal}} \equiv E_M + q\varphi_s + qV_f = q\varphi_b \times \cosh^{-2}(E_{00}/kT) + q(\varphi_s + V_f) \times \tanh^2(E_{00}/kT). \quad (15)$$

Thus $E_{M\text{-metal}}$ varies linearly with applied bias.

For the assumption of a constant tunneling effective mass to hold for n-type semiconductors, $E_b\tanh^2(E_{00}/kT)[= (1-\alpha_M)E_b]$ must be less than $E_g m_v/(m_v+m^*)$ where E_g is the direct semiconductor bandgap and m_v the tunneling effective mass for the valence band. In the remainder of this treatment it will be implicitly assumed that this condition is satisfied.

Representative normalized energy distributions of the emitted carriers are shown in Fig. 4 for two values of E_b/kT and for high, low and intermediate values of kT/E_{00}. It is apparent from these results that an increase in E_b/kT reduces the half-width of the energy distribution expressed in units of E_b. When kT/E_{00} increases, the character of the dominant current flow mechanism shifts from T-F emission towards thermionic emission. These results show that the predicted energy distribution, especially at high and low temperatures, cannot be approximated by a Gaussian distribution. The prediction by PADOVANI and STRATTON[8] of a Gaussian distribution for T-F emission is a direct consequence of their use of the MURPHY and GOOD[19] parabolic approximation to the energy dependence of the exponent in the expression for the carrier transmission coefficient $\tau(E)$ [cf. equation (2) and Appendix].

(b) *Criteria for dominance of T-F emission*

(*i*) *Thermionic to T-F transition* (*forward and reverse bias*). A reasonable criterion for the high temperature limit of validity of the present approach is that E_M must fall at least $\Delta\varphi_b$ below E_b (where $\Delta\varphi_b$ is the image force lowering of the barrier[4]). This limitation is not as serious as that imposed by the departures from Maxwell–Boltzmann statistics at low temperatures.

(*ii*) *Non-degenerate semiconductors: T-F field transition* (*reverse bias*). Under reverse bias conditions the criterion for validity of equation (5), which relates forward and reverse current flux, is that $E_{M\text{-metal}}$ must exceed the Fermi level, ξ_m, in the metal by at least kT. Note that when this condition is fulfilled transverse degrees of freedom add another kT to the average energy of the tunneling electrons. This further ensures the validity of the Maxwell–Boltzmann approximation. Thus the maximum reverse bias voltage, $V_{r\text{-max}}$, for

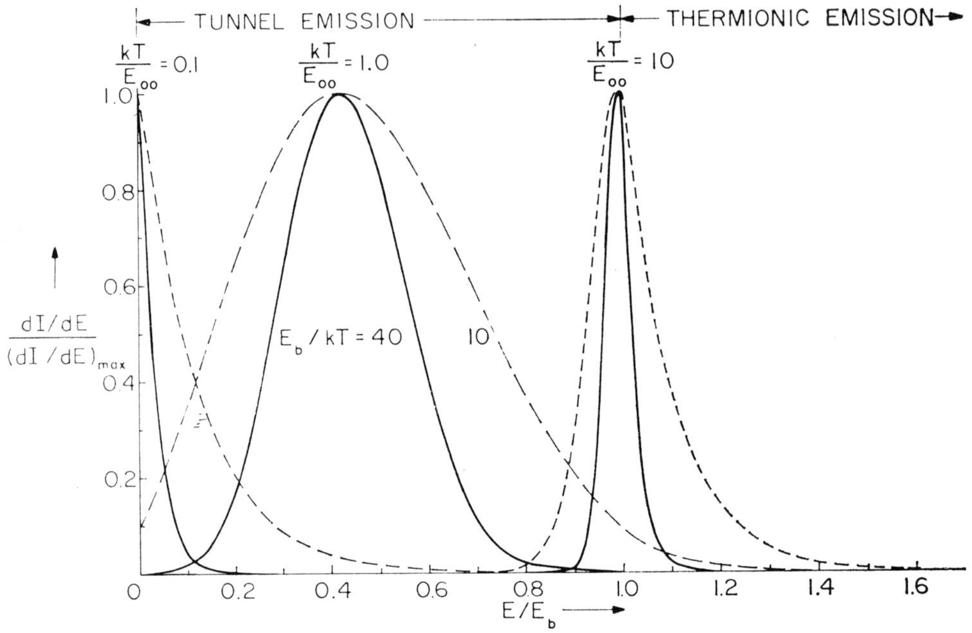

FIG. 4. Normalized energy distributions of emitted carriers for selected values of kT/E_{00} and the semiconductor band bending E_b in units of kT.

application of the present theoretical approach is

$$V_{r\text{-max}} \approx \varphi_s + \varphi_b \sinh^{-2}(E_{00}/kT) - (kT/q)\coth^2(E_{00}/kT). \quad (16)$$

For $-V_f > V_{r\text{-max}}$, Fermi–Dirac statistics must be used to evaluate the reverse current and the current is essentially pure field emission. If applied, the present model with Maxwell–Boltzmann statistics in this case yields, for a given bias voltage, a value of α_M which is too small and a current which is too large. Equation (16) is thus a somewhat conservative criterion for the existence of T-F emission.

(iii) Degenerate semiconductors: T-F to field transition (forward and reverse bias). In degenerate semiconductors the potential energy configuration is still parabolic within the depletion region and, in the present notation, $q\varphi_s$ becomes the negative of the Fermi kinetic energy in the semiconductor. Provided that the model predicts a value of E_M which exceeds $(-q\varphi_s + kT)$ and provided that equation (16) is satisfied, T-F emission will be dominant and our model should provide reasonable accuracy. The first condition requires that V_f be less than a maximum value, $V_{f\text{-max}}$, where

$$V_{f\text{-max}} \approx \varphi_b + \varphi_s \sinh^2(E_{00}/kT) - (kT/q)\cosh^2(E_{00}/kT) = V_{r\text{-max}} \sinh^2(E_{00}/kT). \quad (17)$$

Thus the bias range for existence of T-F emission vanishes when $V_{f\text{-max}}$ reaches zero. When $V_{f\text{-max}} = 0$, equation (17) yields a relationship for determination of the temperature, T_M, below which T-F emission at zero bias becomes field emission;

$$\sinh^2(E_{00}/kT_M) = \left[\frac{\varphi_b - kT_M/q}{-\varphi_s + kT_M/q}\right]. \quad (18)$$

The above criteria are similar to those derived by PADOVANI and STRATTON [equations (16) and (30)][8] from a consideration of the failure of their high and low temperature versions of the MURPHY–GOOD approximation[19] for forward bias. We would obtain more nearly identical results to those of Padovani and Stratton if we chose the criterion to be $E_M = -q\varphi_s$ for $V_{f\text{-max}}$ and assumed that $E_{00} \gg kT_M$. We have been unable, however, to

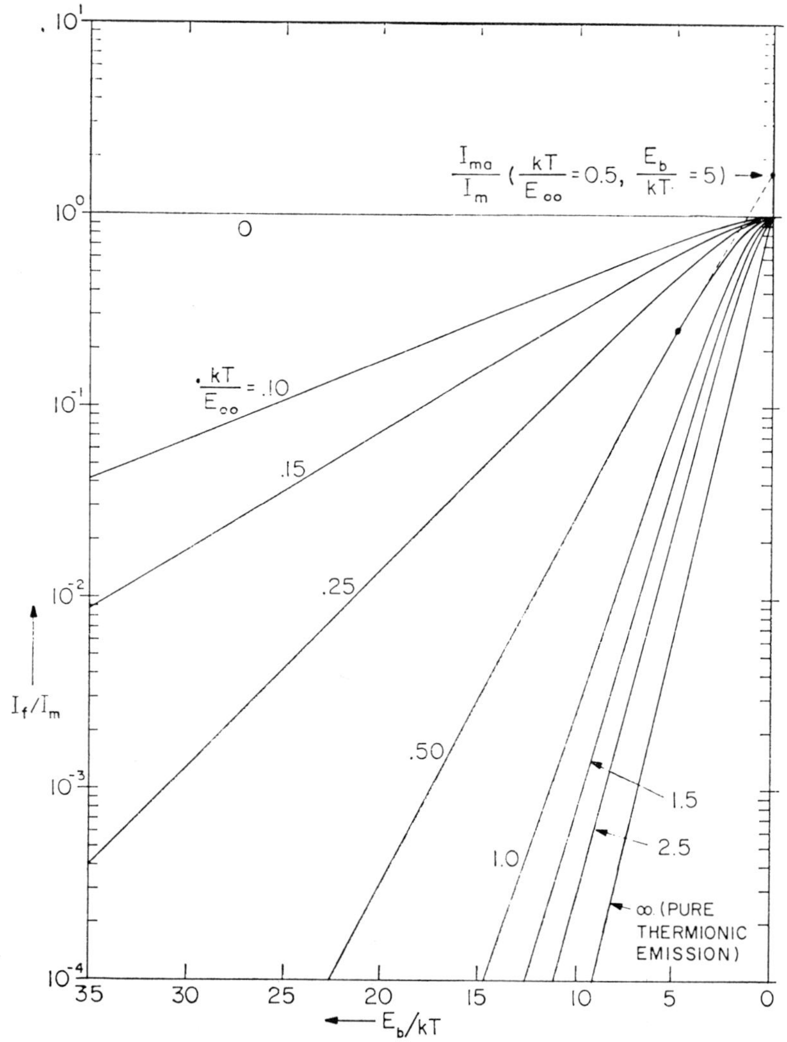

FIG. 5. Forward I–V diode characteristic for thermionic and thermionic-field emission. Normalized forward current density vs. band-bending in units of kT is shown for selected values of kT/E_{00}. A graphical illustration of I_{ma}/I_m ($kT/E_{00} = 0.5$, $E_b/kT = 5$) is shown.

determine an equivalent criterion for $V_{r\text{-max}}$ from their treatment (cf. Appendix).

(*c*) *Thermionic and thermionic-field emission* (*case* $kT/E_{00} \geq 1$)

(*i*) I_f/I_m *as a function of* E_b/kT *and* kT/E_{00}. In the high temperature-low doping range where $kT/E_{00} \geq 1$, thermionic emission of carriers over the barrier becomes dominant as $kT/E_{00} \to \infty$. Then the present model predicts

$$(I_f/I_m)_{\text{thermionic}} = \exp(-E_b/kT). \qquad (19)$$

In Fig. 5, I_f/I_m is shown as a function of E_b/kT for moderate to high values of kT/E_{00}. For applied bias greater than a few times kT/q, Fig. 5 gives the normalized I–V relationship [cf. equations (5) and (10)]. This presentation shows the thermionic limit described by equation (19).

To facilitate correlation with experimental results, and because the semi-log plots are nearly straight lines, it is useful to express the diode characteristics in the form

$$I_f/I_m = (I_{ma}/I_m)\exp(-E_b/nkT), \qquad (20)$$

where the terms (I_{ma}/I_m) and n are slowly varying functions of kT/E_{00} and E_b/kT when kT/E_{00} is greater than unity. The quantity I_{ma}/I_m is an apparent normalized 'flat-band' current density which is never less than unity. A graphical illustration of $I_{ma}/I_m(kT/E_{00} = 0.5,\ E_b/kT = 5)$ is shown in Fig. 5. The semi-log slope of the I_f–E_b characteristic determines the n value;

$$n = n(kT/E_{00}, E_b/kT) \equiv \frac{-\mathrm{d}(E_b/kT)}{\mathrm{d}\ln(I_f/I_m)}$$

$$= \frac{q}{kT}\frac{\mathrm{d}V_f}{\mathrm{d}\ln I_f} = \frac{-I_f}{I_m}\frac{\mathrm{d}(E_b/kT)}{\mathrm{d}(I_f/I_m)}. \qquad (21)$$

Values of $\ln(I_{ma}/I_m)$ and of n are shown in Figs. 6 and 7 respectively as a function of kT/E_{00} for selected values of E_b/kT. Interpolation for other values of E_b/kT is simple to do with an accuracy adequate for most purposes. Note that $n \to 1$ as $kT/E_{00} \to \infty$. The value of n was evaluated from a computation of I_f/I_m and its first derivative with respect to E_b/kT [cf. equation (21)]:

$$\frac{\mathrm{d}(I_f/I_m)}{\mathrm{d}(E_b/kT)} = \int_0^1 \left[1 - \frac{E_b}{kT}(\alpha + kT\,y(\alpha)/E_{00})\right] \times \exp\left[-\frac{E_b}{kT}(\alpha + kT\,y(\alpha)/E_{00})\right]\mathrm{d}\alpha - \exp\left[-\frac{E_b}{kT}\right]. \qquad (22)$$

Subsequently (I_{ma}/I_m) was obtained using equation (20).

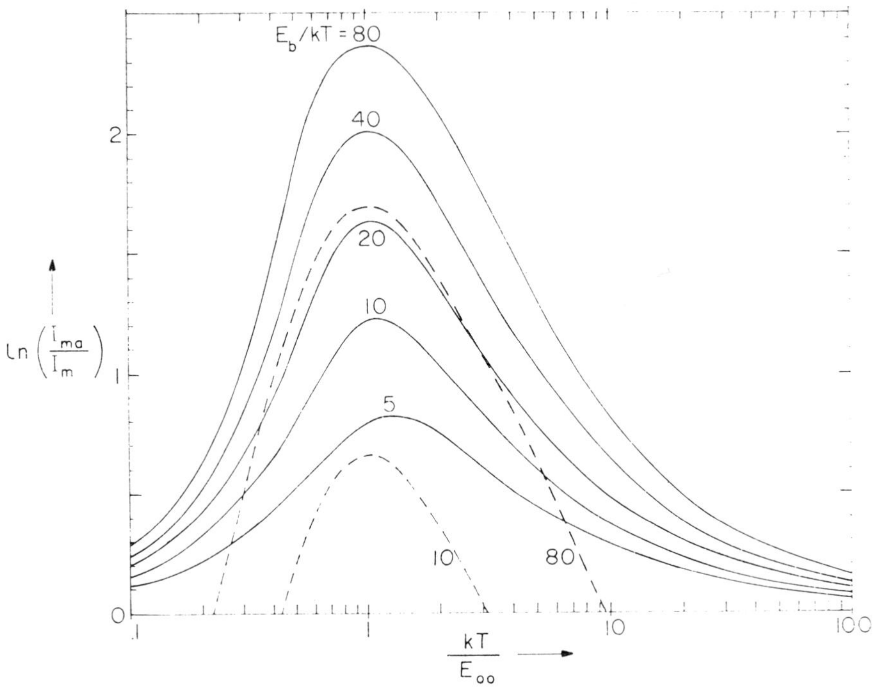

FIG. 6. Normalized apparent saturation ('flat-band') current density vs. kT/E_{00} for selected values of E_b/kT. The dashed lines indicate corresponding solutions for a Gaussian energy distribution of carriers [cf. equation (A.6)].

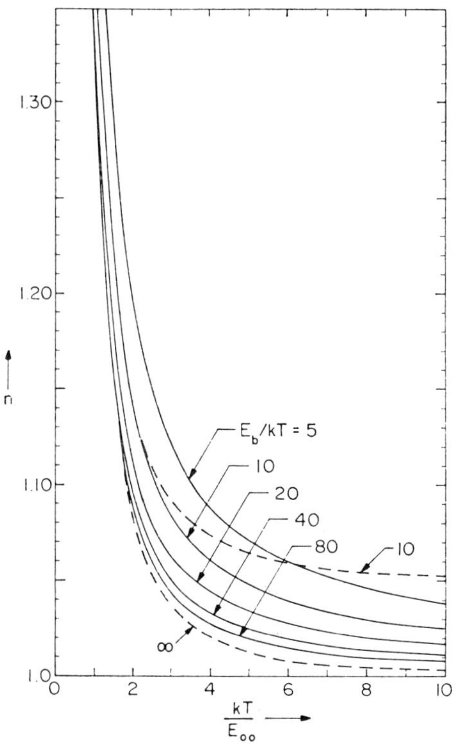

FIG. 7. Diode n value vs. kT/E_{00} for selected values of E_b/kT. The dashed lines indicate corresponding solutions for a Gaussian energy distribution of carriers [cf. equation (A.5)].

As an aid in utilizing experimental I–V data in cases where kT/E_{00} is unknown, $\ln(I_{ma}/I_m)$ is shown in Fig. 8 as a function of n for selected values of E_b/kT. Once again the parameters vary sufficiently slowly with E_b/kT that an accurate interpolation offers few problems. Figure 7 does not permit an accurate reading of kT/E_{00} for large n. The combination of Figs. 6 and 8, however, can be used to correlate kT/E_{00} and n when $kT/E_{00} \approx 1$.

(*ii*) *I–V characteristic*. From equations (4), (5), (7), (10) and (20), the I–V_f characteristic can be expressed in terms of n and I_{ma}/I_m in the form

$$I/I_m = (I_f - I_r)/I_m = (I_s/I_m) \times \exp(qV_f/nkT)[1 - \exp(-qV_f/kT)], \tag{23}$$

where

$$I_s/I_m \equiv (I_{ma}/I_m)\exp[-q(\varphi_b - \varphi_s)/nkT]. \tag{24}$$

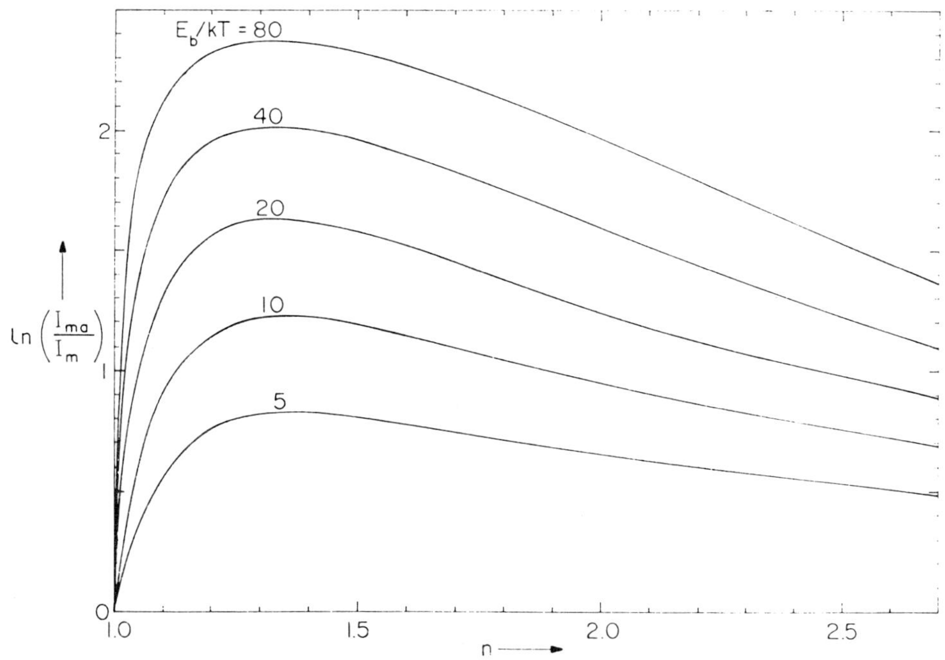

FIG. 8. Normalized apparent saturation current density vs. diode n value for selected values of E_b/kT.

Since n is greater than unity except in the asymptotic case $kT/E_{00} \to \infty$, a true reverse saturation current is not expected, but I_s is an apparent saturation current density when $V_f \gg kT/q$.

The following procedure can be used to calculate φ_b and E_{00} for a given diode. If I, V_f and n are measured at one point on the I–V_f characteristic, then the value of I_s can be obtained from equation (23). Equations (7) and (24) yield

$$\varphi_b = (1-n)\varphi_s + (nkT/q)\ln\left[\frac{A^*T^2}{I_s}\frac{I_{ma}}{I_m}\right]. \quad (25)$$

Note that even for n values near unity this result can vary considerably from that of the thermionic model, which yields $\varphi_b = (kT/q)\ln[A^*T^2/I_s]$.

Padovani and Sumner[20] have used an empirical relationship similar to equation (25) which appeared to determine reasonable values of φ_b. Terms in φ_s and I_{ma}/I_m, however, did not occur in their relationship. To calculate φ_b from equation (25), explicit knowledge of the doping concentration is necessary to obtain φ_s [$\varphi_s = (kT/q)\ln(N_c/N)$ where N_c is the effective density of states in the conduction band]. If the effective mass of carriers in the semiconductor is known, then φ_b, φ_s and N can be calculated from n, T and I_s by a rapidly converging iteration starting with $(E_b/kT) \approx n\ln[A^*T^2/I_s]$. First approximations to $\ln[I_{ma}/I_m]$ (from Fig. 8), to E_{00} (from Fig. 7) and to N [from Fig. 2 or equation (9)] can then be obtained. These values give approximations to φ_b [from equation (25)] and to φ_s, which in turn give a more precise value for E_b/kT from which to repeat the iteration. This procedure converges rapidly if $kT/E_{00} \gg 1$. However, for heavily doped material (small φ_s) and for lightly doped material (n close to unity), φ_b remains relatively independent of the term in φ_s. When $kT/E_{00} \gg 1$, both n and $\ln[I_{ma}/I_m]$ are slowly varying functions of kT/E_{00} and the value of E_{00} is strongly dependent on the accuracy of the experimentally determined value of n. Note that a knowledge solely of kT/E_{00} and n does not suffice to give an accurate value of φ_b because n is relatively insensitive to changes in E_b/kT (cf. Figs. 5 and 7).

(d) Field and thermionic-field emission (case $kT/E_{00} \le 1$)

(i) I_f/I_m as a function of E_b/E_{00} and kT/E_{00}. In the low temperature–high doping range where $kT/E_{00} \le 1$, field emission of carriers through the full width of the semiconductor depletion region becomes dominant as $kT/E_{00} \to 0$. For the present model the asymptotic limit I_f/I_m as $kT/E_{00} \to 0$ is

$$(I_f/I_m)_{\text{field}} = \exp(-E_b/E_{00}) \quad (26)$$

[cf. equation (8)]. In Fig. 9, I_f/I_m is shown as a function of the temperature independent product E_b/E_{00} for moderate to low values of kT/E_{00}. This presentation also shows the field emission limit described by equation (26) and a graphical illustration of I_{ma}/I_m ($kT/E_{00} = 0.5$, $E_b/E_{00} = 5$). For a given value of kT/E_{00} the data of Fig. 9 have counterparts in Fig. 5 since

$$E_b/E_{00} = (E_b/kT) \times (kT/E_{00}).$$

Thus results with a one-to-one correspondence to those in Section *c-i* should be expected.

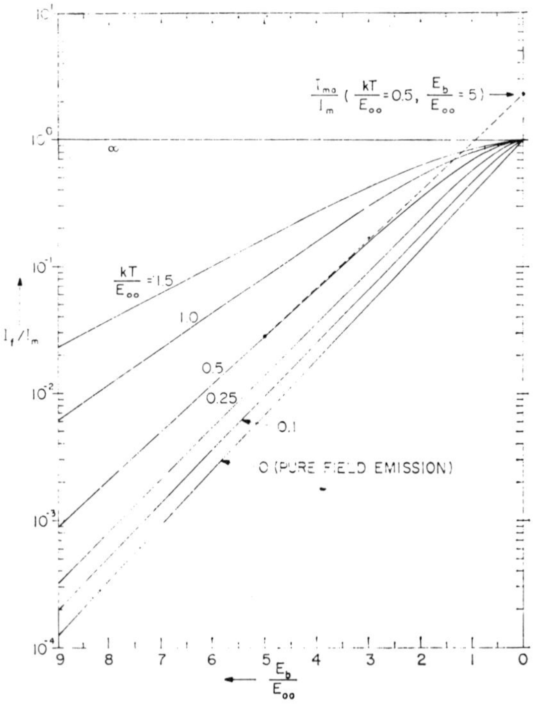

Fig. 9. Forward I–V diode characteristic for field and thermionic-field emission. Normalized forward current density vs. E_b/E_{00} is shown for selected values of kT/E_{00}. A graphical illustration of I_{ma}/I_m ($kT/E_{00} = 0.5$, $E_b/E_{00} = 5$) is shown.

Accordingly the I_f/I_m vs. E_b/E_{00} characteristics can be represented in the form

$$I_f/I_m = (I_{ma}/I_m)\exp(-E_b/E_0), \quad (27)$$

where

$$E_0 \equiv nkT. \quad (28)$$

Note that $E_0/E_{00} \to 1$ as $kT/E_{00} \to 0$.

In Figs. 10 and 11, $\ln(I_{ma}/I_m)$ and E_0/E_{00} are shown as functions of kT/E_{00} for selected values of E_b/E_{00}. In Fig. 12, $\ln(I_{ma}/I_m)$ is shown as a function of E_0/E_{00} for selected values of E_b. This presentation is analogous to that in Fig. 8. The use of the temperature independent product E_b/E_{00} in Figs. 9–12 permits field-emission currents to be more accurately compared with the low temperature asymptotic characteristic.

(*ii*) *I–V characteristic*. Results analogous to those in Section *c-ii* can be obtained by the substitution of terms in E_0 for those in nkT [cf. equations (23) and (28)], thus

$$I/I_m = (I_s/I_m)\exp(qV_f/E_0) \times [1-\exp(-qV_f/kT)], \quad (29)$$

where now

$$I_s/I_m = (I_{ma}/I_m)\exp[-q(\varphi_b-\varphi_s)/E_0] \quad (30)$$

[cf. equation (24)]. In addition

$$\varphi_b = [1-(E_0/kT)]\varphi_s + (E_0/q)\ln\left[\frac{A^*T^2}{I_s}\frac{I_{ma}}{I_m}\right] \quad (31)$$

[cf. equation (25)].

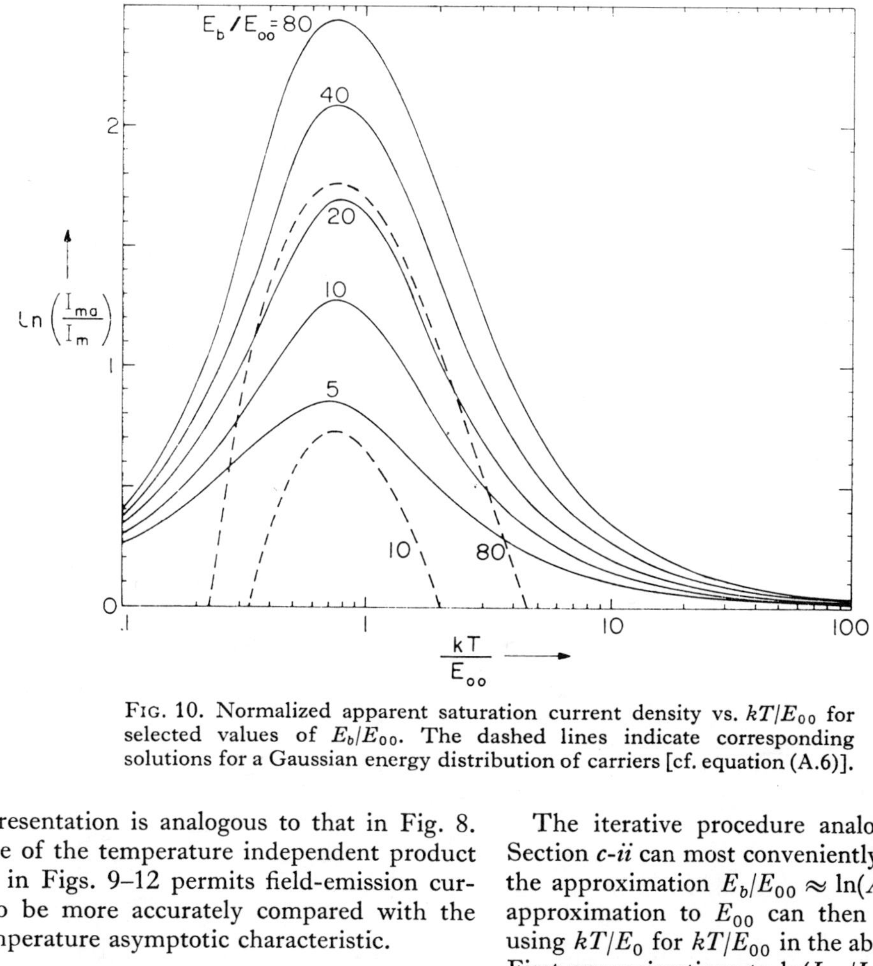

FIG. 10. Normalized apparent saturation current density vs. kT/E_{00} for selected values of E_b/E_{00}. The dashed lines indicate corresponding solutions for a Gaussian energy distribution of carriers [cf. equation (A.6)].

The iterative procedure analogous to that in Section *c-ii* can most conveniently be started from the approximation $E_b/E_{00} \approx \ln(A^*T^2/I_s)$. A first approximation to E_{00} can then be obtained by using kT/E_0 for kT/E_{00} in the abscissa of Fig. 11. First approximations to $\ln(I_{ma}/I_m)$, N, φ_s and φ_b can then be calculated. These in turn lead to a more precise value of E_b/E_{00}. Note that in this iterative procedure the doping concentration N,

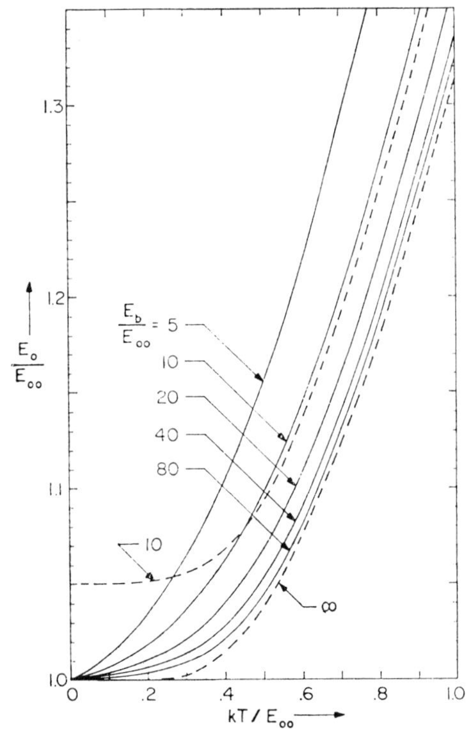

which is determined in terms of E_{00} [cf. equation (9)], can be calculated much more precisely than in the thermionic case. Analogous to the case of the thermionic emission a knowledge solely of kT/E_{00} and E_0 does not give an accurate value of φ_b because E_0 is relatively insensitive to changes in E_b/E_{00} (cf. Figs. 9 and 11).

(e) Excess temperature

Padovani and Sumner[20,21] have claimed that an empirical temperature independent parameter, T_0, can be used to describe the observed I–V characteristic of Schottky barriers over a wide temperature range. The relationship of their results to our present approach is as follows:

$$
\begin{aligned}
I_f &= I_s \exp(qV_f/nkT) \\
&= I_s \exp[qV_f/k(T+T_0)]. \qquad (32)
\end{aligned}
$$

Fig. 11. E_0/E_{00} ($= nkT/E_{00}$) vs. kT/E_{00} for selected values of E_b/E_{00}. The dashed lines indicate corresponding solutions for a Gaussian energy distribution of carriers [cf. equation (A.5)].

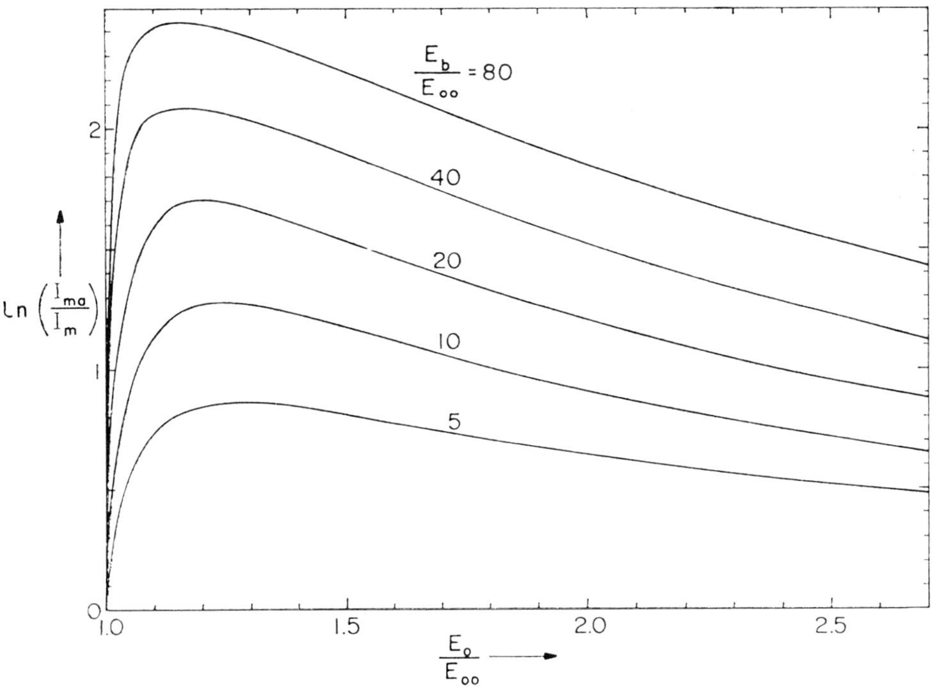

Fig. 12. Normalized apparent saturation ('flat-band') current density vs. E_0/E_{00} ($= nkT/E_{00}$) for selected values of E_b/E_{00}.

Thus the diode n value, expressed in terms of the excess temperature, T_0, is

$$n = 1+(T_0/T). \tag{33}$$

As is apparent from Figs. 7 and 11, our predicted n values are strongly temperature dependent. In Fig. 13 the excess temperature, T_0, vs. temperature

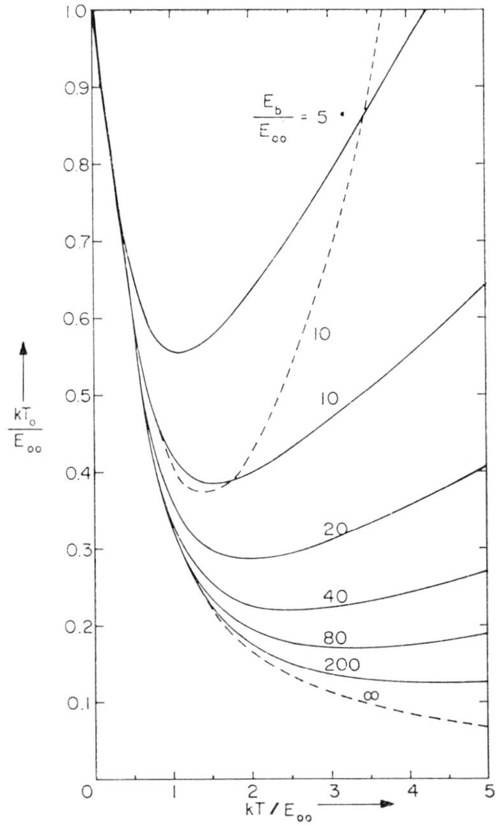

FIG. 13. Normalized excess temperature parameter kT_0/E_{00} vs. kT/E_{00} for selected values of E_b/E_{00}. The dashed lines indicate corresponding solutions for a Gaussian energy distribution of carriers [cf. equation (A.5)].

relationship is shown in normalized form. The excess temperature associated with the present approach is strongly temperature dependent except for the case of large E_b/E_{00} in the higher kT/E_{00} range where thermionic emission is nearly predominant. An empirical temperature-independent T_0 can, however, be a useful parameter for experimental characterization if one is willing to accept a rough fit to the n value for the case of thermionic emission, and if one does not extend the experimental measurements to too low a temperature. We feel that Padovani and Sumner have not used a sufficiently accurate determination of n in the high temperature region to allow a significant comparison between experiment and theory.

4. EXPERIMENTAL CORRELATIONS

(a) W–GaAs *Schottky barriers*

The above theoretical presentation is capable of reconciling a number of discrepancies between Schottky barrier heights deduced from current–voltage (I–V), capacitance–voltage (C–V) and photoresponse data. CROWELL *et al.*[(6)]† have studied W–Si and W–GaAs Schottky barriers at room temperature. The W–Si data are consistent with a purely thermionic origin of the I–V characteristic, a fact which is not surprising since $kT/E_{00} \approx 52$ (cf. Fig. 3) for the Si used in making these barriers ($N \approx 2\times10^{15}$ cm^{-3}). The W–GaAs data, however, showed a considerably lower barrier height, 0.71 eV, deduced from the 'thermionic' I–V characteristic in comparison with 0.77 eV deduced from the C–V and 0.80 eV deduced from the photoresponse (cf. Table 1). Using the n value of 1.04 for the W–GaAs barriers studied by CROWELL *et al.*[(6)] in the converging iterative procedure of section *c-ii* (which starts from $[E_b/kT] \approx n \ln[A^*T^2/I_s]$), a room temperature value for $\ln[I_{ma}/I_m]$ of 1.3 is obtained (cf. Fig. 8). The I–V barrier height of 0.77 eV deduced from equation (24) is in much better agreement with the C–V and the photoresponse measurements than that obtained directly from the pure thermionic model (cf. Table 1). Using the above theoretical analysis, a kT/E_{00} value of 3.9 is deduced from the n value of 1.04 (cf. Fig. 7) which is in fair agreement with the kT/E_{00} value of 5.3 calculated from equation (9) using the known

† We are indebted to Dr. F. A. Padovani for pointing out a discrepancy between the data plotted in Fig. 2 and the results quoted in Table 1 of this reference. The data in the table are used here. These are the results which were obtained before the experimental results were replotted for publication but with an erroneous scale change. The data as plotted in Fig. 5 of the reference are, however, consistent with the results of the table.

Table 1. *Schottky barrier parameters at* 300°K

Metal-semiconductor system	W–Si[6]	W–GaAs[6]	Au–GaAs
n value at 300°K	1.02	1.04	1.08
kT/E_{00} from n value (Section *c.ii* iteration)	—	3.9	2.3
kT/E_{00} from impurity concentration N [cf. equation (9)]	53	5.3	21
Barrier Height (eV) from			
(*a*) I–V (assuming pure thermionic emission)	0.67	0.71	0.74
(*b*) I–V (present theoretical analysis)	—	0.77	0.85
(*c*) C–V	0.65	0.77	0.83
(*d*) Photoresponse	0.65	0.80	0.83

impurity concentration of 5×10^{16} cm^{-3} for the GaAs samples.

(*b*) Au–GaAs *Schottky barriers*

PADOVANI[22]† has made detailed room temperature measurements of the I–V characteristic near zero bias for barriers of Au on lightly doped GaAs, and has correlated the reverse current with photoresponse measurements of the barrier height. He has interpreted his results in terms of pure thermionic emission because the current was proportional to $[\exp(qV_f/kT)-1]$ for $-kT/q \leq V_f \leq +kT/q$, and the reverse current varied as $\exp(q\Delta\varphi/kT)$ where $\Delta\varphi$ was the photoelectrically measured change in the barrier height. An anomaly that Padovani could not explain was a reverse current density of approximately twenty-seven times that expected from the thermionic model. The above observations can be explained in terms of our theoretical presentation. Since Padovani reported an n value of 1.08 from the forward characteristics (for $V_f \gg kT/q$), the I–V characteristic expected from the above theory is given by equation (23) as

$$I = I_s \exp[qV_f(n^{-1}-1)/kT][\exp(qV_f/kT-1)]$$
$$= I_s \exp[-0.074\, qV_f/kT][\exp(qV_f/kT-1)]. \quad (34)$$

† We wish to thank Dr. Padovani for a prepublication copy of this work.

One would not expect the $\exp[-0.074\, qV_f/kT]$ term to be detected in Padovani's graphical presentation. A value of 2.3 for kT/E_{00} corresponds to the n value of 1.08. This places the peak of the electron tunnel current density at an energy of approximately 0.1 eV below the barrier peak [cf. Fig. 3 and equation (14)]. A correlation of the photothreshold lowering with the reverse current is thus expected.

The barrier height deduced using equation (25) is 0.85 eV. This result is in reasonable agreement with Padovani's C–V and photoresponse barrier measurements, both of which are reported as 0.83 eV (cf. Table 1). The discrepancy between the kT/E_{00} value of 2.3 deduced from the n value, and the kT/E_{00} value of 21 deduced from the measured impurity concentration of $N = 4 \times 10^{15}$ cm^{-3} can be understood in terms of a more critical evaluation of the barrier height measurements. In the presence of the anomalous barrier lowering reported by Padovani, one should expect that the photoelectrically measured barrier should be much lower than that measured from the C–V characteristic, which does not include image force effects. In addition a C^{-2} vs. V plot measures the barrier height that would occur at zero electric field if the derivative of the barrier height with respect to the electric field were to continue unchanged when a diode is biased towards the flat band condition.[7] Thus for true 'agreement', the capacitively measured barrier height should have been approximately 0.88 eV instead of the 0.83 eV reported, since approximately 0.05 eV lowering is anticipated when $(V_D - V - kT/q)^{1/4}$ is extrapolated to zero (cf. Padovani's Fig. 5). This is also strong evidence for the existence of a hyperabrupt junction caused by charged traps near the surface of the GaAs, the effects of which Padovani observed. Thus an enhanced effective doping at the surface can explain all of the features of Padovani's observations.[22]

(*c*) *Temperature sensitivity*

From our criteria for the existence of T-F emission, $T > T_M$ [cf. equation (18)] and $-V_{r\text{-max}} < V_f < V_{f\text{-max}}$ [cf. equation (17)], we see that on either side of zero bias there is a region where the I–V_f characteristic is thermally sensitive. The temperature sensitivity becomes negligible as pure field emission is approached.

The greatest temperature sensitivity actually occurs at zero bias where $\mathrm{d}I/\mathrm{d}V_f = qI_s/kT$ [cf. equations (23) and (29)]. This is the same result as one gets from thermionic emission. Note that this temperature sensitivity extends to lower temperatures as the semiconductor approaches nondegeneracy.

5. DISCUSSION

The theoretical treatment presented here has been of assistance in the interpretation of the I–V characteristic of metal–semiconductor (Schottky) barrier diodes. At the same time it should be noted that there are several additional factors which it would be desirable to consider, namely; surface-state and interface effects in nonideal diodes, image force barrier-lowering, and the effects of the potential energy configuration in the immediate neighborhood of the metal surface.

In a limited sense the present theoretical treatment can give us some idea of the effects mentioned above. We can analyze portions of the I–V characteristic to obtain φ_b and N and thus detect interface effects which cause the barrier height to appear to be bias dependent.(7) While image force effects have not been incorporated into the present treatment, the deduced barrier height should fall between the barrier height determined from photoresponse measurements (i.e., with image force lowering) and the barrier height which would occur in the absence of image force lowering. Effects of the band structure in the metal are only included implicitly and in an approximate form. It would be desirable to use a more precise expression for the transmission coefficient both below and above the barrier energy,(16) but such a correction should not change the essential nature of our theoretical results.

REFERENCES

1. D. Kahng, *Solid-St. Electron.* **6,** 281 (1963).
2. M. M. Atalla and R. W. Soshea, Scientific Report No. 1, Contract No. AF19 (628)-1637, Hewlett-Packard Associates (1962).
3. C. R. Crowell, *Solid-St. Electron.* **8,** 395 (1965).
4. S. M. Sze, C. R. Crowell and D. Kahng, *J. appl. Phys.* **35,** 2534 (1964).
5. D. Kahng, *Bell. Syst. tech. J.* **43,** 215 (1964).
6. C. R. Crowell, J. C. Sarace and S. M. Sze, *Trans. Metal. Soc. AIME* **233,** 478 (1965).
7. C. R. Crowell and S. M. Sze, *Solid-St. Electron.* **9,** 1035 (1966).
8. F. A. Padovani and R. Stratton, *Solid-St. Electron.* **9,** 965 (1966).
9. F. A. Padovani, and R. Stratton, *Phys. Rev. Lett.* **16,** 1202 (1966).
10. J. W. Conley, C. B. Duke, G. D. Mahan and J. J. Tiemann, *Phys. Rev.* **150,** 466 (1966).
11. J. W. Conley and G. D. Mahan, *Phys. Rev.* **161,** 681 (1967).
12. J. L. Moll, *Physics of Semiconductors*, equations (12–28) and (12–30), p. 246. McGraw-Hill, New York (1964).
13. C. R. Crowell and S. M. Sze, *Physics of Thin Films*, Vol. 4, pp. 338–341. Academic Press, New York (1967).
14. C. R. Crowell, *Solid-St. Electron.*, **12,** 55 (1969).
15. W. A. Harrison, *Phys. Rev.* **123,** 85 (1961).
16. C. R. Crowell, and S. M. Sze, *J. appl. Phys.* **37,** 2683 (1966).
17. C. R. Crowell and S. M. Sze, *Phys. Rev. Lett.* **15,** 659 (1965).
18. S. M. Sze, C. R. Crowell, G. P. Carey and E. E. LaBate, *J. appl. Phys.* **37,** 2690 (1966).
19. E. L. Murphy and R. H. Good, Jr., *Phys. Rev.* **102,** 1464 (1956).
20. F. A. Padovani and G. G. Sumner, *J. appl. Phys.* **36,** 3744 (1965).
21. F. A. Padovani, *J. appl. Phys.* **37,** 921 (1966).
22. F. A. Padovani, *Solid-St. Electron.* **11,** 193 (1968).

APPENDIX

T-F I–V characteristics assuming a Gaussian energy distribution for tunneling charge carriers

From equation (13), the logarithm of the energy distribution of the forward carrier flux I_f in the tunneling range of energies is

$$\ln \frac{\mathrm{d}(I_f/I_m)}{\mathrm{d}\alpha} = \ln(E_b/kT) - (E_b/kT) \times (\alpha + kTy(\alpha)/E_{00}). \quad (A.1)$$

A Guassian distribution which matches the maximum and curvature of this distribution will be characterized by a maximum of

$$\left.\frac{\mathrm{d}(I_f/I_m)}{\mathrm{d}\alpha}\right|_{\alpha_M} = (E_b/kT) \times \exp[-(E_b/E_{00}) \tanh(E_{00}/kT)] \quad (A.2)$$

at α_M [cf. equation (14)], and by

$$\left.\frac{\mathrm{d}^2}{\mathrm{d}\alpha^2} \ln\left[\frac{\mathrm{d}(I_f/I_m)}{\mathrm{d}\alpha}\right]\right|_{\alpha_M} = \frac{E_b \cosh^2(E_{00}/kT)}{E_{00} 2 . \tanh(E_{00}/kT)}. \quad (A.3)$$

If qV_f/kT is subtracted from the right-hand side of equation (A.1) and if the right-hand side of equation (A.2) is multiplied by $\exp(-qV_f/kT)$, then the same expressions hold for the reverse flux I_r. If we integrate over the Gaussian distribution and use equation (5) to relate I_f and I_r,

$$\frac{I}{I_m} = \frac{[\pi E_{00} E_b \tanh(E_{00}/kT)]^{1/2}}{kT \cosh(E_{00}/kT)} \times \exp - [(E_b/E_{00}) \tanh(E_{00}/kT)] \times [1 - \exp(-qV_f/kT)]. \quad \text{(A.4)}$$

The parameters T_0, E_0 and n defined in the present treatment are then related by

$$k(T+T_0) \equiv E_0 \equiv nkT \equiv \frac{-\mathrm{d}E_b}{\mathrm{d}\ln(I_f/I_m)} = \left[\frac{\tanh(E_{00}/kT)}{E_{00}} - \frac{1}{2E_b}\right]^{-1}. \quad \text{(A.5)}$$

In addition

$$\ln(I_{ma}/I_m) = \tfrac{1}{2} \ln \left[\frac{\pi E_b E_{00} \tanh(E_{00}/kT)}{k^2 T^2 \cosh^2(E_{00}/kT)}\right] - \tfrac{1}{2}. \quad \text{(A.6)}$$

The half-width, δE_M, of the Gaussian distribution is

$$\delta E_M = \frac{2[E_b E_{00} \ln(2) \tanh(E_{00}/kT)]^{\frac{1}{2}}}{\cosh(E_{00}/kT)}. \quad \text{(A.7)}$$

The above results differ from those of PADOVANI and STRATTON[8] (P–S) in the following ways.

(*i*) The terms in $\tanh(E_{00}/kT)$ in equations (A.3), (A.6), (A.7) and in the pre-exponential term in equation (A.4) are omitted in the 'equivalent' treatment of P–S. Note that these terms are needed at high temperatures to explain the increased curvature (relative to the case of $kT/E_{00} = 1$) of the peak of the energy distribution (cf. Fig. 4).

(*ii*) P–S used different values of the Richardson constant for the forward and reverse characteristics and hence were unable to formulate an expression [equivalent to equation (A.4)] which gave zero current at zero bias.

(*iii*) Equation (A.5) shows that there is an implicit curvature in the ln I_f vs. V_f plot (cf. Fig. 5). The P–S treatment has neglected the term in E_b on the right-hand side of equation (A.5). The reverse I–V_f characteristic has the same discrepancy from the P–S treatment. The ϵ' of P–S [equation (47)] for the reverse characteristic is equivalent to $-[(1/E_0)-(1/kT)]^{-1}$. Note that the term in E_b of equation (A.5) is relatively successful in representing the results of the exact calculation (see Figs. 7, 11, and 13).

(*iv*) The expressions for the pre-exponential term in the P–S reverse bias treatment are not consistent with the functional form for their forward characteristic. This plus the fact that their values of $V_{r\text{-max}}$ are obviously not consistent with equation (16) suggest that there is an algebraic error in the P–S reverse bias calculation equivalent to equation (A.3).

The Gaussian distribution appears to be an excellent assumption for determining the pre-exponential term for cases where $E_{00}/kT \approx 1$ (see Figs. 6 and 10), but for practical barrier heights, the prediction of the magnitude of I inherent in the distribution in equation (A.1) becomes rapidly in error as E_{00}/kT deviates appreciably from unity. The fact that equation (A.4) does not have the appropriate functional form for thermionic emission when $kT/E_{00} \gg 1$ [rather than errors in the expression for $\tau(E)$] made it imperative for P–S to derive a criterion for limiting the high temperature application of their expression for T–F emission. Their omission of the pre-exponential term in $\tanh(E_{00}/kT)$ in the equivalent to equation (A.4) helps to decrease the high temperature discrepancy, but refinements of the MURPHY–GOOD[19] approximation such as truncating the Gaussian distribution [P–S equation (17)] at zero energy in the semiconductor would only increase the error. In contrast our computed solution contains a gradual transition between T-F and thermionic emission just as CROWELL and SZE's[7] more accurate treatment of the perturbation of thermionic emission by tunneling demonstrates a fairly large range of operating conditions where appreciable tunneling occurs but is compensated by reflections of electrons which would classically be emitted.

3.2.2 metallurgy

The metallurgy of the barrier contact formed at low temperature is simple because the metal does not interact with the semiconductor. But when forming an ohmic contact or an environmentally stable barrier the structure is sintered at high temperature and the interaction is considerable. The requirements for a low resistance contact are:

1./The contact metal should uniformly wet and react with the semiconductor.
2./A uniform highly doped region should be formed with a low contact resistance resulting in ohmic behaviour.
3./The contact system should be compatible with other metals used and stable throughout subsequent processing procedures.

The contact metals commonly used involve a dopant such as Sn, In or Ge for n-type and Zn or Mn for p-type, alloyed with Au, Ag, or Ni. The latter provide a surface suitable for wire bonding, and hold the low melting point dopants in intimate contact with the semiconductor during alloying.

The alloying process involves as many as five elements; thus, the precise reactions that happen are often the subject of controversy. No definitive study has been made of the metallurgy, apart from the reaction of Au and Au-Ge with GaAs in the paper by Gyulai et al[88]. The Au-GaAs eutectic occurs at a temperature around 600°C, but visual inspection during alloying shows considerable reaction at 400°C, which cannot be between Au and As because their eutectic is at 665°C. It is thus likely that the reaction is between Au and Ga, a fact confirmed by observing the effect of a Au/Ga contact during heating which indicated no further reaction. In the case of the Au-Ge eutectic contact (which melts at 356°C), the reaction with the GaAs takes place at a lower temperature.

The proposed model postulated that gallium vacancies were created after the alloying of the Au with the GaAs, with Ge going into these vacancies to act as a donor. The RRE group studied the penetration of the Au using Rutherford scattering methods. The phase diagram of GaAs-Ge shows that the metallurgical reaction is negligible at these temperatures[59].

Silver reacts with Ga and As in the 500-600°C temperature range, although no eutectic forms. In the silver based contacts the dominant reaction occurs between the GaAs and the low melting point constituent of the contact alloy (either tin or In-Ge). The pseudobinary diagram of GaAs-Sn in Figure 3-5 shows that the liquidus for temperatures in the range 400-600°C occurs with approximately 93-97% tin. A reasonable supposition is that GaAs is taken into solution and regrows upon cooling heavily doped with tin. The same effect might well occur with In-Ge in place of the tin.

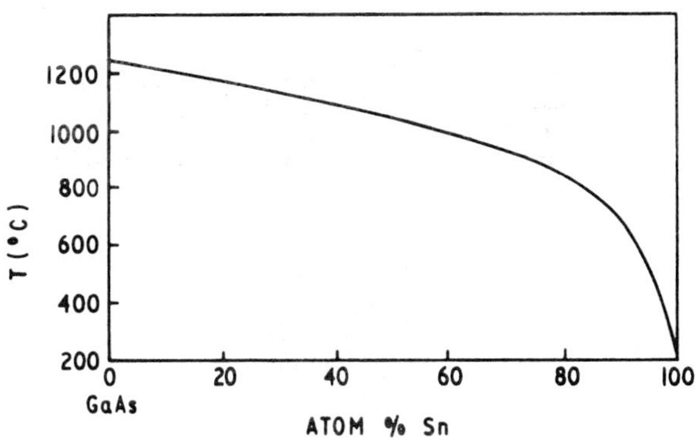

Figure 3-5/GaAs-Sn Pseudobinary

It is curious that the presence of a regrowth region has not been positively identified in contacts using these materials, although it would be extremely narrow and difficult to identify, even with the aid of scanning electron microscopy.

The physical nature of the alloyed films has been noted for Au and Au-Ge, Ag-In-Ge[60], and Ag-Sn.

The latter is illustrated in the key paper by Basterfield et al. All have shown the existence of metal spikes into the semiconductor, cracking around contact periphery, and non uniform alloy wetting; our experience has supported these observations for a wide variety of contact alloys and alloying conditions. The attack of the metal on GaAs is of the form of rectangular pitting, bounded largely by < 111 > crystallographic facets (the dimensions of which vary with the quantity and type of alloy used). They can penetrate up to 2 μm into the surface, although limiting the quantity of metal can reduce the degree of this effect to around 0.5 μm in the practical case. A number of angle sections of contacts on InP under different alloying temperatures are shown in Figure 3-6. Note that the spike diameter increases with temperature. Also shown is a section of a device in which a deep crack has appeared around the contact periphery.

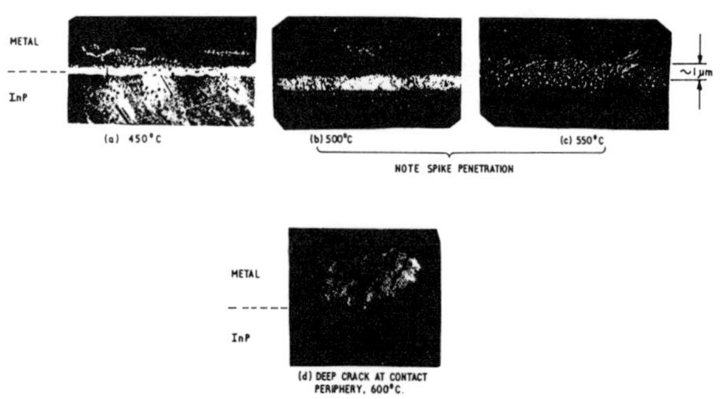

Figure 3-6/Metallurgical Sections of Ag-Sn Contacts to InP. Contacts Alloyed for 1 Min at Temperature Shown.

All of these various effects to some degree determine the electrical characteristics of the metal-semiconductor junction and hence those of the completed device. Until the metallurgical interactions are understood and can be controlled, the performance of devices will be variable. With new analytical processes becoming available (such as the ion probe spectrometer), it should now be possible to analyse these metal semiconductor interfaces as a function of depth and to obtain valuable quantitative information about contact metallurgy.

One method of reducing contact problems is to use an epitaxially grown, highly doped layer as a buffer between the contact and the active region of the device. Unfortunately sharp doping transitions are difficult to achieve. and the resulting finite buffer layer thickness makes a significant contribution to the thermal impedance of the device. Directly applied metallic contacts therefore have advantages if the metallurgical problems are overcome.

Stable Schottky barriers also present problems in fabrication. Although the contact is usually elemental, many phases can exist after sintering and the long term stability depends on the correct choice of materials and process conditions, just as for low resistance contact technology.

key paper

Structure Features Of Alloyed Contacts To GaAs

J. Basterfield, M.J. Josh, and M.R. Burgess
(Mullard Research Laboratories, Redhill, Surrey, UK)

Acta Electronica, Volume 15 (1972)
Pages 83-87

The contact technology has a great influence on successful manufacture of GaAs Gunn devices. The present paper describes one aspect of the metallurgical structure of the GaAs-metal interface after alloying.
The most significant aspect of the results presented is the appearance of a metal-GaAs interface which is not planar but punctuated on the GaAs side by a series of small holes bounded by (111) facets. The topography of the interface is dependent on the alloying cycle and the nature of the contact metal.
The spiked nature of the interface is likely to influence device behaviour but it is only one of several contact features which might affect contact properties.

La technologie des contacts a une importance énorme dans la réalisation fructueuse de dispositifs Gunn à l'arséniure de gallium. Cet article décrit un aspect de la structure, du point de vue métallurgique, de l'interface GaAs-métal après alliage.
Le plus important résultat mis en évidence concerne l'interface métal-GaAs qui n'est pas uniforme mais qui présente sur le côté GaAs une suite de petits trous délimités par des facettes (111). La topographie de l'interface dépend du cycle d'alliage et de la nature du contact métallique. La présence de « pointes » sur cet interface a vraisemblablement une influence sur le comportement du dispositif mais c'est seulement l'un des nombreux paramètres qui peuvent jouer un rôle dans l'étude des propriétés de contact.

The successful manufacture of gallium arsenide Gunn devices depends to a large extent on contact technology. Contacts are often formed by alloying, and a number of investigations into the electrical behaviour of the contacts as a functio~ of processing conditions have been made [1, 2]. The metallurgical structure of the contact was studied by R.H. Cox et al. [1, 3] who demonstrated the existence of metal spikes at the metal/GaAs interface. Gunn suggested that a thin layer of high resistivity material may exist beneath the contact or alternatively that the current is carried only by a large number of small patches [4]. Strain detected at the periphery of the contact by R.H. Cox and T.E. Hasty was attributed to quenched-in vacancies and dislocations [3], while M. Aven and R.K. Swank indicated that the creation of a large number of sharp spikes beneath the contact might improve contact properties [5]. The present paper which is part of an overall structural study, describes one aspect of the metallurgical structure of the GaAs/metal interface after alloying.

EXPERIMENTAL PROCEDURE

Alloyed contacts were made by evaporating metal films, namely silver, tin, silver-tin alloys and indium, through metal masks on the surface of a slice of GaAs polished and etched in 1 % by volume bromine in methanol prior to alloying the composite in an atmosphere of hydrogen. Because of the large amount of material required, melt-grown N-type GaAs doped with 10^{18} atoms·cm^{-3} of silicon was used for part of the structural investigation, but comparison checks showed that similar results were obtained in the epitaxial layers used for devices.
Prior to removal of the alloyed metal contact using the technique described by R. Tijburg and D. de Nobel [6] the surface topography was examined by optical microscopy and when necessary in a JEOL JSM-U3 scanning electron microscope (SEM). After removal, SEM examination of the interface was essential due to the sub-micron dimensions of the structure.

RESULTS

Topography of the metal/GaAs interface

The structure of Ag-Sn contacts illustrates the nature of the metal/GaAs interface. A film consisting of 500 Å of Sn and 5 000 Å of Ag evaporated in turn

Fig. 1a. *Optical micrograph of Ag-Sn contact alloyed at 600 °C.*

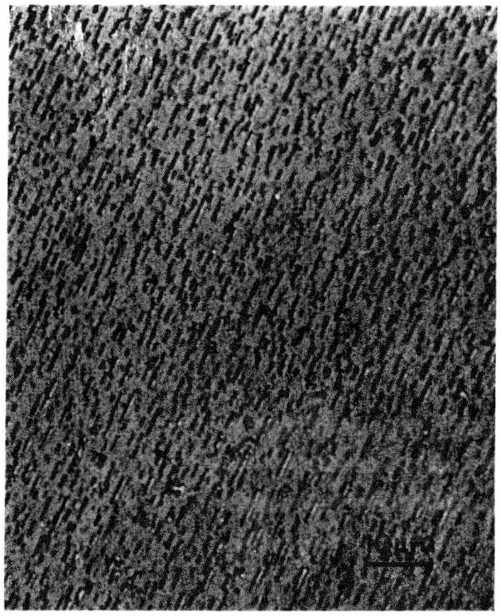

Fig. 1b. *Optical micrograph of GaAs surface after removal of contact metal.*

was alloyed at 600 °C for 10 minutes with heating and cooling rates fixed at 70 °C/minute. The surface of the alloyed metal film shown in figure 1a was continuous and fairly uniform indicating that the tin had melted and taken Ag and GaAs into solution, but the top Ag surface was largely unmelted. After removal of the metal film the topography of the GaAs at the interface was revealed as shown in figure 1b. Although a uniform overall metal film thickness had been apparently maintained during alloying the reaction at the interface was not uniform and a fine scale distribution of short lines was observed over the GaAs surface. On examination in the SEM these acicular regions were found to be long shallow holes varying from 0.1 μm up to 5.0 μm in length and from 0.05 μm up to 0.7 μm in width. A typical area is shown in figure 2 which demonstrates the increased detail revealed by the SEM compared with the optical photomicrograph of figure 1b. Since comparison of figures 1a-b showed no obvious correlation between the surface topography of the contact metal and the holes in the GaAs beneath it, removal of the contact metal or metallographic sectioning is essential to observe the nature of the reaction at the interface.

The exact shape of the holes was determined by tilting experiments in the SEM on some large holes exposed when the contact metal coalesced into balls during alloying as shown in figure 3. Each of the four faces comprising the sides of the hole intersected the (100) surface of the specimen along the [011] and [01$\bar{1}$] directions, and measurement of the angles between the facets and the (100) surface unambiguously assigned {111} crystallographic planes to each face. The elongation of the holes results from the different rates of dissolution from opposite pairs of {111} facets which correspond to planes ending in arsenic or gallium atoms respectively. The depth of the holes can be readily calculated from their width and is approximately 0.5 μm maximum for the holes shown in figure 2. The crystallographic nature of the holes suggests that the orientation of the GaAs will influence the interface topography.

Occasionally some limited information on the reaction at the interface could be gained before the contact metal was removed. If the metal film broke up during alloying, parts of the GaAs surface were exposed as shown by the alloyed Ag-Sn (4 500 Å Sn - 500 Å Ag) contact shown in figure 4. The large white areas are lumps of Ag-Sn alloy corresponding to agglomeration of the film, while the rectangular areas are holes filled with Ag-Sn left behind by the retreating film. Alternatively if the metal film was sufficiently thin its surface topography after alloying may give some indication of hole formation at the interface. Figure 5 shows the surface of a 100 Å thick Ag film alloyed to GaAs at 600 °C with acicular depressions, which on removal of the Ag were found to correspond to holes similar to those shown in figure 2.

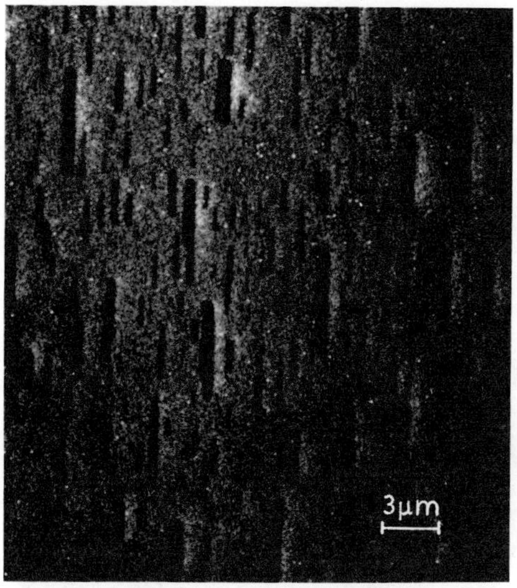

Fig. 2. *Scanning electron micrograph of specimen shown in 1b.*

Influence of processing on interface topography

If the alloying temperature was high enough such that large molten globules of metal were formed, large holes were dissolved beneath them, since, for a given equilibrium solubility of GaAs in the metal,

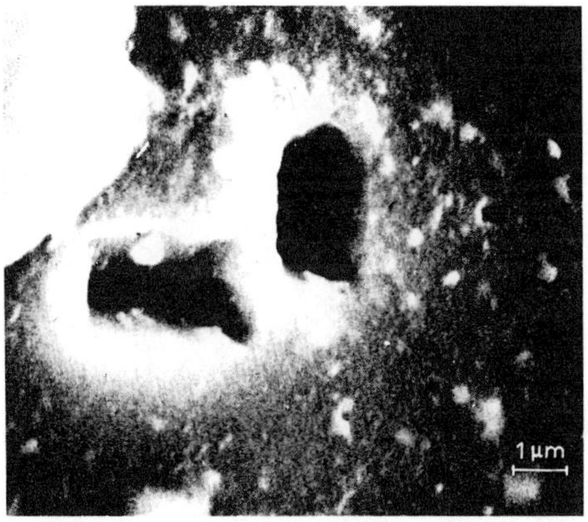

Fig. 3. *Scanning electron micrograph showing deep holes formed in GaAs surface by contact metal.*

the amount of GaAs dissolved increases with metal film thickness. An example of this general behaviour is shown in figure 6 which is a SEM photograph of a GaAs surface after removal of a 2 000 Å thick tin film alloyed at 600 °C. The largest holes correspond to areas previously covered by the largest tin globules.

Abrasion of the GaAs surface prior to evaporation produced scratch grooves which were found to act as anchor points for globules of tin whose size was of the same order as the scratch width and a pattern of holes corresponding to the scratch distribution was formed. Hence removal before contacting of the surface layers formed during the shutdown period of the chloride process for epitaxial growth of GaAs was best carried out by etching rather than by mechanical polishing [7].

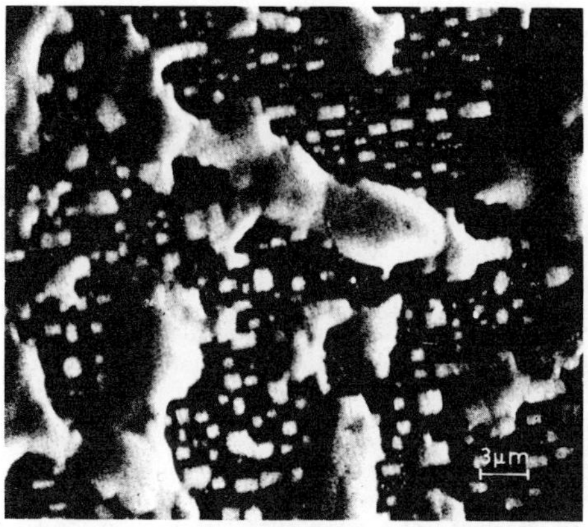

Fig. 4. *Scanning electron micrograph showing contact structure after globule formation.*

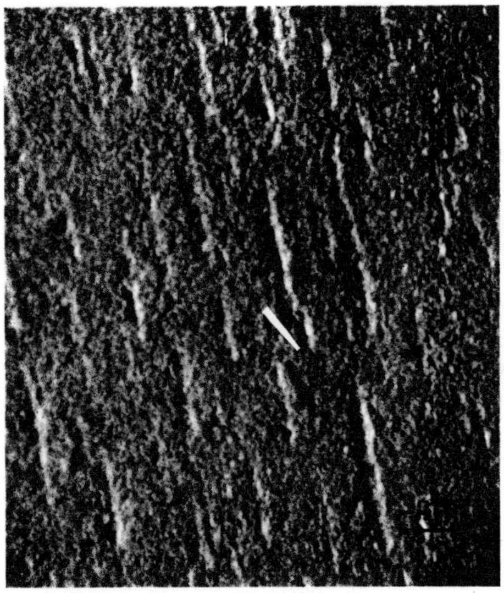

Fig. 5. *Scanning electron micrograph of silver contact showing depressions caused by hole formation.*

It was observed that if the metal contact was evaporated into circular holes defined by photoresist, alloying was not always uniform over the contact area, and the contact metal in the outer regions showed a greater tendency to form globules than in the centre of the area. However, when similar contacts were formed by evaporation through metal masks this effect was not observed and may be ascribed to photoresist contamination possibly causing wetting problems. This might introduce an element of irreproducibility into the contact properties and care should be taken to ensure complete removal of the photoresist.

Influence of the contact metal on hole formation

Holes were formed when alloy contacts were made with a number of metals and alloys such as Sn, Ag, Ag-Sn, Au-Sn and In, and in all cases dissolution of the GaAs took place by the formation of holes for the limited range of alloying conditions and metal film thicknesses used. There were some differences in the shape, size and distribution of the holes. For example Ag and Ag-rich Ag-Sn alloys show strongly facetted holes, whereas with Sn alone the facets are less pronounced and with longer alloying times large (up to 10 μm across) flat-bottomed holes were observed. Further, for a given alloy composition the size of the holes increased with the thickness of contact metal above the GaAs.

DISCUSSION

The most significant aspect of the results presented here is the appearance of a metal/GaAs interface which is not planar, but punctuated on the GaAs side by a series of small holes bounded largely by {111} crystallographic facets. Hole formation appears to be general to many metal and alloy systems, but at present we have no experimental evidence on the electrical effects of such an interface structure in Gunn devices, although it is expected that there will be some. The interface structure may for example be compared with the assumptions on contact structure made by other workers in the field already mentioned in the introduction to this paper [4, 5, 8]. It is also a close practical example of the spiky interface put forward by M. Aven and R.K. Swank as a means of improving contacts to wide band gap semiconductors [5]. It could give rise to an effective high field region beneath the contact in the manner described by J.B. Gunn [4], and finally the thermal strain distribution at the interface which will be modified by the spiked structure may have an electrical effect.

Hole formation produces a local variation in the thickness of the active device layer. Provided the holes are small this is probably not important, but if large holes are present, say greater than 1-2 μm deep, they could punch through a substantial part of the active layer. This would be particularly serious

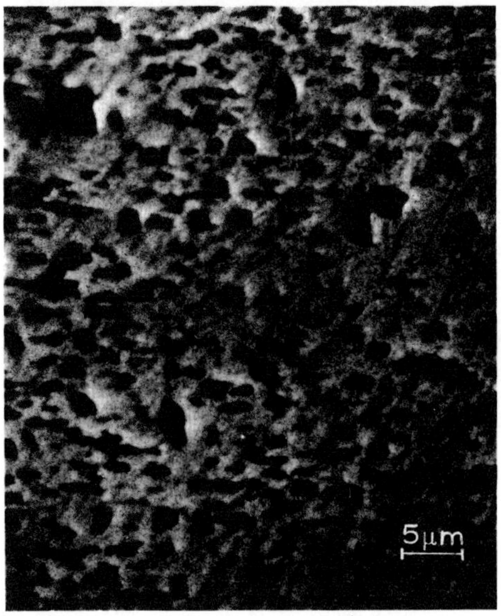

Fig. 6. *Scanning electron micrograph showing GaAs surface after removal of alloyed tin film.*

on material used for Q-band operation which is only about 3 μm thick. For Ag-Sn alloys where alloy globules have formed due to the use of thick Sn layers ($\sim$5 000 Å) holes deeper than 2 μm have been observed.

We have confined ourselves in this paper to some results obtained on the topography of the metal/GaAs interface, but of course other structural features have been observed such as the formation of recrystallised GaAs and other compounds, and the existence of other layers at the interface due to diffusion, strain, imperfections or the original GaAs surface. The electrical behaviour of the device due to the spiky nature of the interface cannot be considered in isolation from these and work is proceeding to determine a definitive metallurgical structure of the metal/GaAs interface.

CONCLUSIONS

1. In typical alloyed metallic contacts the metal/GaAs interface is not planar, but is interrupted on the GaAs side by holes which are often bounded by {111} facets.

2. The topography of the interface is dependent on the alloying cycle and the nature of the contact metal.

3. The spiked nature of the interface is likely to influence device behaviour.

4. The topography of the interface is only one of several structural features which might affect contact properties.

ACKNOWLEDGMENTS

The authors are grateful for much experimental assistance from Mr. R. Harbott and for discussions with Dr. W. Gurney, and Mr. R. Tijburg (Philips Research Laboratories, Eindhoven, Holland).

Manuscript received in June 1971.

Bibliography

1 COX (R.H.) and STRACK (H.A.). — **Ohmic contacts for GaAs devices.** Solid-State Electron., GB, vol. 10, 1967, pp. 1213-1218.

2 KNIGHT (S.) and PAOLA (C.). — **Ohmic contacts for gallium arsenide bulk effect devices.** In : **Ohmic contacts to semiconductors,** New York, B. Schwartz, Electrochemical Soc., pp. 102-114.

3 COX (R.H.) and HASTY (T.E.). — **Metallurgy of alloyed ohmic contacts for the Gunn oscillator.** In : **Ohmic contacts to semiconductors,** New York, B. Schwartz, Electrochemical Soc., pp. 88-101.

4 GUNN (J.B.). — **Properties of a free, steadily travelling electrical domain in GaAs.** IBM Journal, vol. 10, 1966, pp. 300-309.

5 AVEN (M.) and SWANK (R.K.). — **Ohmic contacts to wide-band-gap semiconductors.** In : **Ohmic contacts to semiconductors,** New York, B. Schwartz, Electrochemical Soc., pp. 69-81.

6 TIJBURG (R.P.) and DE NOBEL (D.). — **Improved method for the fabrication of GaAs microwave devices.** MOGA 70, 8th International Conference on Microwave and Optical Generation and Amplification (summaries), Amsterdam, 7-11th September 1970, pp. 91-93.

7 JOSH (M.J.) and BASTERFIELD (J.). To be published.

3.2.3 technology

Vacuum evaporation and sputtering are the most important techniques for deposition of the metallic contact layers; they are used in production by all of the commercial organisations involved with compound semiconductors. Chronologically, the first devices were constructed using alloy bead contacts; as these units are still commonly used in material assessment, a description of the technique is relevant.

Alloy-bead Contacts

Dice of adequate size are cut from the semiconductor wafer, cleaned, and placed on the hot stage of a silica alloying enclosure. (See Figure 3-7). The preform alloy bead is placed on the epitaxial surface, and the enclosure is sealed from the atmosphere. Forming gas, which can be bubbled through HCl, is passed continuously through the enclosure during the heating and cooling cycle. HCl is included in the heating cycle from room temperature up to about 300°C for GaAs (250°C for InP); for the remainder of the heating and cooling cycles, only forming gas is passed through the enclosure.

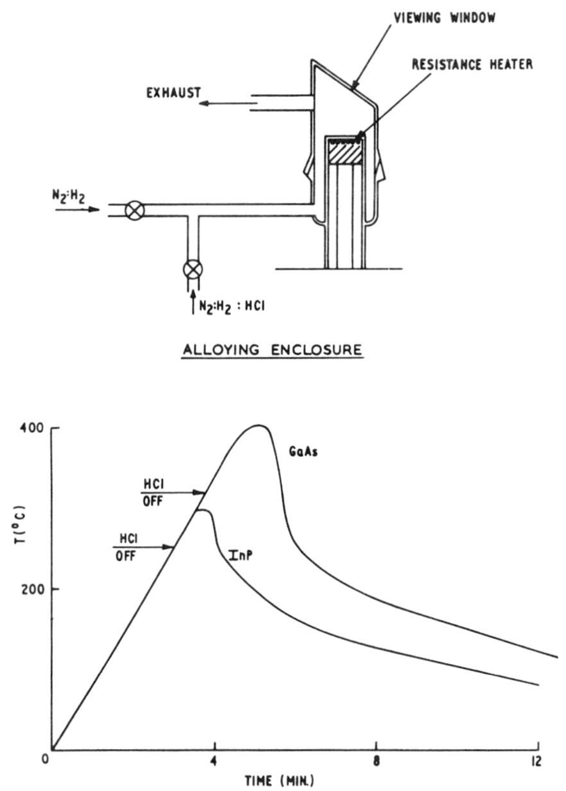

Figure 3-7/Alloying Cycles, Sn Bead

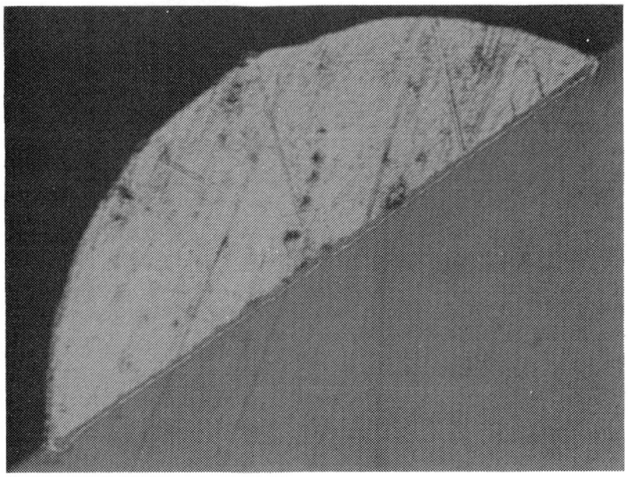

Figure 3-8/Section of Alloy Bead Contact Showing Stained Epi-Layer

The alloy materials, based on tin, indium, or lead, can be the pure element or doped with impurities. Cd, Zn, Sb, P, As, Al, Bi, Ga, Ge, Te, and Ag can all be incorporated into the basic alloy metal at concentrations in the 1-5% range, without affecting the alloying thermal cycle. After contacting the layer with the alloy bead the chip is soldered into a suitable mount, and a contact wire is melted into the bead, if necessary.

Variations of the alloyed bead process include the use of radiant heating (as opposed to the conductive heating described) and heater strip assemblies where the chip is placed directly on a resistance heated metal platform. The schedules are basically similar, although the thermal cycle varies from laboratory to laboratory. A section through an alloyed structure is shown in Figure 3-8.

Thin-film Contacts

An important aspect of thin-film contacting is the preparatory cleaning schedule. The simplest procedure uses chemical cleaning, followed by a short free etch immediately prior to insertion into the vacuum chamber. The chemical cleaning procedures have been described; relevant etches are $NH_4OH:H_2O_2:H_2O$ for GaAs, and $HCl:HNO_3:H_2O$ for InP (both in the proportions listed in the table of etches).

The etching time is approximately 1-20 secs, such that a few thousand angstroms can be removed from the surface. Occasionally, a quick immersion of the wafer into neat HF has been found to improve contact adhesion, although, as with most of the pretreatment procedures, the reasons for the improvement are unclear.

Ion bombardment cleaning has been studied with GaAs[61] and found to be beneficial providing that the ion energy is kept below 50 eV. Above this level, the effects of ion damage were evident in the characteristics of Schottky barrier tunnelling contacts. The quality of the contacts deposited on the low energy bombarded surface is claimed to have been equivalent to that of contacts on vacuum cleaved surfaces. It is not known whether this work was extended to ohmic contacts or to lower doped material.

Conventional oil diffusion pump evaporation systems working down to about 10^{-7} Torr are used for the metal evaporation. They are fitted with cold traps; the pressure during the actual evaporation process is approximately 10^{-6} Torr. A number of different configurations and procedures for deposition are used in the RRE laboratory; the substrate is maintained above the evaporation source and each of the alloy constituents is evaporated separately, with facilities for controlling substrate temperature. The evaporation can be executed by electron beam or straightforward heated filament. The disadvantage of the electron beam system is that the evaporation rate is difficult to control. Since the final quantity of each metal deposited on the wafer is critical, control is achieved more easily by complete evaporation of a known charge from a heated boat while maintaining the identical dimensional arrangement from run to run.

IMPATT devices, which require stable Schottky barriers to operate at high power densities, use materials such as platinum for the barrier metal. Sputtering systems are necessary for the deposition of these refractory metals. Despite the success of the sputtering technique for barrier contacts, it has not been applied frequently to the deposition of ohmic contacts. One might assume that sputter cleaning *in situ* would improve contact reproducibility; however, experiments in this field have been inconclusive. The simplicity of the filament evaporator remains very attractive.

The following points are relevant to thin metallic contact deposition for low resistance contacts.

1./The use of UHV systems does not appear to result in any significant improvement in contact behaviour.
2./Sophisticated cleaning using other than chemical means has not given conclusive results.
3./No significant differences seem to exist between filament and electron beam evaporation or sputter deposition.
4./The effects of heat cleaning *in vacuo* have not been fully investigated, although this process is standard for barrier contacts.
5./Variability in metal adhesion, structure, and appearance has not been explained.

The production of some devices requires that the contact layer cover the whole wafer surface. for others, some area definition is necessary. Metal evaporation masks clamped over the wafer surface are perfectly adequate for a relatively crude definition. The form of these masks is that of thin foil with the pattern photo-etched through it; hole diameters of about 6 mils are possible using this procedure. For precision arrays, photo-lithography is necessary; again, there are two techniques. The metal can be deposited all over the wafer, which is subsequently coated with photoresist and the pattern exposed. Excess metal is then removed by etching. Alternatively, the wafer can have the pattern exposed *before* the evaporation process. After removal from the coating unit, the resist pattern is dissolved away. The metal in contact with the wafer surface remains, while that in contact with the photoresist floats off.

The latter technique is often favoured, primarily because of difficulties in controlling the metal etchants required in the former procedure. The tendency is for the etch to attack quickly along the metal-semiconductor interface, necessitating that the etch be controlled to minimise the degree of undercutting. Yet it must be ensured that *all* of the unwanted metal is removed, leaving no metallic ions on the free surface. Control of the "float-off" process is necessary at the photoresist stage to produce a resist film 1.5-2 times thicker than the contact metal and of high edge definition. If the procedure is executed properly, the metal breaks away at the pattern edges during "float-off" and a high yield results.

Other metal deposition processes have been adapted for semiconductor contacts. One utilizes a vapour plating process; the metal is transported (as in epitaxial wafer growth) and converted onto the substrate surface. This process has been performed with tin onto GaAs[62] by the transport of tin from stannous chloride. After further processing, the results were good ohmic contacts on n-type material of $> 10^{15}$ cm^{-3} doping. This "stannising" method has been developed further in the UK[63], but has not been accepted as a routine process.

Chemical plating has also been employed, but with minimal success (judging from the dearth of reporting) in the literature. Certainly, experiments at the RRE on plating to otherwise untreated semiconductor surfaces have resulted in non uniform coverage. Presumably this fact is due to small differences in surface potential, resulting in preferential deposition. It is not probable that any production alternative to vacuum evaporation of metallic contacts will be developed. Any improvements on contacting technology most likely will result from better control in semiconductor surface preparation and improved vacuum techniques.

3.2.4 rectifying contacts

Schottky barriers are formed by the evaporation of elemental metals directly onto the semiconductor surface, using appropriate masking procedures. The wafer, heated to around 200°C under vacuum to outgas, is then cooled to 150°C and maintained at that temperature throughout the evaporation of the metal. Metals are cleaned and evaporated from a molybdenum source boat, again after outgassing when molten. Nickel can be evaporated from a tungsten spiral, and chromium from plated spirals; refractory materials must generally be deposited in a sputtering system.

Barrier height data for various metals on GaAs is well documented[64,65]. Gold or nickel are normally used to form junctions for C-V profiling; varactors and mixers may well require layered metal structures or partially alloyed complexes[66]. IMPATT's generally use platinum contacts sintered at 450°C.

The data on InP is sufficiently limited; Mead[65] gives a barrier height of 0.5 eV for Au and 0.55 eV for Ag on n-type vacuum-cleaved material. The corresponding figure for p-type material is 0.76 eV for Au. Smith[68] measured Au on epitaxial n-type InP, and found a barrier height of 0.4 ± 0.02 eV from I-V, C-V, and I_s-T measurements. In his calculations, the modified Richardson constant was found to lie in the range 1-2 A cm^{-2} °K^{-2}. RRE measurements on epitaxial layers confirm that the barrier height is around 0.4 eV, independent of the metal used.

While these InP diodes adhere to the thermionic emission theory at low bias levels, all reveal low reverse breakdown voltages and high reverse current due to tunnelling. Thus, the evaporated diodes are unsuitable for C-V profiling. To overcome this limitation the free surface of the material is treated so that mercury contacts produce the required characteristics[69]. An insulating layer is formed on the surface by immersion in H_2O_2: H_2O until the optimum characteristic is obtained. The resulting MIS structure must be of sufficient conduction to prevent inversion of the semiconductor surface, yet be sufficiently insulating to modify the barrier height properties to prevent exess tunnel currents. It has not been established whether the same process could be applied to evaporated metal contacts, although it seems improbable. The precise nature of the junction is not yet known.

To make a permanent junction on InP, the only satisfactory procedure that currently exists is epitaxy. Again, diffusion effects can prevent the formation of step junctions; however, the transient-growth method has been used to grow 2-3 μm thick, heavily doped p^+ layers from a zinc doped melt[70]. Although the growth time is only 10 seconds at 650°C, it appears this duration is sufficient to modify the profile of the origjnal n-layer. However, the current-voltage characteristics reveal good breakdown properties; with greater control of the doping in the p^+ region and the growth parameters, it should be possible to produce ideal step junction characteristics by this method. (This technique may well be the *only* one which forms good junctions on n-type material; this fact should be remembered when considering the fabrication of FET's or IMPATT's from n-InP).

In reference to Section 3.2.2 on contact metallurgy, it may be concluded that an evaporated and alloyed contact containing a specific p-type impurity would make a good rectifier on n-type material. Experiments with Ag-In-Zn and Ag-Mn onto both GaAs and InP produced non-ohmic characteristics, but very poor rectifiers. Presumably, this result was due to any heavily doped p-region being extremely narrow, allowing easy tunnelling of electrons.

3.2.5 ohmic contacts

Any discussion of ohmic contacts must start with the definition of "ohmic." We have already seen that an I-V characteristic which is linear at one temperature may not be so when the device is cooled. But for our purposes here, an ohmic contact is defined as one which has an I-V characteristic which is *linear through the origin at room temperature (300° K)*, and has low specific resistance. Departures from linearity at higher fields may occur, but it is the low field region which defines the contact property.

A great number of contacts have been developed for GaAs. Gold based alloys include Au-Sn, Au-In, Au-Ge, Au-Ge-Ni, and Au-Zn; silver based alloys include Ag-Sn, Ag-In-Ge, Ag-In-Zn, and Ag-Mn. The choice is therefore very wide, but what makes it more difficult is the lack of established criteria for selection. Each laboratory has developed its technology around one particular contact system; thus, in their experience, that system gives the best results. It is possible, though, to clarify the most obvious features:

1./Each alloy system seems capable of producing equally good contacts.
2./The thermal alloying cycle of the gold based alloys is lower than that required for silver.
3./The gold based alloys result in a more stable metal surface for wire bonding than does silver.
4./The silver based alloys appear to give more reproducible results to low doped material.

We have investigated Ag-In-Ge, Ag-Sn, Ag-Mn (for p-type), Au-Ge, and Au-Ge-Ni. There seems to be very little to choose electrically between the two n-type silver alloys, but we have favoured Ag-Sn because it is easier to evaporate. For the gold alloys, Au-Ge de-wetted from the surface unless the nickel overlay was included. With the nickel overlay, the physical appearance of the contact after heat treatment varies, both from wafer to wafer and across the surface of a single wafer. The p-type contact Ag-Mn[71] has been shown to give ohmic contacts on heavily doped p-type GaAs ($\sim 10^{18}$ cm^{-3}). Silver-tin has therefore become the standard n-type contact in the RRE laboratory.

The normal ratio of the two alloys is 10% tin by volume, (or 7% by weight) with deposited thicknesses of 600Å tin and 6000Å silver. After evaporation, the contacts are alloyed in a silica furnace under an atmosphere of hydrogen at 600°C for one minute; the temperature-time cycle is dependent on the thermal time constant of the resistance heated furnace. This silver-tin contact gives contact resistances as low as 3.10^{-5} Ω-cm^{-2} on n-type GaAs doped at the 3.10^{15} cm^{-3} level[72].

InP contacting technology has profited from the work on GaAs, and again many of the same alloy compositions have been investigated. Ag-Sn, in the same ratio as for GaAs, provides good ohmic contacts when alloyed at temperatures around 450°C. The gold alloys, like Au-In-Ge, give best results around 350°C. At higher temperatures the metals start to penetrate the semiconductor excessively.

The important parameters in determining the contact properties are:

1./Choice of alloy type
2./Composition of the alloy
3./Sintering temperature (peak)
4./Sintering temperature cycle

The effect each of these parameters on GaAs has not been fully studied, because each system has been empirically optimised for minimum contact resistance. Only through the discovery of the improved performance achieved from InP TED's having 'anomalous' contacts[55] did the full range of contact phenomena attract attention. To investigate these phenomena with InP, only the particular alloy system (Ag-Sn) was fixed and the composition was changed while maintaining constant sintering conditions. Both experiments yielded similar results. Figure 3-9 shows the variation of microwave power and efficiency of TED's, as a function of the alloy composition, from pure silver to 30% tin in silver. The contact behaviour is quite different at the two extremes, although nominally "ohmic." Decreasing the sintering temperature further results in a "rectifying" characteristic; eventually it behaves as a Schottky barrier. Thus, the tunnelling current is controlled by the contact technology. In fact, the processes are so critical that the control of the technology becomes quite difficult. It is, however, possible to produce any desired contact property through control of the technology.

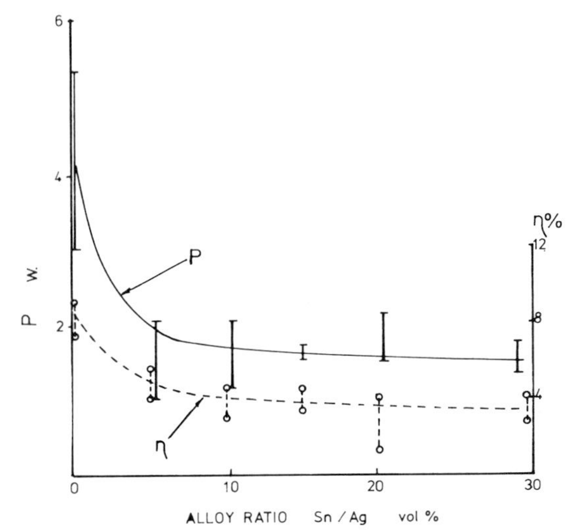

Figure 3-9/Microwave Power & Efficiency of InP TED as a Function of Contact Alloy Ratio

Ohmic contacts to heavily doped material (> 10^{17}) are less problematic, so that there are a wide variety of techniques for contacting to substrates. One method is to evaporate 1 μm of Au/Ge eutectic onto substrates held at 200°C during deposition. This method is compatible with the use of Au/Ge solder preforms for bonding chips into packages.

3.2.6 contact characterisation

Rectifying Contacts

Barrier contacts are characterised by their current-voltage, capacitance-voltage, and photoemission properties. We have already seen that:

$I = I_s\,(\exp(qV/nKT) - 1)$

so that a plot of log I versus V is linear, with a slope of q/nKT and an intercept on the current axis of I_s. It is therefore possible to obtain the value of the non-ideality factor n (which is temperature dependent) and A*ST exp $(-\phi_b/n\,KT)$, which yields an estimate of the barrier height (ϕ_b). To obtain ϕ_b more accurately requires an activation energy plot of log I/T^2 versus $1/n(T)kT$; forward bias can perform this task. Providing that $qV > 3\,kT$, the slope is $-(\phi_b - qV)$. In reverse bias, providing $-qV >$ 3 kT, the slope yields ϕ_b directly.

The current-voltage measurement is relatively straightforward; however, currents can be as low as 10^{-10} Amps, and care must be taken with the input impedance of measuring equipment. When measuring barrier height by the activation energy method, the value of n(T) at each measurement temperature is determined from the forward characteristic.

The capacitance-voltage determination of barrier height is more direct:

since $C \propto (1/Vd - V)^{1/2}$

where Vd is known as the diffusion voltage (see Figure 3-1(c)).

A plot of C^{-2} versus V is extrapolated back to the voltage axis, intercepting at Vd. The barrier height now can be deduced from:

$\phi_b = Vd + (E_C - E_F)$.

A powerful method for measurement of barrier height uses photoemission. The sample is illuminated, by either a transparent contact or transmission through the semiconductor to the barrier interface. Two excitation processes can occur — photoemission of excited electrons in the metal over the barrier or band to band excitation within the space charge region. By using transmission of the illumination through the semiconductor (when $h\nu > E_g$), the light is strongly absorbed. The photocurrent per absorbed photon (R) as a function of the photon energy (h) is given by the Fowler theory[73]. For $(h\nu - \phi_b) >$ 3 kT, the theory reduces to the form:

$\sqrt{R} \sim (h\nu - \phi_b)$

When the square root of the photoresponse is plotted as a function of photon energy, a straight line should be obtained with an extrapolated value to the energy axis, thereby giving a direct measure of barrier height. Figure 3-10 shows the photoresponse of a W-GaAs diode.

A positive feature of photoemission measurements is that they can determine either the barrier lowering with reverse bias voltage or the variation of barrier height with temperature independently.

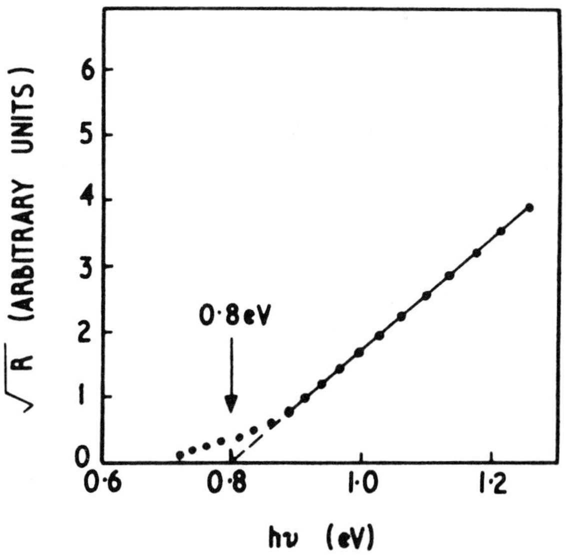

Figure 3-10/Photoresponse of W-GaAs Barrier

"Ohmic" Contacts

The most identifiable parameter of an ohmic contact is its contact resistance, although this measurement is often not easily made. Let us assume the metal contact of known diameter d is made to an n-type epitaxial layer of thickness ℓ and resistivity ρ grown on a heavily conducting substrate. The total resistance through this structure is given by:

$$R_T = \frac{\rho}{d\pi} \arctan \frac{4}{d/\ell} + \frac{4R_c}{\pi d^2} + R_o$$

where R_c is the specific contact resistance in $\Omega\text{-cm}^2$ and R_o is the resistance of substrate.

In the limit $d \gg \ell$:

$$R_T = \frac{\rho\ell}{A} + \frac{R_c}{A} + R_o$$

$\rho\ell/A$ is the resistance of the volume of epitaxial layer; it can be determined providing ρ is known.

Values of R_c around 10^{-4} Ω cm^2 can be achieved on GaAs and InP doped to the 10^{15} level, resulting in a 1Ω series resistance for a device 100 μm square.

The anomalous type of contact has what can be a much higher resistance; but to comprehend the nature of the contact, we need to know more than its resistance. Figure 3-11 shows the I-V of such a contact after cooling to 77°K. In (a) are shown the characteristics of a Schottky barrier with a very small reverse current; in (b) there is considerable reverse current which increases with applied voltage, indicative of tunnelling. Further confirmation of tunnelling is provided by looking at the second differential d^2I/dV^2 at low temperature, which shows additional periodic fluctuations due to phonon effects[74]. In each case there is a barrier involved; however, the current transport is dominated by thermionic emission in (a) and by field emission in (b). A region between where thermally assisted tunnelling takes place also exists. These results are discussed in the key paper by Colliver et al.

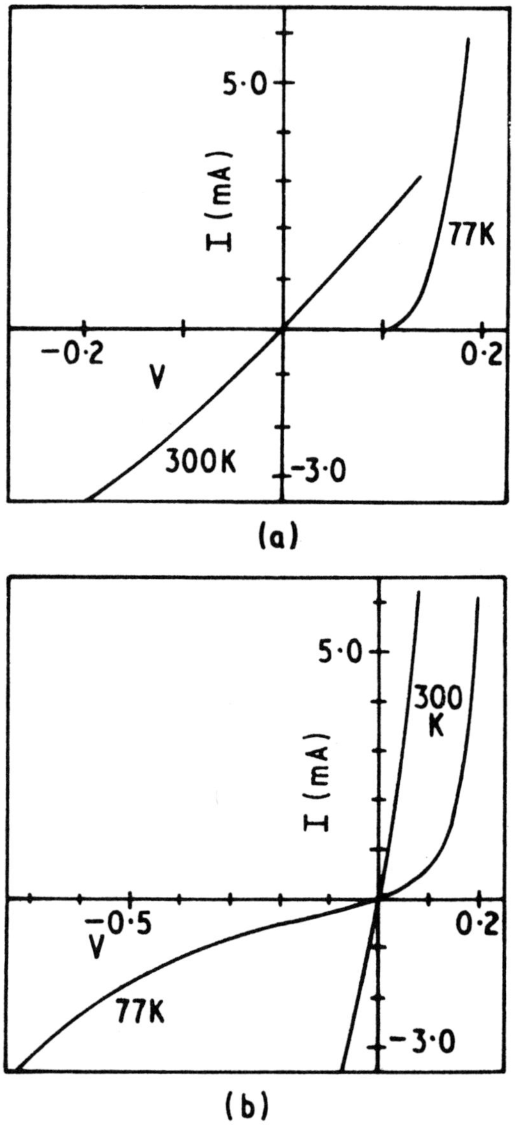

Figure 3-11/Current Voltage Characteristics of InP Devices at 300°K and 77°K

There has been little investigation of these contacts (which appear ohmic at room temperature) other than to identify the current voltage behaviour. Problems occur in the production of such contact reproducibility; the question is raised of how to specify the contact more precisely. One possible parameter is the barrier height. With InP devices, the effective barrier height of the Schottky barrier (estimated from the temperature dependence of the current at fixed reverse bias) lies between 0.15 and 0.2 eV[75], for various devices with an I-V similar to Figure 3-11(a). Comparable measurements have not been made on the devices as in Figure 3-11(b).

Both I-V and C-V methods of measuring the barrier height are difficult with these devices; it is not clear whether one can apply Schottky theory to the I-V characteristic when the current is so much higher than theory would predict, and the reverse characteristic is generally not good enough to measure meaningfully capacitance. Photoemission has not yet been tried, but it could prove the best way of measuring these very low barrier energies.

key paper

Cathode Contact Effects In InP Transferred-Electron Oscillators

D.J. Colliver, K.W. Gray, D. Jones, H.D. Rees, G. Gibbons, and P.M. White
(Royal Radar Establishment, Great Malvern, Worcs., UK)

Institute of Physics and Physical Society Symposium on GaAs (1972)
Pages 286-293

Abstract. Experimental and theoretical evidence is presented indicating that two types of current-limiting contacts are responsible for the recently observed high DC-microwave conversion efficiency in InP transferred electron oscillators.

1. Introduction

We have previously observed high DC to microwave conversion efficiency in short high frequency InP devices (Colliver *et al* 1972, White and Gibbons 1972). In this paper we wish to present both experimental and theoretical evidence showing that a current-limiting cathode contact is responsible for this behaviour.

Many studies have been made on the effect of cathode boundary conditions on the space-charge dynamics of microwave devices (Boer and Dohler 1969, Conwell 1970, Grubin *et al* 1972, Kroemer 1968, McCumber and Chynoweth 1966, Shaw *et al* 1969, Yu *et al* 1971). However no work has been published on the effect of boundary conditions on high frequency microwave device characteristics, notably efficiency.

The current–voltage dependences of the cathode contacts of high efficiency devices have been extracted and are reported in §2, followed by a description of device microwave properties in §3. The characteristics of short length devices with barrier contacts have been computed using a new simulation method and are described in §4.

2. Device construction and low voltage characteristics

The present experimental investigations pertain to devices fabricated from vapour epitaxially grown n-InP layers with ⟨100⟩ orientation (Hales *et al* 1971, Joyce and Williams 1971). The substrates are high conductivity tin-doped InP and layers have been grown with and without sulphur-doped buffer layers between the substrate and the n layer. Active layer free carrier concentrations range from 7×10^{14} to $2{\cdot}6 \times 10^{15}$ cm^{-3} with room temperature mobilities around 4000 cm^2 V^{-1} s^{-1}; both estimates pertain to layers grown on adjacent semi-insulating InP substrates but resistivity measurements on devices yield compatible values. Three basic types of top (cathode) contact have been used, alloyed bead, evaporated and alloyed film and evaporated and alloyed film with a subsequent electroplated thick gold layer (Gibbons and White 1971). Optical and scanning electron micrographs indicate uniform and reproducible alloy fronts with alloy spikes penetrating a few hundred nanometres from the metal into the semiconductor at very high density.

Measurements of the device *I–V* characteristics at low voltages over a temperature range from 77 to 450 K give linear characteristics only occasionally. The dominant forms of characteristic are indicated by figures 1(*a*) and 1(*b*) and we refer to these two forms as type 1 and type 2 respectively. The type 1 characteristic is strongly temperature dependent, becoming more asymmetrical with decreasing temperature. We attribute this behaviour to a Schottky barrier metal–semiconductor

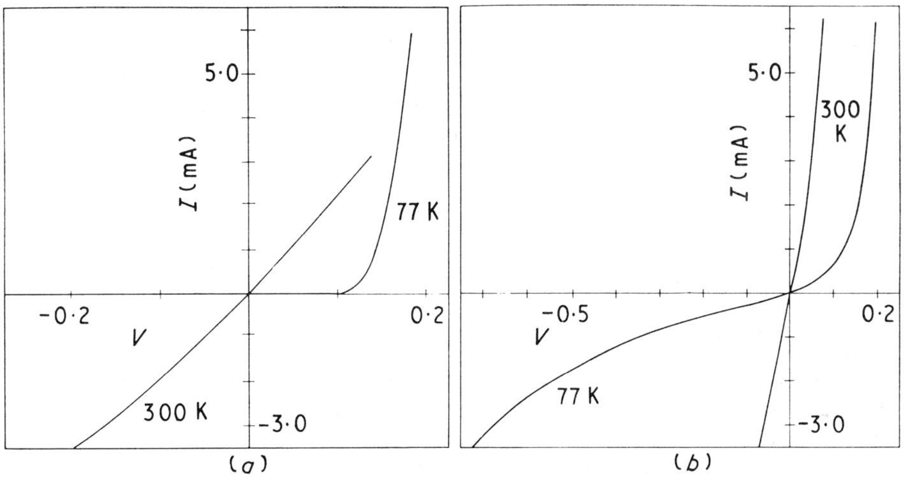

Figure 1. Low voltage *I–V* characteristics at 77 and 300 K for two devices on two different layers. An example of a type 1 device is shown in (*a*) and a type 2 device in (*b*).

contact (in which the current is always controlled by thermionic emission) in series with the n-layer resistance. The effective barrier height of the Schottky barrier estimated from the temperature dependence of the current at fixed reverse bias (Padovani 1971, Yu *et al* 1971) varies at low voltage from 0·15 to 0·25 eV for the various devices with the type 1 characteristic.† The type 2 characteristic is similar for forward bias and low reverse bias, but at larger reverse voltages the current is considerably greater than for the type 1 curve and shows no more than a weak temperature dependence. We attribute this behaviour to a barrier contact with a narrower depletion region in which tunnelling from the metal to the semiconductor conduction band (with little or no thermal activation) is important.

A steady progression of the *I–V* characteristic from type 1 through type 2 to linear can be readily understood in terms of an increasing donor doping level immediately under the metal. For relatively light doping a wide depletion region

† The barrier height is voltage dependent, decreasing from its low voltage value by 20% at bias values of 1·5 V or larger.

adjacent to the barrier leads to a characteristic dominated by thermionic emission. For greater doping levels the narrower depletion region permits appreciable tunnelling under reverse bias and for very high donor concentrations any barrier is too narrow to limit current and the contact is effectively an ohmic n^+ contact. The expected correlation of I–V characteristic with contact doping is supported by the observed dependence of barrier properties on alloying temperature and the composition of the contact metal. However, a more surprising observation is the significant increase in the asymmetry of the characteristic which invariably follows the electroplating of an evaporated and alloyed contact. It has been occasionally observed that the InP cleaves during the plating process and the introduction of strain may be a possible cause of the changes, but it is difficult to give a straightforward explanation of the detailed physical mechanism. Large dislocation densities near the metal contact are thought to be possibly responsible.

In normal operation as a microwave source the barrier contact is reverse biased. The mode of operation of the devices will therefore be influenced by the contact type since continuous progression from a low field injection (ohmic) condition to high field (barrier) is possible. In addition the temperature dependence of the high field cathode device characteristics may be expected to vary from a very high sensitivity (thermionic injection) to a negligible sensitivity (tunnel injection).

3. Microwave characteristics

Tests of the properties of the devices as microwave sources were usually carried out with the devices encapsulated in S4 packages. The packaged devices were mounted adjacent to the broad wall of a rectangular waveguide and contact to the top was made by a variable diameter post. Tuning and matching were effected by a sliding short and by an E–H tuner between the source and detector. The standard drive waveform comprised 300 ns pulses with a duty cycle of 10^{-3}.

There is a pronounced difference in the characteristics of devices with ohmic and injection-limited cathodes as indicated by the curves of current, power and efficiency against voltage for the two cases illustrated in figures 2(*a*) and 2(*b*). Both sets of curves pertain to devices made from the same 10 μm layer and oscillation frequencies are similar, 15–22 GHz. In addition to the differences summarized by figures 2(*a*) and 2(*b*), the injection-limited devices were distinctive in showing no tendency to generate low frequency bias circuit oscillations and in having significantly higher breakdown voltages, typically 60 V compared with 30 V for ohmic devices.

Differences between type 1 and type 2 devices can be attributed entirely to their different temperature sensitivities for electron injection from the cathode. No characteristic differences are apparent if type 1 and type 2 devices are compared at similar current levels. Figure 3 shows the temperature dependence of the relative oscillator efficiency for two type 1 devices (with estimated barrier height at low voltage of 0·15 and 0·25 eV) and a type 2 device. Ohmic contact devices show a similar fall-off with increasing temperature and therefore this fall-off is associated with the bulk InP. The higher barrier type 1 device has a slightly lower peak efficiency which reflects this bulk property. From a practical device viewpoint the high efficiency and relatively weak temperature sensitivity for the type 2 device are of obvious significance.

An investigation of a large number of type 2 devices, for which the typical operating current densities varied from 1×10^3 A cm^{-2} downwards, and of several type 1 devices over their operating temperature ranges, showed device behaviour

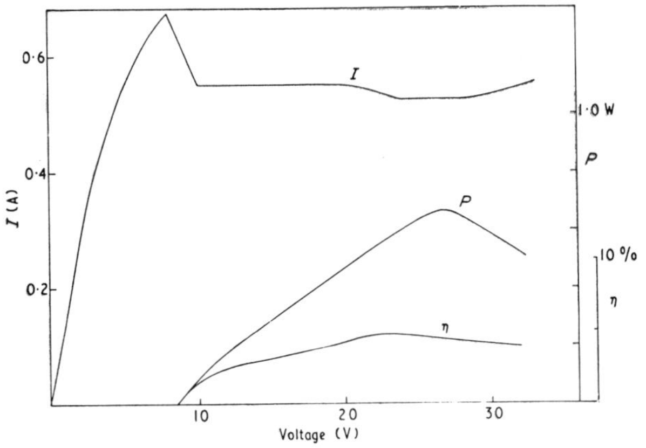

(a)

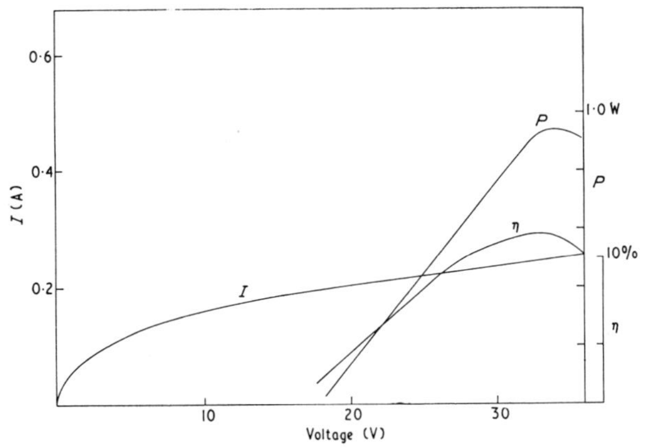

(b)

Figure 2. The dependence of current, microwave power and conversion efficiency on voltage for (a) an ohmic contact device and (b) a device of similar area on the same layer with a type 2 cathode contact.

varying continuously from that of ohmic contact devices to virtually blocking diodes. For type 1 devices the maximum oscillator efficiency was obtained for operating current densities about 25% of that for high temperature reverse operation or with the diode forward biased (see figure 1 of White and Gibbons 1972). In the case of type 2 devices on the same layer and nominally the same area the maximum conversion efficiency occurred when the cathode contact reduced the ohmic contact saturation current by around 15%. This difference is not fully understood.

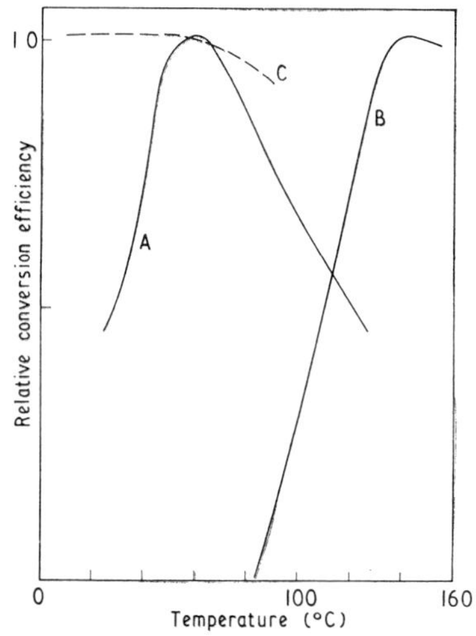

Figure 3. Dependence of the relative conversion efficiency on temperature for various devices. The full curves (A) and (B) are for type 1 devices with low voltage barriers of 0·15 and 0·20 eV respectively. The dotted curve (C) is for a type 2 device which is similar to an ohmic contact device.

4. Interpretation of device microwave characteristics

The experimentally determined properties of the devices indicate that considerably increased oscillator efficiency can be obtained with appropriate cathode characteristics. The present theoretical analysis is restricted to considerations of the means by which the overall oscillator characteristics are influenced by changes in the electron injection mechanism at the cathode. Interpretation of the experimental results in terms of a theoretical model proposed by Yu *et al* (1971) has many attractive features, but direct experimental evidence for this model pertained to bulk GaAs devices with oscillation frequencies near 1 GHz corresponding to effective transit velocities of about $0{\cdot}8 \times 10^7$ cm s^{-1}. Among the present InP results, high efficiencies have been observed only for frequencies above 12 GHz and effective transit velocities are about $1{\cdot}5 \times 10^7$ cm s^{-1}. In this regime electron relaxation effects are certainly important and any analysis omitting them, such as that of Yu *et al*, could be misleading.

We recently described a number of electron relaxation effects revealed in a series of device simulations which took proper account of the electron relaxation processes within the device (Jones and Rees 1972). These simulations have been extended to incorporate the effects of high field cathode boundary conditions, which in many ways represent the physical properties of the experimental cathode contacts. For the present purposes of analysing the basic oscillator modes a model of electron transfer appropriate to GaAs has been used, but the main device characteristics are expected to be valid for most materials exhibiting electron transfer effects.

An oscillation mode intermediate between an accumulation transit mode and a normal domain mode and exhibiting key features of the experimental results has been observed in the simulations. Its relationship to a pure accumulation mode is illustrated by figures 4(*a*), (*b*) and (*c*) which show the field profiles during an oscillation cycle for 4 μm GaAs devices in series with parallel resonant circuits. In each case a donor doping concentration of 2×10^{15} cm^{-3} and an n$^+$ graded anode contact are simulated.

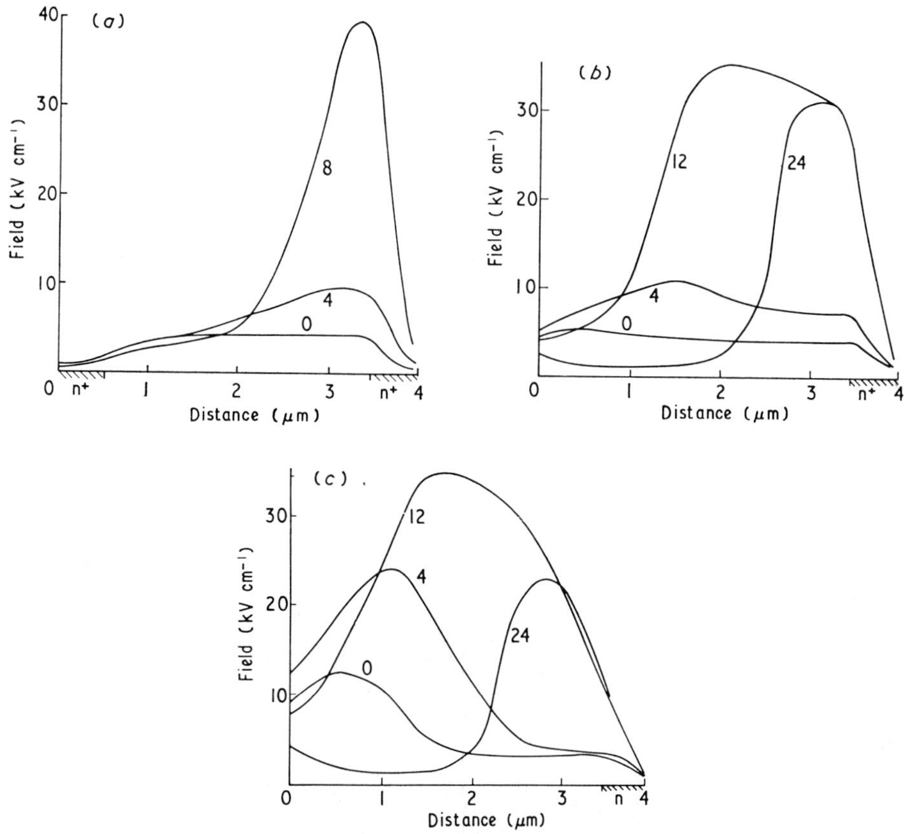

Figure 4. Computed electric field profile during an oscillation cycle for a 4 μm GaAs device. The relative time between profiles is given in ps. The behaviour for three cathode field conditions are for (*a*) an ohmic contact showing an accumulation transit mode, (*b*) a simulated 0·18 eV Schottky barrier (thermionic emission) showing a modified accumulation transit mode and (*c*) a simulated 0·20 eV barrier showing a short device domain mode.

Figure 4(*a*) pertains to a simulation where the cathode is also an n$^+$ graded contact. Nucleation of the accumulation instability occurs in a zone about 2 μm from the cathode contact; the displacement reflects the fact that low energy electrons injected from the n$^+$ contact need to gain from the field an energy equal to the intervalley separation before transfer is possible. The early stages of growth of the accumulation layer are illustrated by the progression from the 0 to the 4 ps curves. The subsequent growth of the layer is rapid, leading to an anode field after 8 ps of 40 kV cm^{-1}. Later, the falling voltage imposed by the circuit response is accompanied by movement of the accumulation layer into the anode contact and the device returns to its initial state. The overall efficiency, about 2% at 36 GHz, is low, largely because the displaced nucleation point of the accumulation layer confines the high field region to a narrow zone near the anode and the voltage excursion of the device is consequently low.

This drawback can be reduced by a low barrier contact which raises the cathode field to a value comparable with the threshold field. This situation is illustrated in figure 4(*b*). A weak depletion forms initially at the cathode, but its growth is inhibited by the rapid rise of device voltage (Jones and Rees 1972) and it is of little significance. In effect the accumulation layer is nucleated close to the cathode. The oscillation mode remains an accumulation transit mode but the efficiency is

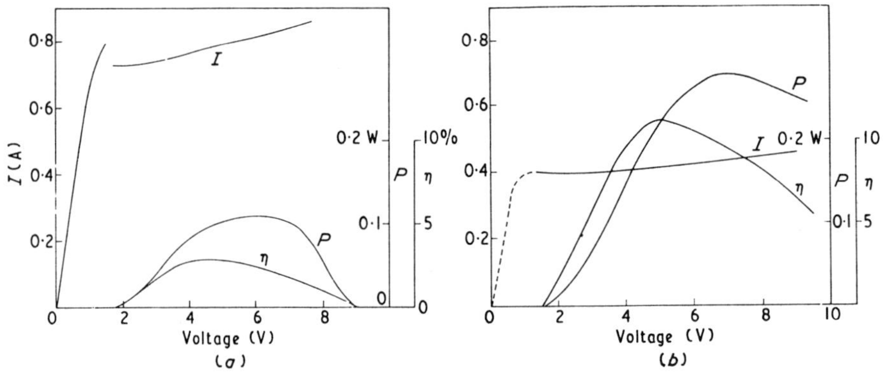

Figure 5. Computed characteristics of current, microwave power and conversion efficiency against voltage for (*a*) an ohmic contact device and (*b*) a 0·20 eV barrier device.

improved, in this case to 8·5% at 25 GHz. A barrier height for thermionic injection of 0·18 eV would correspond to the cathode conditions appropriate to this simulation.

The result of further raising the effective barrier height by just 10% to 0·20 eV is shown by figure 4(*c*). With a higher cathode field the nucleated instability is essentially a dipole domain, the anode field remaining below threshold. However, after 12 ps the domain grows to fill the entire n region of the device leading to a similar field configuration to that attained in the previous case. The subsequent transit of the accumulation region from cathode to anode is also similar. This mode is referred to as a short device domain mode and in this example an efficiency of 11% at 25 GHz is obtained. It can be noted that the peak anode fields in figures 4(*b*) and 4(*c*) are less than figure 4(*a*) which probably accounts for the increased breakdown voltages of barrier cathode devices.

The DC current falls as the cathode barrier is increased; its value is 0·75 A for the n^{+} cathode and 0·50 and 0·41 A for the configurations of figures 5(*b*) and 5(*c*) With a further increase of barrier height the DC current continues to fall and the efficiency degrades from the 11% maximum. The AC voltage and current waveforms are almost sinusoidal and are insensitive to the precise barrier height. From a practical standpoint there is, therefore, little distinction between the accumulation transit and the short device domain modes.

The main characteristics of the short device domain mode are only weakly dependent on equilibrium carrier concentration; only small changes are observed as the donor concentration is varied from 1 to 6×10^{15} cm^{-3}. The essential requirement is that the high field region should expand to fill the whole device, or at least most of it, at the voltage peak. Only a small reduction in domain size can be induced by increasing n since the fractional depletion is simultaneously reduced. However, for greater device lengths, over about 10 μm for GaAs, degeneration to a lower efficiency domain transit mode readily occurs unless more critical conditions on the nl product and barrier height are satisfied. Consequently the high efficiency results can be associated primarily with short devices and high frequencies, but the precise limits on l and f will naturally be somewhat dependent on material. Using a model of electron transfer thought to be appropriate to InP (incorporating weaker intervalley scattering), similar effects have been seen in simulations of devices with lengths up to 10 μm. In this case the preferred oscillation frequencies are centred on 15 GHz, which is in broad agreement with experiment.

Figures 5(*a*) and 5(*b*) summarize the computed characteristics of current, microwave power and microwave efficiency against voltage for the n^+ and barrier cathode devices appropriate to figures 4(*a*) and 4(*c*). The structure of these curves, particularly the reduced current and increased efficiency of the barrier cathode compared with the n^+ cathode device, agrees reasonably with that for the experimental InP results shown in figure 3.

5. Summary

The experimental results reported in this paper indicate that high microwave oscillator efficiencies can be achieved with short devices if fairly precise conditions on the cathode injection characteristics are satisfied. In practice various types of Schottky barrier contact have been used. The experimental observations are interpreted with reference to computer simulations which show that increased efficiency can be obtained with cathode fields of the order of the threshold field. The mode of operation, characterized by a field configuration in which the high field region substantially fills the device at the peak voltage excursion, is viewed as intermediate between an accumulation transit and a domain mode and is descriptively termed a short device domain mode. Broad agreement is obtained with the experimental I–V characteristics, effective transit velocities and increased breakdown voltages as well as with the increased oscillator efficiency. The experimental results pertain to device lengths less than 12 μm and frequencies above 12 GHz, which is consistent with the inference from simulations that less critical conditions for high efficiency prevail in the case of short devices. However, it is just this frequency range for which electron relaxation effects restrict the LSA relaxation mode, making the two modes to some extent complementary.

Alternative ways of generating a relatively temperature-independent high field cathode, such as a cathode-doping notch are under developmental study.

Acknowledgements

The authors wish to acknowledge the experimental assistance of L Irving, A McConachie and B Nightingale and the continuing support throughout this work of Dr C Hilsum.

This paper is published with the permission of the Director of the Royal Radar Establishment.

References

Boer K W and Dohler G 1969 *Phys. Rev.* **186** 793
Colliver D J, Gray K W and Joyce B D 1972 *Electron. Lett.* **8** 11
Conwell E M 1970 *IEEE Trans. Electron. Devices* **ED-17** 262
Gibbons G and White P M 1971 *Proc.* 1971 *European Microwave Conf., Stockholm* (Stockholm: Royal Swedish Academy of Engineering Science) vol 1 paper A2/4
Grubin H L, Shaw M P and Solomon P R 1972 *IEEE Trans. Electron. Devices* to be published
Hales M C, Knight J R and Wilkins C W 1971 *Proc. 3rd Int. Symp. on GaAs: Inst. Phys. Conf. Ser. No 9* p50
Jones D and Rees H D 1972 *Electron. Lett.* **8** 363
Joyce B D and Williams E W 1971 *Proc. 3rd Int. Symp. on GaAs: Inst. Phys. Conf. Ser. No 9* p57
Kroemer H 1968 *IEEE Trans. Electron. Devices* **ED-15** 819
McCumber D E and Chynoweth A G 1966 *IEEE Trans. Electron. Devices,* **ED-13** 4
Padovani F A 1971 *Semiconductors and Semimetals* ed R K Willardson and A C Beer (New York: Academic Press) vol. 7A p75
Shaw M P, Solomon P R and Grubin H L 1969 *IBM J. Res. Develop.* **13** 587
White P M and Gibbons G 1972 *Electron. Lett.* **8** 166
Yu S P, Tantraporn W and Young J D 1971 *IEEE Trans. Electron. Devices* **ED-18** 88

3.2.7 summary

Metal to semiconductor contacts are only understood in a general phenomenological way, except for isolated exceptions. For the majority of devices this view suffices, providing the contact does not radically affect or control the ultimate performance or degradation of the particular device. But where the contact is made directly to the active region of the device (particularly in TED's; to a lesser extend in IMPATT's), it has been shown with both GaAs and InP that the performance can be dominated by the contact properties. If these characteristics could be understood and controlled, improved performance should result. For controlled barriers, it is possible that heterojunction structures are more suitable than metal-semiconductor techniques.

Of significant importance is the long term behaviour of the contact. The majority of the microwave devices operate under high electric fields at elevated temperatures — perfect conditions for metallurgical problems. Unfortunately, the reliability of any structure takes a long time to establish; therefore, new contact structures require long development times.

The characterisation of contact properties usually is executed at low fields, with empirical extrapolation to high fields. There is much to be learnt about the metal-semiconductor interactions under operating conditions.

chapter four

the heat sink

Efficient heat removal is problematic with any semiconductor device, particularly with those designed for generation or amplification in the microwave region. In a planar structure, the major thermal impedance is the semiconductor substrate. Since it is impractical technologically to invert the device on the heat sink, the only way to reduce the thermal impedance is to thin the substrate to its absolute minimum consistent with chip handling and bonding. For two-terminal devices, where the field is applied normally to the plane of the device, considerable gains are achieved by inverting the structure to get the active region of the device as close as possible to the heat sink. The power density in TED's and IMPATT's can be as high as 10^8W cm^{-3}. To remove this heat the thermal impedance from the active region must be reduced to a minimum.

4.1 theory

The structure of a two-terminal device and the associated thermal resistances are illustrated in Figure 4-1. Let us assume that heat is generated uniformly in the active region of the device — a valid assumption for thin active layers.

The Contact Layer, R_1

R_1 is the heavily doped p^+ or n^+ layer, epitaxially grown, which forms one contact to the active region. Thermally, it reacts as a simple series impedance, such that:

$$R_1 = \frac{A}{K\ell} \ {}^\circ CW^{-1}$$

where A is the area (cm^2), ℓ is the layer thickness (cm), and K is the thermal conductivity (W $^\circ K^{-1}$ cm^{-1}).

It should be noted that the value of K for heavily doped material is some 20% less than for the high purity active region. To reduce R_1, we need to minimise or remove the contact layer altogether.

The Metal Contact, R_2 and R_6

For an "ideal" contact, this region makes only a small contribution to the thermal impedance,

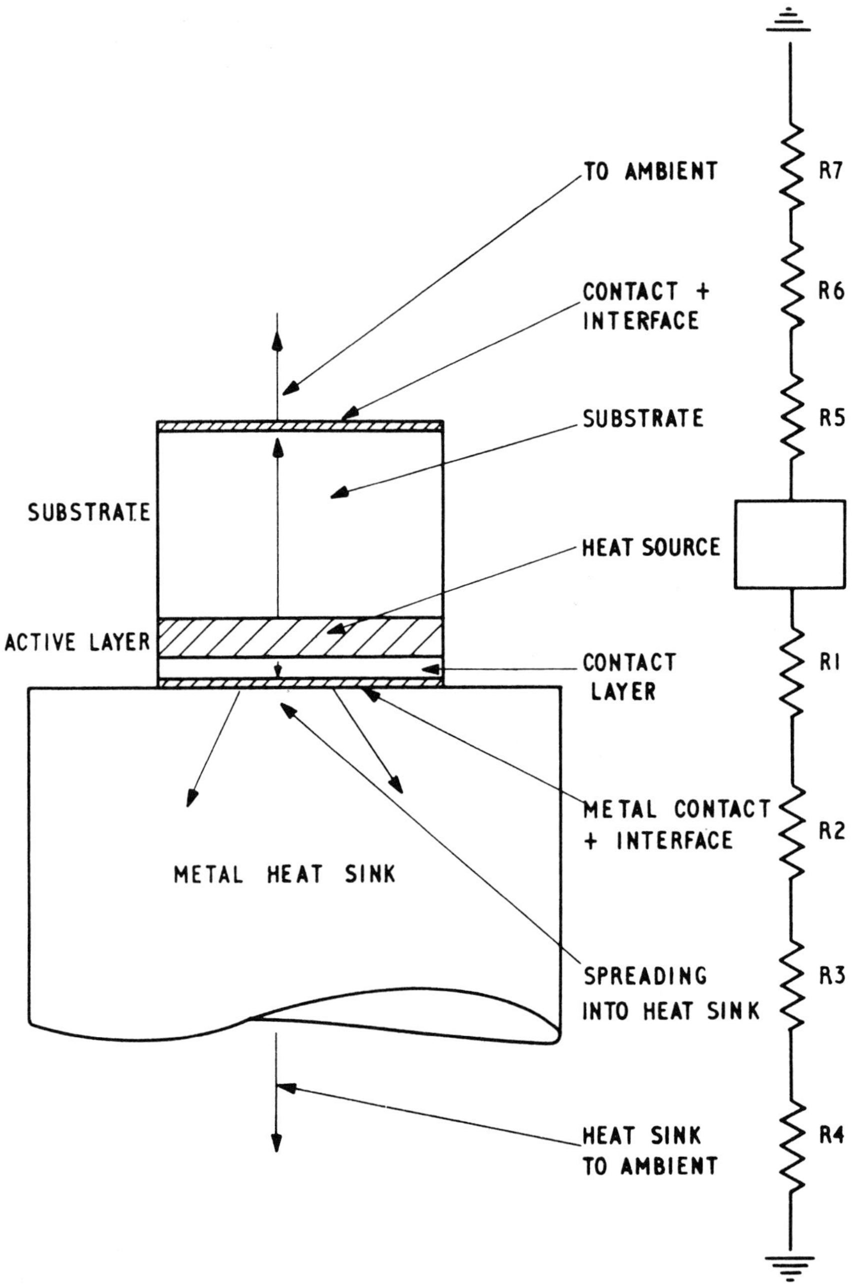

Figure 4-1/Thermal Structure of Two-Terminal Device.

providing that the contact/heat sink interface is free of voids or inclusions and that the bond is good. But, if the contact has a significant electrical resistance, it will generate heat.

The Substrate, R_5

The substrate thickness is usually greater than 25 μm, making R_5 large enough for the device structure to be considered as a single sided heat sink. Efforts have been made recently to reduce the substrate thickness and heat sink from both sides. The problems that then arise are ones of encapsulation and thermal stress relief.

Resistance to Ambient, R_7

The electrical contact is usually a gold wire or tape which has a high thermal impedance; coupled with R_5 this contact makes the structure one-sided.

Spreading Resistance, R_3

The spreading resistance to heat flow from the semiconductor into the heat sink represents one of the major thermal impedances. The value is naturally dependent on the boundary conditions considered, with a worst case solution given by[93]:

$$R_3 = 8/3\pi^2 K_p r \ ^\circ\text{C/W}$$

Obviously, there is a case for making the thermal conductivity of the heat sink, (K_p), as large as possible. As a result, type II diamond has been used; it has a thermal conductivity five times greater than copper.

Heat Sink to Ambient, R_4

Heat is normally carried away by conduction; into the waveguide, for microwave devices. The process can be improved by forced air or liquid cooling; generally, though, R_4 is a relatively small factor in the overall thermal resistance. There is little to be gained through its individual reduction.

Active Layer

The effects of thermal resistance within the active layer have not yet been considered. This layer causes undesirable temperature gradients; the profile depends on whether single or double-sided heat sink structures are used. Consider the structure shown in Figure 4-1. Let:

active layer thickness = ℓ cm
contact layer thickness = ℓ_2 cm
(with thermal conductivities of K_1 and K_2, respectively)
power generated in the active layer = P watts

The temperature at x — x is measured from substrate to layer boundary — is then given by the equation in the key paper by Perlman:

$$T_x = \frac{P\ell}{A K_1}\left\{\frac{1}{2}\left[1-\left(\frac{x}{\ell}\right)^2\right]+\frac{\ell_2 \cdot k_1}{\ell k_2}\right\}+\frac{8P}{3\pi^2 K_p r} + T_A$$

This equation can be extended to the double heat sink structure, by assuming that a fraction (δ) of the heat flux flows through the bottom contact. The T_{max} occurs $[(1-\delta)\ \ell]$ from the substrate interface; therefore, the structure is equivalent to a single sided structure of length $\delta\ell$ with a generated power of δP. From the equation above, it is evident that the temperature across the active region is provided by:

$$T = \frac{P\ell}{2\,AK_1}$$

Because the power generated at constant bias conditions in these devices is proportional to device area, the thermal resistance through the semiconductor is independent of area. However, the spreading resistance term increases proportionally with diameter. This effect is shown in Figure 4-2. For the particular geometry selected the spreading resistance term is the major item in determining device temperature when the diameter exceeds 100 μm.

In pulsed devices, analysis is necessarily more rigorous, as the pulse length, duty factor, and thermal diffusivity of the various components become the determining factors. Using Perlman's analysis for GaAs and InP pulsed TED's gives calculated figures which compare reasonably well with the experiment.

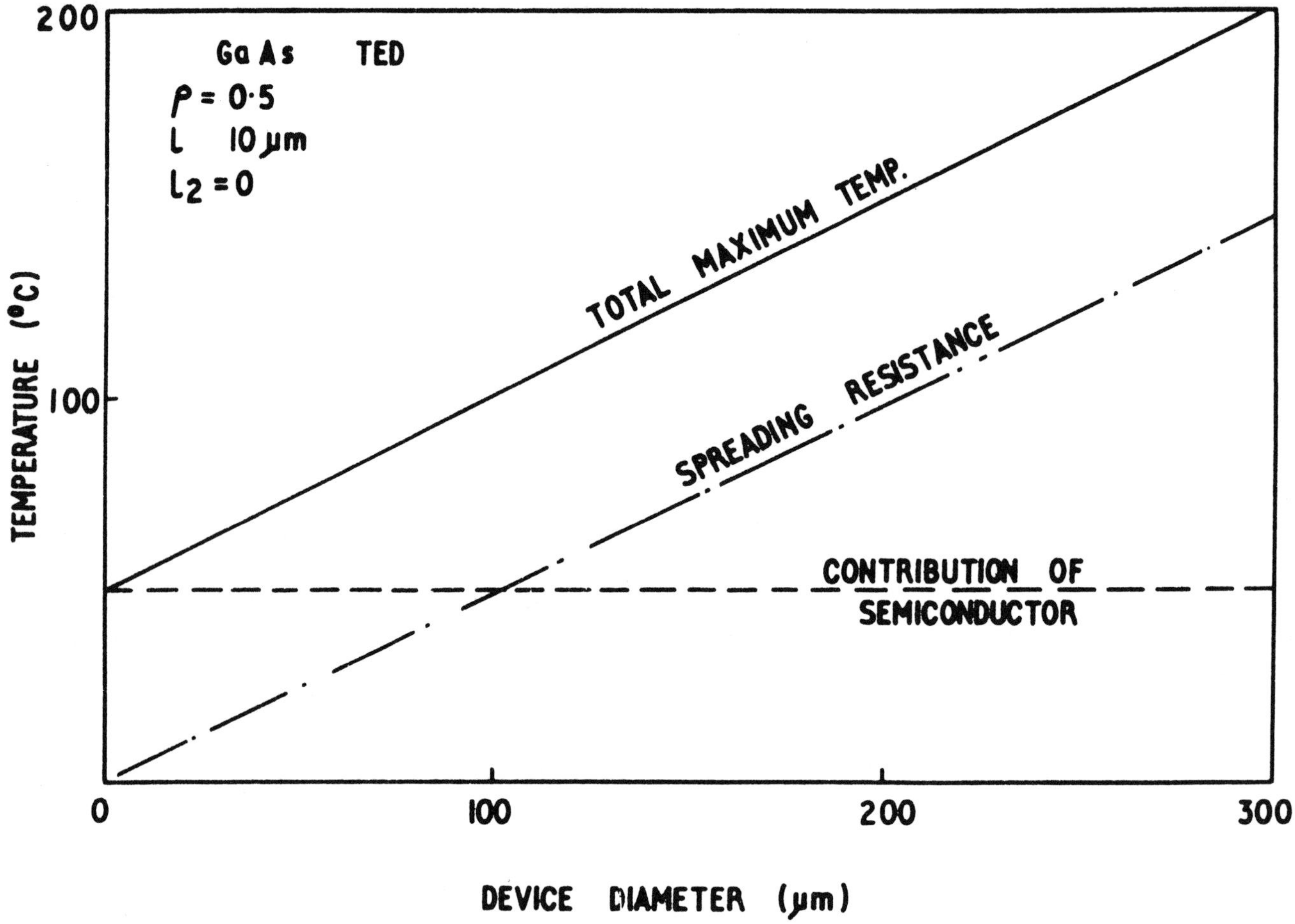

Figure 4-2/Maximum Device Temperature as a Function of Diameter

In order to minimise the thermal impedance of two-terminal structures, we need to:

1./Fabricate small diameter devices, preferably less than 100 μm.
2./Use multiple arrays of small diameter devices for high power operation to minimise spreading resistance.
3./Achieve good mechanical and thermal bonds through all metal-metal and metal-semiconductor interfaces.
4./Fabricate double-heat-sink structures if other parameters permit.

key paper

Pulsed Heat Conduction In A Layered Semiconductor-Metal Transferred Electron Oscillator Geometry

B.S. Perlman
(RCA Electronic Components, Princeton, New Jersey, US)

RCA Review (December 1969)
Pages 637-667

Summary—A method of analyzing the temperature gradient within the active layer of a bulk-effect GaAs transferred-electron device during pulsed excitation is presented. The effects of material and source parameters are considered with respect to the thermal transient response and associated time constants for specific device geometries. The results have been presented in graphical form using normalized variables. The increase in average device temperature due to spreading-resistance effects has been calculated. The thermal time constants for devices using either GaAs or Si substrates have been computed using the active-layer length and relative capping-layer thickness as parameters.

When this device is characterized by an optimum product of carrier concentration and active-layer length, the heat density is inversely related to this length. The asymptotic steady-state temperature is linearly dependent on the n-region thickness, while the total dissipated power remains independent of the n-region thickness and is proportional to the cross-sectional area. The limiting thermal response is imposed by the active layer, where the time constants are related to the square of the different relative lengths. The asymptotic temperature has been evaluated from consideration of the inverse temperature dependence of the thermal conductivity of the semiconductor material.

INTRODUCTION

A CONVENIENT METHOD for analyzing the variation of temperature throughout composite multilayered media is the use of the Laplace transformation.[1] The heat-conducting properties of a complicated thermal geometry are closely analogous to an equivalent electrical noninductive lossy transmission line. The

[1]Carslaw, H. S. and J. C. Jaeger, *Conduction of Heat in Solids*, 2nd. Ed., Oxford University Press, New York, 1959.

thermal flux temperature distribution may be visualized as a current-density–voltage distribution in an equivalent electrical transmission system.

The problem of particular interest concerns the uniform generation of heat in a pulsed layer of active semiconducting material, i.e., the n-layer of a transferred-electron oscillator (TEO), and the resulting transient temperature distribution. Several analyses have provided significant insight into the thermal properties of similar materials in the steady state.[2-5] However, in many important high-power applications, where steady flow is virtually impossible, a pulsed mode of operation is used to maintain a safe heating limit for the device. During a typical pulse, a thermal transient is developed that results in many interesting, and usually undesirable, rf electrical and material phenomena (e.g., FM, jitter, redistribution of carrier concentration, thermally induced avalanche, etc.).[6,7] The ultimate rise and distribution of temperature within the active layer, which directly affects the electrical performance of the device, is related to the heat-sink thermal constants and spreading resistance, the source interpulse spacing, the layer thickness, and the input power density.

The primary aim of this analysis is to develop a quantitative picture of the transient temperature distribution in the layered geometry associated with a transferred-electron semiconductor device. Only one-dimensional heat transfer by conduction from the bulk active region representing a uniform source is considered, although the analytical methods used are readily applicable to other source conditions.

By establishing a close analogy between the thermal geometry and an equivalent electrical network, this problem may be more clearly interpreted and solved using several techniques of circuit analysis.*

---

[2] Knight, S., "Heat Flow in n^{++}-n-n^{+} Expitaxial GaAs Bulk Effect Devices," *Proc. IEEE (Letters)*, Vol. 55, p. 112, Jan. 1967.

[3] Swan, C. B., et. al, "Composite Avalanche Diode Structures for Increased Power Capability," *IEEE Trans. ED*, Vol. 14, p. 584, Sept. 1967.

[4] Clorfeine, A. S., "Power Dissipation Limits in Solid State Oscillator Diodes," *Microwave Jour.*, Vol. 11, p. 93, March 1968.

[5] Gooch, C. H., "The Thermal Properties of GaAs Laser Structures," *IEEE Jour. Quantum Elec.*, Vol. QE-4, p. 140, April 1968.

[6] Haitz, R. H., "Nonuniform Thermal Conductance of an Avalanche Microwave Oscillator," *IEEE Trans. ED*, Vol. 15, p. 350, June 1968.

[7] Brown, R. F., "Transient Temperature Distribution in Diode Lasers and the Time Duration of the Output Pulse at 300°K," *IEEE Jour. Quantum Elec.*, Vol. QE-4, p. 135, April 1968.

* The analogy to cascaded transmission lines used here in a one-dimensional analysis should be applicable to heat transfer in the active layer of this semiconductor device if the active layer is very thin compared with its lateral dimensions, the heat generation is uniform, and the contact layer does not have a larger surface area than the active area.

In addition to the use of the Laplace transform, we introduce a thermal input impedance, using matrix techniques, at the source–sink interface and utilize the principle of superposition to provide a solution for variable sink temperatures, V_A.

The heat diffusion through the sink geometry may be analyzed in terms of several thermal time constants, although a first-order approximation of a single constant is satisfactory. The limiting response is imposed by the relatively poor thermal characteristics associated with the active layer of GaAs material.

When this particular device is characterized by an optimum product of carrier concentration and active layer length, the resulting heat density is affected by the length used, with the asymptotic transient temperature is *linearly* related to this length. The equilibrium temperature associated with an infinite half space spreading resistance is independent of this length and linearly related to the device radius.

A possible complication in this analysis would be imposed by an inverse temperature dependence of the thermal conductivity of the semiconductor material. The transient solution therefore ignores this problem, although a steady-state solution is found for variable conductivity so as to evaluate the magnitude of error introduced by the linear analysis.

The resulting temperature distribution within the active layer for both single-pulse and periodic excitation has been derived for several geometries. The results are presented in graphical form using normalized variables with respect to both the thermal time constants and steady-state temperatures. The time constants associated with specific geometries, using both GaAs and Si substrates, are computed using the active layer length and relative capping layer thickness as parameters.

A more thorough familiarity with the problem, with particular regard for both material and geometrical considerations, should provide an understanding of certain device limitations that result from the temperature distribution within the active region and make it possible to minimize these effects. An improvement of the device electrical efficiency is directly associated with the package thermal efficiency and, consequently, the overall heat-sink design.

The TEO (Gunn Oscillator) as a Heat Source

A. Electrical Characteristics

The application of a uniform electric field to a bulk semiconductor results in the generation of a uniform heat density. In the active n-layer of a GaAs epitaxial device that is suitably doped to exhibit the "two-valley" conduction band-effect,[8] a differential negative resistance appears, thus permitting this material to sustain microwave oscillations. The rf power that is removed from the material represents only a small portion of the dc power that must be supplied to the device. The remaining energy is consumed by the material, resulting in uniform heating throughout the active region.

The heat density may be represented by

$$A(s) = \frac{A_o}{s} = \frac{JE}{s}(1-\eta), \qquad (1)$$

where we have introduced the Laplace transform $A(s)$ of the power density. A_o. The dc-to-rf conversion efficiency is given by η. Although this analysis considers only uniform heating, the introduction of non-uniform values of $A(s)$ could be used to analyze localized heating effects and other devices (i.e., avalanche diodes, etc.).

The usual geometry for a layered device with a rectangular cross section is shown in Figure 1. The thickness l of the active region is grown epitaxially to establish the electrical length necessary for efficient oscillator performance at a specific frequency. The transit-time frequency f_T is inversely related to this length (i.e., $f_T \simeq 10^7/l$ Hz). The lower limit on frequency is determined by both the increased temperature and the difficulties in homogeneous growth associated with thick layers. A 1 GHz oscillator operating in a resonant mode near f_T would require as much as 100 microns of active material, with a carrier concentration near 1.5×10^{14} cm^{-3}. The optimum efficiency for an oscillator tuned into a resonant mode near f_T is obtained when the nl product is a constant (i.e., $nl \simeq 1.5 \times 10^{12}$ cm^{-2}).[9]

At the threshold field, $E_T \simeq 3000$ V/cm, the current density $J_T = E_T/\rho_o$, where ρ_o is the volume resistivity ($1/ne\mu$ ohm-cm) of the active layer, which has a carrier concentration n and electron mobility μ. The operating field, which is usually several times E_T, may be specified as βE_T, where $\beta \geq 1$. The valley current density becomes γJ_T, where $\frac{1}{2} \leq \gamma \leq 1$.

We may now express the heat power density A_o from Equation (1), in terms of the material constants* and the constant nl product specified.

---

[8] McCumber, D. E., and A. G. Chynoweth, "Theory of Negative-Conductance Amplification and of Gunn Instabilities in 'Two-Valley' Semiconductors," *IEEE Trans. ED*, Vol. 13, p. 4, Jan. 1966.

[9] Copeland, J. A., "Theoretical Study of a Gunn Diode in a Resonant Circuit," *IEEE Trans. ED*, Vol. 14, p. 55, Feb. 1967.

* In this analysis we shall neglect the temperature susceptibility of the resistivity ρ_o which is evaluated at an ambient temperature $T_o \simeq 300$°K.

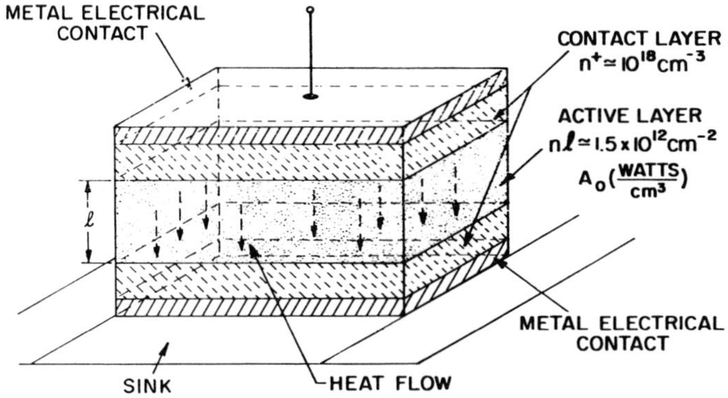

Fig. 1—Layered geometry of a TEO Gunn bulk-effect device.

$$A_o = \frac{\beta\gamma E_T{}^2}{\rho_o}(1-\eta) = \frac{\beta\gamma}{l}[1.28 \times 10^4] \text{ watt/cm}^3. \qquad (2)$$

Here l is in cm and a nominal value of $\eta \simeq .05$ has been used. Figure 2 shows the variation of power density A_o versus length l with $\beta\gamma$ as a parameter, obtained with Equation (2). Although the power density

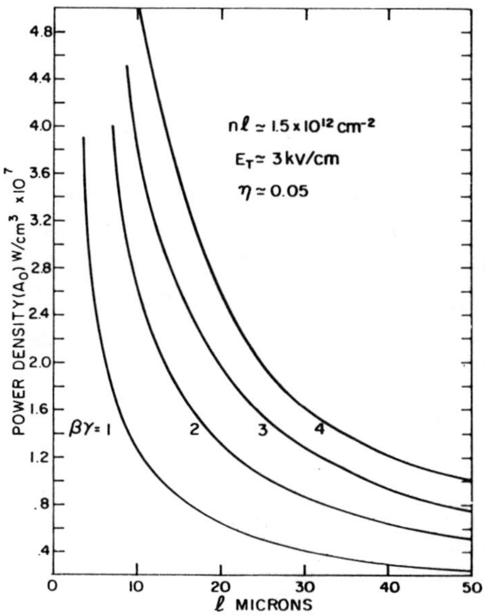

Fig. 2—Variation of input power density with layer thickness.

increases with decreasing layer thickness, the actual dissipated power is independent of l and is proportional to the cross-sectional area. The temperature rise within the active layer is affected by l and determines the limitations on performance for this device.

B. Material Considerations—Thermal Characteristics

The properties of several well known materials,[10-14] that could conceivably be used in the sink geometry are indicated in Table I.

The appearance of different thermal properties for high-resistivity and highly doped single-crystal semiconductors creates a thermal problem. The thermal conductivity of highly doped material ($n > 10^{18}$ cm^{-3}), which is necessary for good ohmic constants, is actually lower than that of lightly doped material, even though more electrons are available for conduction of heat. The decreased thermal conductance appears to result from a phonon–electron scattering process.[11, 12] In this transient analysis, we have disregarded the resulting $1/T$ conduction dependence. This inverse temperature effect on K has been used in the Appendix to calculate the steady-state behavior.

The most common geometry for a TEO consists of n/n+ epitaxial GaAs, where the difficulty is that the alloying of metallic contacts directly to the already thermally poor n layer necessitates the use of highly doped capping material. The relatively poor thermal response of GaAs limits the effectiveness of heat removal through the n+ contact material, thereby requiring a minimum layer thickness. Even if a direct alloy to the active n layer were achieved (i.e., Sn, Au-Ge, etc.), a significant improvement in thermal performance would not occur. Unfortunately, the high-conductivity materials (i.e., Ag, Au) do not effectively alloy with n-GaAs ($n \leq 10^{16}\,cm^{-3}$). The use of a planar geometry would separate the conduction processes (electric versus thermal) but still retain a semi-insulating GaAs thermal barrier unless another single-crystal substrate material were used. When the required electrical length l becomes large ($> 30\,\mu$), the temperature rise in a layered geometry can become excessive and a planar structure may be more desirable. The most effective method for removing heat

[10] Berman, R., *Physical Properties of Diamond,* Clarendon Press, Oxford, 1965.

[11] Maycock, P. D., "Thermal Conductivity of S_i, G_e, III-V Compounds and Alloys," *Solid State Elect.*, Vol. 10, p. 161, March 1967.

[12] Holland, M. G., "Photon Scattering in Semiconductors from Thermal Conductivity Studies," *Phys. Rev.*, Vol. 134, p. A471, April 1964.

[13] Willardson, R. K. and Beer, A. C. *Semiconductors and Semimetals,* Vol. 2, "Physics of III-V Compounds," Academic Press, New York, 1966.

[14] Drabble, J. R., and Goldsmith, H. J., *Thermal Conduction in Semiconductors,* Pergamon Press, New York, 1961.

from thin n-layer material (e.g., $l \leq 20\ \mu$) is to flip-chip bond the epitaxial side to the heat sink. A minimum thickness of n+ material is desirable.

The most interesting, as well as effective, material for use in the heat-sink geometry is diamond Type IIa.[10] This material can be used to spread the heat efficiently from the concentrated area of the device over a broader base, thereby minimizing thermal flux entering the heat

Table I—Thermal Constants for Selected Materials

Material	Thermal Conductivity K (Watts/°C − cm)	Diffusivity α (cm²/sec)
GaAs (n)	0.5	.29
GaAs (n+)	0.4	.232
Si (undoped)	1.45	.825
Si (high doping)	1.15	.650
Be_2O_3	2.45	.572
Diamond (IIa)	22.0	1.01
Diamond (IIb)	13.0	.592
Silver	4.27	1.71
Gold	3.16	1.18
Copper (OFHC)	4.0	1.14
Aluminum	2.36	.86
Molybdenum	1.46	.58
Tin	0.625	.38
Brass (70:30)	1.04	.33
Lead	0.34	.25

sink. The relatively high thermal conductance of type IIa diamond is very effective in reducing the spreading resistance associated with the device-to-metal contact.[15] The allowable dissipation of the resulting package is thereby improved.

Although the termal conductivities of the various materials vary considerably, a less significant difference exists between their diffusivities. The higher specific heat and density of diamond II, Be_2O_3, etc. limit their thermal response, since the associated thermal time con-

[15] Kennedy, D. P., "Spreading Resistance in Cylindrical Semiconductor Devices," *Jour. Appl. Phys.*, Vol. 31, p. 1495, Aug. 1960.

stants are inversely proportional to the diffusivity, α. Diamond IIa has the minimum thermal resistance and therefore yields the least rise in temperature, but *silver* has the smallest time constant.

Transient Temperature Response for Several Geometries

The transient temperature build-up within the active n layer $(0 < x < l)$ of a transferred-electron-oscillator geometry is readily found as a solution to a one-dimensional thermal-conduction equation, which is described in Appendix I. Using electrical analogies, the layered geometry may be represented as a thermal network, as shown in Figure 3, consisting of a set of cascaded transmission lines. Using

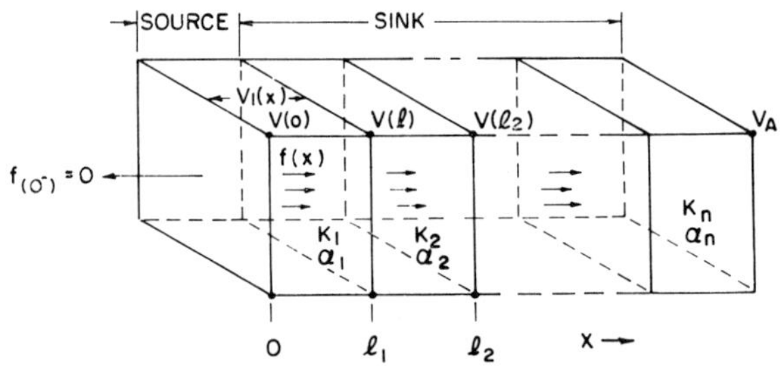

Fig. 3—Cascaded thermal network.

Laplace notation, each line is characterized by a transfer matrix relating the flux and temperature distribution between layers.

Separating the source region (i.e., active layer) from the sink geometry and introducing an input "sink" impedance, $Z(l) = V(l)/f(l)$ where $x = l_1 = l$, we may relate the passive sink parameters to a termination on the source. The sink is terminated by a finite controllable source, V_A. The use of a fixed termination V_A neglects the spreading resistance normally associated with steady spherical flow $(\partial/\partial t = 0)$ from the device into a larger sink. This model is justified during the initial transient, where the temperature rise within the device will be shown to be considerably faster than that in the sink, which has a considerably larger thermal mass. The contribution of the spreading resistance will be determined for the case of a repetitive pulse source where the system has reached an equilibrium condition and the continuity of an average thermal flux is maintained through the sink geometry.

The temperature distribution within the active n layer is derived in Appendix I and may be expressed in Laplace space as

$$V_1(x) = \bar{A}\left[1 - \frac{C_1(x)}{C_1(l) + \sigma S_1(l)}\right] + \frac{V_A}{s}\left[\frac{C_1(x)}{AC_1(l) + \sigma S_1(l)}\right], \quad (3)$$

where

$V_1(x)$ = transformed temperature in active layer as a function of the distance l measured into the device from the top of the source region.

V_A = sink-load-temperature boundary condition.

$\bar{A} = A(s)\alpha_1/(sK_1)$ = normalized equivalent source temperature for a zero initial material temperature (i.e., $t = 0$, $V_o = 0$).

$A(s)$ = source power density (watts/cm^3), with time dependance explicit in Laplace operator s.

A = variable in ABCD combined transfer matrix for cascaded sink geometry (see Equation (4)).

α_n = thermal diffusivity (cm^2/sec for n^{th} layer (e.g., $n = 1$ for source region).

K_n = thermal conductivity (watts/ ^{o}C — cm) for n^{th} layer.

$C_n(x) = \cosh q_n x$

$S_n(x) = \sinh q_n x$

$q_n = \sqrt{s/\alpha_n}$ = normalized Laplace operator.

$\sigma = Z(l)/Z_1$ = dimensionless variable.

Z_n = characteristic impedance of nth region.

The impedance transfer matrix for each layer may be expressed, as shown in Appendix I, as,

$$\begin{bmatrix} A_n B_n \\ C_n D_n \end{bmatrix} = \begin{bmatrix} C_n(l_n) & -Z_n S_n(l_n) \\ Z_n^{-1} S_n(l_n) & C_n(l_n) \end{bmatrix} \quad (4)$$

where $l_n = x_n - x_{n-1}$ corresponds to the length of the nth subregion. The composite matrix, $\begin{bmatrix} AB \\ CD \end{bmatrix}$ for a series of cascaded lines is obtained by direct matrix multiplication. The primary-source driving point impedance $Z(l)$ may be shown to be,

$$Z(l) = \frac{V(l)}{f(l)} = -B/A, \quad (5)$$

where A, B are found from the composite matrix.

UNIT-PULSE HEATING OF A SEMI-INFINITE HOMOGENOUS MATERIAL

The results derived in Appendix I and summarized above are used to examine the problem in which heat is generated in an active layer $0 < x < l$ and dissipated in similar material that extends infinitely in the positive x direction. This geometry is shown in Figure 4(a), with

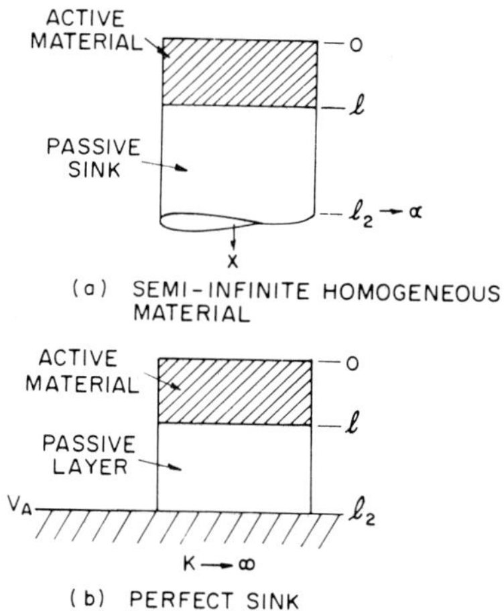

Fig. 4—Finite composite media.

$V_A \to 0$ as $l_2 \to \infty$. From Equations (3), (4), and (5), the normalized sink impedance σ is found to be

$$\sigma = \frac{Z(l)}{Z_1} = \frac{Z_2}{Z_1}\left(\frac{S_2(l_2)}{C_2(l_2)}\right)\Bigg|_{\substack{l_2 \to \infty \\ V_A \to 0}} = \frac{K_1}{K_2}\left(\frac{\alpha_2}{\alpha_1}\right)^{1/2}. \tag{6}$$

For homogeneous media, both source and sink are characterized by the same material constants, and $\sigma = 1$. For the active-layer, temperature distribution, Equation (3), becomes

$$V_1(x) = \bar{A}\left[1 - \frac{1}{2}\left(\exp\{-q_1(l-x)\} + \exp\{-q_1(l+x)\}\right)\right]. \tag{7}$$

A solution for $V_1(x, t)$ is obtained by replacing $\bar{A}$ by the suitable Laplace transform of a single pulse of width T, e.g., $\bar{A} = (\alpha/s^2)A_o(1 - e^{-sT_1})$. The resulting inverse transform of Equation (7) may be normalized to a temperature V_N

$$V_N = \frac{A_0 l^2}{K_1}. \tag{8}$$

Then,

$$\frac{V_1(x,t)}{V_N} = \frac{\alpha_1 t}{l^2}\left[1 - 2i^2 \operatorname{erfc}\frac{l-x}{2\sqrt{\alpha t}} - 2i^2 \operatorname{erfc}\frac{l+x}{2\sqrt{\alpha t}}\right]$$

$$\text{with } 0 < x < l \text{ and } 0 \leq t < T_1, \tag{9}$$

and

$$\frac{V_1(x,t)}{V_N} = \frac{\alpha_1}{l^2}\left\{T_1 - 2t\left[i^2 \operatorname{erfc}\frac{(l-x)}{2\sqrt{\alpha t}} + i^2 \operatorname{erfc}\frac{(l+x)}{2\sqrt{\alpha t}}\right]\right.$$

$$\left. + 2(t-T_1)\left[i^2 \operatorname{erfc}\frac{(l-x)}{2\sqrt{\alpha}(t-T_1)} + i^2 \operatorname{erfc}\frac{(l+x)}{2\sqrt{\alpha}(t-T_1)}\right]\right\} \text{ with } t > T_1. \tag{10}$$

Similarly, the average solution becomes

$$\frac{V_{av}(t)}{V_N} = \frac{\alpha_1 t}{l^2}\left[1 - \frac{2(\alpha_1 t)^{1/2}}{l}\left(\frac{1}{3\sqrt{\pi}} - 2i^3 \operatorname{erfc}\frac{l}{\sqrt{\alpha t}}\right)\right] \text{ with } 0 < t - T_1, \tag{11}$$

and

$$\frac{V_{av}(t)}{V_N} = \frac{\alpha_1}{l^2}\left\{T_1 - \frac{2}{3l}\left(\frac{\alpha_1}{\pi}\right)^{1/2}\left(t^{3/2} - (t-T_1)^{3/2}\right)\right.$$

$$\left. + \frac{4\alpha_1{}^{1/2}}{l}\left(t^{3/2} i^3 \operatorname{erfc}\frac{l}{\sqrt{\alpha t}} - (t-T_1)^{3/2} i^3 \operatorname{erfc}\frac{l}{\sqrt{\alpha(t-T_1)}}\right)\right\} \text{ with } t > T_1. \tag{12}$$

The normalized temperature $V_1(x,t)/V_N$ is plotted in Figure 5, using Equations (8)—(12), for $x = 0$ and l. Figure 5 also shows the average value V_{av} and the differential $V(0) - V(L)$ normalized to V_N, as functions of the variable $\alpha_1 t/l^2$. The pulse width T_1 has been introduced as a normalized parameter $T_1' = \alpha_1 T_1/l^2$.

Also indicated are the results for a TEO with an active source-region length $l = 20\ \mu$m, corresponding to a transit-time frequency near 5 GHz. The diffusivity averaged over n type and n+ type GaAs material has been used here ($\alpha_1 \simeq 0.265$ cm²/sec), with $\alpha/l^2 \simeq 0.067 \times 10^6$ sec^{-1} and the appropriate material constants inserted as discussed previously (e.g., $nl \simeq 1.5 \times 10^{12}$ cm^{-2}).

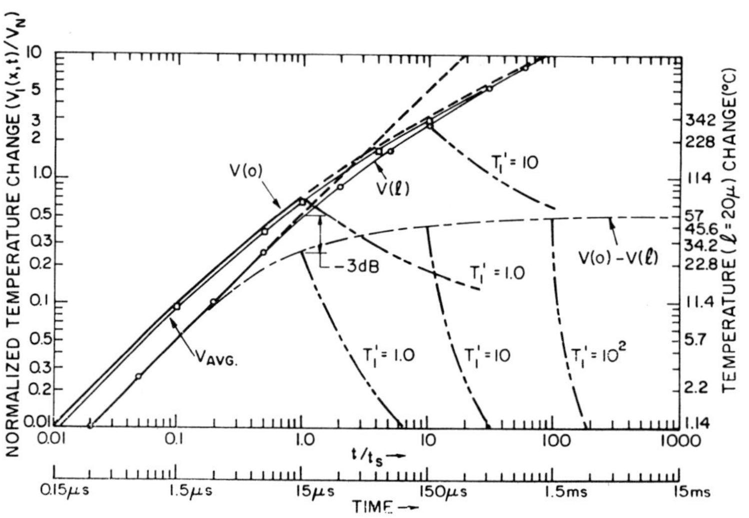

Fig. 5—Transient temperature response of n-layer terminated in infinite n+ material due to unit pulse.

From Equation (2) the normalization temperature V_N may be found for $\beta = 3$, $\gamma = 2/3$, and $K_1 \simeq 0.45$ watt/°C-cm; with a conservative device efficiency, $\eta = 0.05$, $V_N \simeq 5.70l = 114$°C. It should be noted that the thermal properties of GaAs actually vary with temperature[12, 13] (e.g., $K_1 \simeq 1/T$). In addition, the thermal conductivity of GaAs is a function of the electron concentration: $K(\text{n}) \simeq 0.5$ at 300°K for $n = 10^{15}$ cm^{-3} and $\rho = 1$ ohm-cm, and $K(\text{n+}) \simeq 0.4$ at 300°K for $n = 10^{18}$ cm^{-3} and $\rho = 0.001$ ohm-cm.

The results of this example are intended primarily for illustrative purposes. The significance of the n+-layer thickness on effective heat removal is discussed more thoroughly when the GaAs material is referred to a fixed-temperature surface in the examples that follow. It is most interesting to note the "steady-state" temperature difference $V(0)$-$V(L) = 0.5V_N$, as indicated in Figure 5. The steady-state condition may be defined to occur after a time $t_s = l^2/\alpha_1$, when the temperature is within 3 dB of its steady-state value. In this example $t_s \simeq 15$

μsec. This parameter may be interpreted as the thermal time constant for this material geometry, where spreading resistance effects have been neglected since a one-dimensional heat flow is maintained.

Spreading Resistance Effect for Heat Flow into a Semi-Infinite Half-Plane

In the practical case, a small device is usually flip-chip bonded to a large post of thermal conductivity K_p as shown in Figure 4(b). The post geometry may be interpreted as a semi-infinite half-space with a thermal response time considerably larger than the time constants associated with the small device. The temperature build-up in the post may therefore be assumed to be negligible during initial heating transients. When the heat flux entering the post reaches an equilibrium value ($f(t+T) = f(t)$, where T is the period of repetitive heating pulse), an average steady-state thermal resistance to flow from the device cross-sectional area into the half-space of the post may be calculated. Several solutions for this problem have been found for a cylindrical device with a radius r. The assumptions of a variety of boundary conditions result in several values for the spreading resistance R_{sp} with a worst-case solution given by[1],

$$R_{sp} = \frac{V_{pav}}{F_p} = \frac{8}{3\pi^2 K_p r} \ (°\text{C/watt}) \tag{13}$$

where,

V_{pav} = average rise in temperature at infinite plane of the post.

$F_p = \pi r^2 f_p(t)$—is the rate of heat flow across a device area of radius r into infinite half-space.

The effective average heating during pulse conditions is obtained by multiplying the heat density A_o by a duty factor a. For GaAs material with a constant doping length product for the active layer (e.g., $nl \simeq 1.5 \times 10^{12}$), the average temperature produced at the post surface, using Equations (2) and (13), is

$$V_{\text{pav}} \simeq \frac{a\beta\gamma r}{K_p} \times 10^4 = \frac{a\beta\gamma}{K_p} (25.4r \text{ mils}). \tag{14}$$

The normalized temperature is plotted in Figure 6 as a function of r, with K_p as a parameter.

PULSE HEATING OF COMPOSITE MEDIA

Unit-Step Single-Pulse Heating

The thermal time constants and temperature response for the semiconductor device shown in Figure 4(b) may be found as a function of the layer parameters. Media with different thermal characteristics are considered in this case. The source region, $0 < x < l$, with thermal

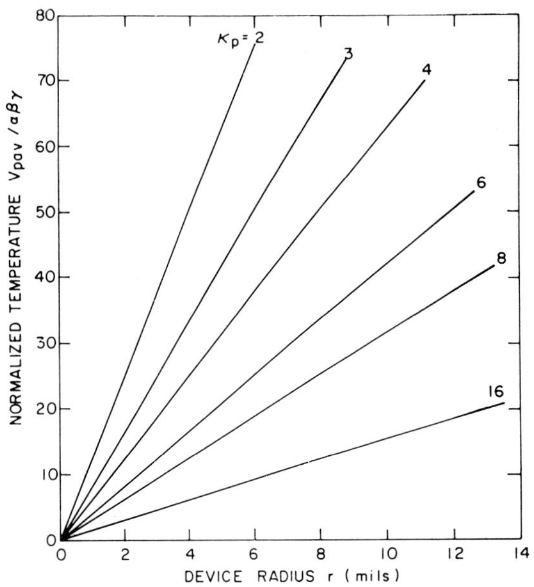

Fig. 6—Average temperature rise due to spreading resistance, with the post thermal conductivity as a parameter.

constants K_1, α_1, etc., passes heat through a subregion $l \leq x \leq l_2$, with constants $K_2\alpha_2$, etc., to an isothermal sink maintained at a temperature V_A.. In this case, we shall neglect the contribution of the spreading resistance during a single pulse transient and use a super-imposed source V_A to provide a suitable isothermal reference plane. As before, the transient temperature rise is referred to zero initial conditions (i.e., $V_o(t) = 0$).

The normalized sink impedance for this geometry is found, using Equations (4)-(6), to be

$$\sigma = \frac{Z_2}{Z_1} \frac{S_2(l_2)}{C_2(l_2)}. \tag{15}$$

The resulting active-layer temperature distribution for a single pulse source function A is found, by substituting Equations (8) and (15) into Equation (3), to be

$$V_1(x) = \frac{\alpha_1}{s^2 l^2} V_N \left[1 - \frac{N(s)}{D(s)} C_2(l_2) \right] + \frac{V_A}{s} \frac{N(s)}{D(s)}, \; 0 < t < T_1. \quad (16)$$

The rational function $N(s)/D(s)$ is given by

$$\frac{N(s)}{D(s)} = \frac{C_1(x)}{C_1(l_1)C_2(l_2) + \sigma' S_1(l_1) S_2(l_2)}, \quad (17)$$

where

$$\sigma' = \frac{Z_2}{Z_1} = \left(\frac{\alpha_2}{\alpha_1}\right)^{1/2} \left(\frac{K_1}{K_2}\right) \quad (18)$$

Equation (17) is not readily transformed into real time by using tabulated results. The most direct approach is to apply the theorem of residues by expanding Equation (17) into partial fractions and using the Laplace inversion theorem. The poles of Equation (16) occur at $s = 0$ and $s = -\alpha_1 \beta_m{}^2$, where the $\pm \beta_m$ ($m = 1, 2, \ldots.$) are the roots of one of the two following equations:

$$\cos\beta l \cos k\beta l_2 - \sigma' \sin\beta l \sin k\beta l_2 = 0 \quad (19a)$$

$$\cot\beta l = \sigma' \tan k\beta l_2 \quad (19b)$$

with

$$k = \frac{q_2}{q_1} = \frac{1}{\sigma'} \left(\frac{K_1}{K_2}\right) = \left(\frac{\alpha_1}{\alpha_2}\right)^{1/2}. \quad (20)$$

When $l \neq kl_2$, there is no explicit analytic method for computing the roots of Equation (19a). The only feasible methods are either graphical construction or numerical computation. With $l = kl_2$, the roots are found from $\tan\beta l = \pm\sqrt{1/\sigma'}$.

The residues at $s = 0$ for the $\bar{A}$ source term in Equation (16) are found for the multiple poles $1/s^2$ and $1/s$ to be zero and $[(l^2 - x^2) + 2l_2 l(K_1/K_2)]/2\alpha_1$, respectively. The $s = 0$ residue in the V_A source term is unity. The remaining residues are found by solving Equation

(19) for the roots and substituting into the conventional expansion term,

$$\frac{1}{s^n}\left[\frac{N(s)}{dD(s)/ds}\right]_{s=-\alpha_1\beta_m}$$

using the correct multiplicity factor n. From Equation (16) and the appropriate tabulated transforms,

$$V_1(x,t) = V_s + V_A - \sum_{m=1}^{\infty} (V_{sm} + V_{Am})\, e^{-t/\tau_m},\ 0 < x < l,\ 0 < t < T_1 \tag{21}$$

with the steady-state primary source response, as $t \to \infty$, given as

$$V_s = V_N\left[\frac{1}{2}\left(1 - \left(\frac{x}{l}\right)^2\right) + \frac{l_2}{l}\frac{K_1}{K_2}\right], \tag{22}$$

which is essentially reached for $t \simeq 3\tau_m$, neglecting the contribution of spreading resistance during this short response time.

The steady-state temperature gradient, $\Delta V_s = V_s(0) - V_s(l)$, may be found from Equation (2) to be $V_N/2$, which is plotted in Figure 7 as a function of the length l with $\beta\gamma$ as a parameter. The average dissipated power P_D versus the device area is also plotted in this figure, again with $\beta\gamma$ as a parameter. Thus, a knowledge of P_D and the device area can be used to find an effective value of the drive parameter $\beta\gamma$, which, in turn, may be used to relate V_N to l.

The thermal time constants are

$$\tau_m = \frac{1}{\alpha_1\beta_m^2} = \frac{l^2}{\alpha_1(\beta l)^2}; \tag{23}$$

the transient term coefficients are

$$V_{sm} = \frac{V_N\left(\dfrac{2}{l^2\beta_m^3}\right)\cos k\beta_m l_2 \cos\beta_m x}{(l + \sigma' k l_2)\sin\beta_m l \cos k\beta_m l_2 + (k l_2 + \sigma' l)\cos\beta_m l \sin k\beta_m l_2} \tag{24}$$

$$V_{Am} = V_A \left[\frac{\left(\frac{2}{\beta_m}\right)\cos\beta_m x}{(l+\sigma' k l_2)\sin\beta_m l \cos k\beta_m l_2 + (kl_2+\sigma' l)\cos\beta_m l \sin k\beta_m l_2} \right], \tag{25}$$

which may be slightly simplified using Equation (19). These co-

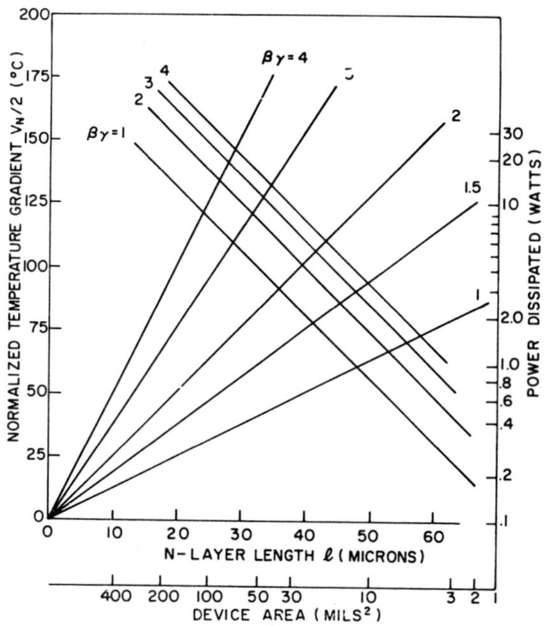

Fig. 7—Steady-state temperature gradient and average power dissipated in active n-layer with a thermal conductivity $K_1 = 0.5$ watts/°C-cm.

efficients may be related to the steady state response V_s, using Equation (21) for $t = 0$,

$$V_s = \sum_{m=1}^{\infty} V_{sm} \tag{26}$$

$$V_A = \sum_{m=1}^{\infty} V_{Am}. \tag{27}$$

The single-pulse thermal transient build-up normalized to $V_A + V_s$

is shown in Figure 8. In a similar manner, the "off" pulse result for the primary source removed at T_1 *is*

$$V_1(x,t) = V_A + \sum_{m=1}^{\infty} \left[\left(V_{sm} \exp\left\{ -\left(\frac{t - T_1}{\tau_m} \right) \right\} \right) \left(1 - \exp\left\{ -\frac{T_1}{\tau_m} \right\} \right) - V_{Am} \exp\left\{ \frac{-t}{\tau_m} \right\} \right], t > T_1, \tag{28}$$

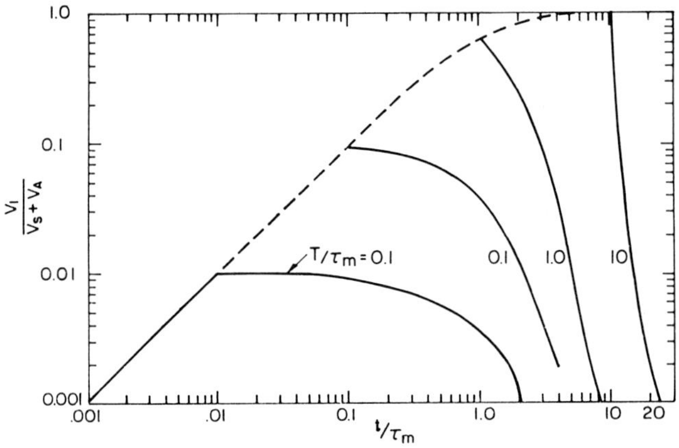

Fig. 8—Single pulse response.

wnere the isothermal surface at l_2 is maintained at V_A (i.e., unit step at $t > 0$). The use of a negative sink V_A (i.e., a thermoelectric device) would reduce the net rise in temperature, assuming sufficient compensation is made for the inevitable contribution of the spreading resistance.

The steady-state asymptotic limit of Equation (22) has been derived by assuming a constant value of thermal conductivity with respect to temperature for the different semiccnductor layers. For the temperatures that would be encountered in practice, $K \sim 1/T$. A comparison may be made between the results obtained for V_s in Equation (22) and those obtained for the inverse temperature value of K as derived in Appendix II.

We may relate V_s to an actual temperature V_{so} with a reference temperature $T_o \simeq 300°K$. These results are indicated in Figure 9 for $x = 0$ as a function of the layer thickness l. Various drive conditions are indicated with $\beta\gamma$ as a parameter (i.e., $A_o \simeq 1.28 \times 10^7$ W/cm^3 for

$\beta\gamma = 2$ and $l = 20\ \mu$m). The error involved in neglecting the inverse temperature dependance of K, relative to the steady state result, is $\leqq 20\%$ for material with $l < 20\ \mu$m and $A_o \leqq 2 \times 10^7$ W/cm^3. In most cases the error is significant, but in thicker layers, where the calculated results indicate abnormally high temperatures, the errors become considerable. These high temperature results are primarily illustrative and have been derived assuming perfectly linear material

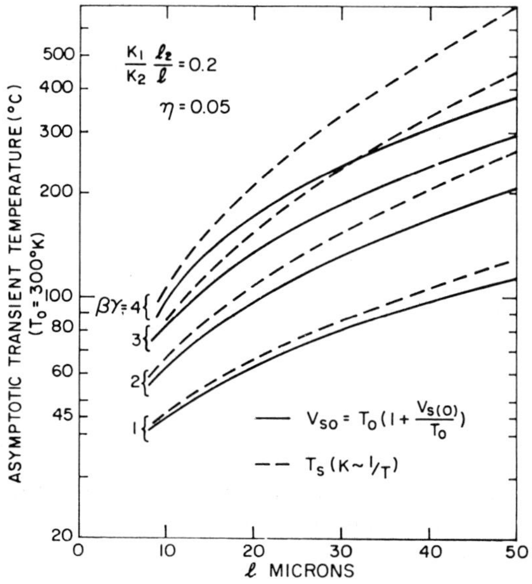

Fig. 9—Steady-state temperature rise at $x = 0$ (maximum temperature location) for optimum $nl = 1.5 \times 10^{12}$ cm^{-2} in a TEO.

parameters and are *not valid* when these parameters are greatly affected by thermal effects (i.e., n or $\rho_o = f(T)$). During cw operation, the temperature V_{pav} from Equation (14) may be added to V_s to give an estimate of the actual device temperature at $x = 0$.

Steady Periodic Heating

The primary source solution for a repetitive pulse[16] applied to the geometry as shown in Figure 4(b) is found by using the appropriate pulse transform for $A(s)$ in Equation (3) and following the procedure for inversion that lead to Equations (21) and (28). The additional

[16] Jaeger, J. C., "Pulsed Surface Heating of SemiInfinite Solid," *Quart Appl. Math.*, Vol. 11, p. 132, 1953.

poles resulting from the periodic source term are at,

$$s = 2n\pi iT,\ n = \pm 1, \pm 2, \ldots,$$

which gives a series of oscillations of periods T, $T/2, \ldots$. These new roots lead to a steady periodic temperature variation, while the $s = 0$ roots provide the steady state average value, $V_s T_1/T$. Both roots lead to a steady periodic term V_{sp}, whereas the roots $s = -\alpha_1 \beta_m^2 = -1/\tau_m$ provide the transient part. The V_A unit-step-source solution has already been found in Equations (21) and (25). The transient result for a periodic source V_A is analogous to the following result for the primary source. The effect of spreading resistance can be added to the steady-state terms by replacing V_A by V_{pav} from Equation (14).

An evaluation of the residue at each pole leads to the combining result

$$V_1(x, t) = V_{sp} - \sum_{m=1}^{\infty} \left(V_{sm} \left[\frac{\exp\{-t/\tau_m\} - \exp\{-(t - T_1)/\tau_m\}}{1 - \exp\{T/\tau_m\}} \right] + V_{Am} \exp\left\{ \frac{-t}{\tau_m} \right\} \right). \quad (29)$$

The previous results for a single pulse may be used to find a convenient solution for V_{sp}. Since Equation (29) is valid for all time, it must agree with Equation (21) for $0 < t < T_1$, and with Equation (28) for $T_1 < t < T$. Thus, equating Equations (29) and (21), V_{sp} for $0 < t < T_1$ is

$$V_{sp} = V_s + V_A - \sum_{m=1}^{\infty} V_{sm} \exp\{-t/\tau_m\} \left(\frac{1 - \exp\{-(T - T_1)/\tau_m\}}{1 - \exp\{-T/\tau_m\}} \right). \quad (30)$$

The periodic temperature V_{sp} given in Equation (30) is the result for each "on" pulse at any time $nT + t$ where n is an integer.

Again using Equation (28), the steady periodic temperature at a time $nT + t$, where $T_1 < t < T$, is

$$V_{sp} = V_A + \sum_{m=1}^{\infty} V_{sm} \exp\{-(t - T_1)/\tau_m\} \left(\frac{1 - \exp\{-T_1/\tau_m\}}{1 - \exp\{-T/\tau_m\}} \right). \quad (31)$$

Since the heating time T_1 is usually considerably smaller than the thermal constants τ_m, we may simplify Equation (30) by using Equation (26) and a small argument exponential expansion for low-duty-cycle operation:

$$V_{sp} \simeq V_A + a \sum_{m=1}^{\infty} V_{sm} \left(\frac{b}{\lambda_m} \right), T_1 \gg \tau_m. \tag{32}$$

Similarly, for high duty cycle (i.e., $T_1/T \geq 0.1$, $T_1 < \tau_m$,

$$V_{sp} \simeq V_A + a \sum_{m=1}^{\infty} V_{sm} \left[1 + \frac{b}{\lambda_m} (1 - a) \right], \quad T \ll \tau_m, \tag{33}$$

and for medium duty ($T \simeq \tau_m, \lambda_m \simeq 1$)

$$V_{sp} \simeq V_A + a \sum_{m=1}^{\infty} \frac{V_{sm}}{\lambda_m} \left[\frac{1}{(e^{1/\lambda_m} - 1)} + b \left(1 - \frac{a}{\lambda_m} \frac{1}{(e^{1/\lambda_m} - 1)} \right) \right], T \simeq \tau_m \tag{34}$$

where $a = T_1/T$ (duty factor), $b = t/T_1$ (normalized heating time), and $\lambda_m = \tau_m/T$ (relative time constants). These approximate expressions have been used to plot the normalized curves in Figures 10(a), (b), and (c), which indicate the periodic build-up of the "average" temperature as a function of the parameters a, b, and λ_m. Similar results could be readily obtained from Equation (30) for wide pulse widths (i.e., $T_1 > \tau_m$). With the inclusion of an additional pulse-modulation term in the source $A(s)$, this analysis may be extended to cover the burst-heating mode, in which short fast pulses at a high duty cycle occur periodically during a gated interval.

In a practical situation, where the external source V_A is not available, one may substitute V_{pav}, the steady-state effect of the spreading resistance from Equation (14), such that $V_A \rightarrow V_{pav}$ in Equations (30) through (34). The validity of this substitution is based upon the fact that the thermal time constant of the post geometry is considerably larger than any τ_m, thereby resulting in a rising pulse transient response for the device (Figure 10) superimposed upon a *constant* temperature, $V_A = V_{pav}$.

The equations derived in this section are very general in form. In many cases considerable simplification is possible after the roots of Equation (19) are found. Their relative magnitude often permits a first-order approximation for the constant τ_m. These roots and the associated thermal-transient response for this layered geometry are functions of the material thermal constants and the length associated with the sink interface layer $l \leq x \leq l_2$.

A computation of the thermal constants τ_m for all GaAs n/n+ material, as well as for GaAs n on Si n+ substrate, based on Equation (19) and the appropriate constants from Table I is shown in Table II.

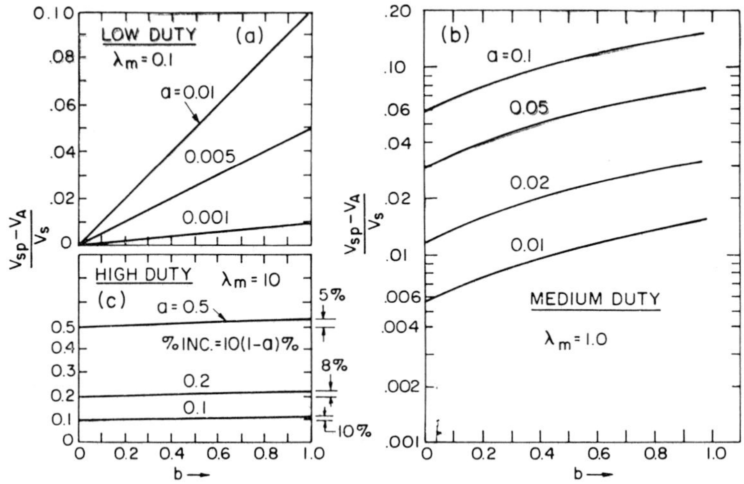

Fig. 10—Temperature rise during pulse for steady periodic heating.

Several relative lengths kl_2/l have been used as parameters. The limiting result for $l = 0$ (i.e., $\beta_m l = m\pi/2$) is also indicated. The roots extracted from Equation (19) require an active-layer length l to be converted into the time constants in Equation (23). The τ_m for both 10 μm and 20 μm samples have been computed. The result for any length is readily obtained by multiplying the 10 μm result by the square ratio of the new length relative to 10 μm. Similarly, in the case of symmetrical heat sinking from both sides of the active layer, the effective length l is halved and τ_m thereby quartered.

The presence of a contact layer ($l_2 \neq 0$) is seen to increase τ_m, although insignificantly, for $kl_2/l < 1/5$ relative to the ideal condition where $l_2 = 0$. The use of Si rather than GaAs(n+) substrate improves the relative transient response when $kl_2/l > 1/5$, but has only a small effect for thin layers. The most practical scheme for reducing the thermal rise in the contact n+ layer is to bond the thin epi-layer to the sink using suitable flip-chip techniques.

kl_2/l	τ_1 μsec $l = 10$ μm	τ_1 μsec $l = 20$ μm	τ_2 μsec $l = 10$ μm	τ_2 μsec $l = 20$ μm	τ_3 μsec $l = 10$ μm	τ_3 μsec $l = 20$ μm	τ_4 μsec $l = 10$ μm	τ_4 μsec $l = 20$ μm
GaAs n/n+, $k = \sigma' = 1.12$								
1	6.10	24.4	.610	2.44	.230	.921	.112	.450
1/2	3.36	13.5	.354	1.52	.125	.5	—	—
1/5	2.1	8.4	.23	.925	.082	.328	—	—
1/10	1.72	6.9	.192	.77	.07	.28	—	—
GaAs n, Si substrate, $k = .670$, $\sigma' = .650$								
1	4.4	17.6	.68	2.72	.213	.86	.12	.48
1/2	2.44	9.8	.354	1.52	.134	.54	—	—
1/5	1.77	7.1	.203	.82	.078	.312	—	—
1/10	1.58	6.3	.178	.71	.064	.256	—	—
GaAs n ($l_2 = 0$)								
0	1.40	5.6	.156	.62	.056	.224	—	—

Table II—Thermal Time Constants for Specific TEO Geometries

Conclusion

A method of analyzing the temperature gradient within the active layer of a bulk-effect GaAs device during pulsed excitation has been presented. The close analogy between the layered thermal geometry of this and other semiconductor devices and a cascaded electrical transmission line permits the use of electrical duals to the thermal parameters. The Laplace transform, matrix techniques, and superposition of sources may readily be applied in analyzing the thermal network.

When a TEO device is characterized by an optimum product of the electron concentration and the active layer length, the heat density is inversely related to this length. The asymptotic steady-state temperature is *linearly* dependent on n-region thickness, while the total dissipated power remains independent of the n-region thickness and proportional to the cross-sectional area. The error involved in neglecting the inverse temperature dependence of the semiconductor thermal conductivity when calculating its steady-state asymptotic value is insignificant in most cases where $l \leqq 20$ and the heat density $A_o \leqq 2 \times 10^7$ W/cm^3.

The thermal transient response and associated time constants are most seriously affected by the contact-layer thickness. The poor thermal characteristics of GaAs require both minimum active-layer and contact-layer thickness to reduce thermal rise time. The limiting response is imposed by the active layer, where the time constants are related to the square of the different relative lengths. The thickness of the contact layer becomes insignificant when it is less than 20% of the active-layer lengths. Since the semiconductor material (primarily because of its relative volume) is actually faster thermally than the larger volume of metallic sink, the transient temperatures derived in this analysis, may be referred to an isothermal surface. The spreading resistance is usually significant and the resulting equilibrium temperature may be added directly to the steady periodic or cw solutions.

Although this analysis has used the TEO as a model for studying the transient temperatures associated with a pulsed device, the mathematical approach may be readily applied to evaluating other layered devices, possibly including those with nonuniform heating.

Acknowledgment

The author is grateful to T. E. Walsh for his encouragement and independent research, which provided the basis for several constructive suggestions.

Appendix I—Heat Conduction in One Dimension

A. Fundamental Relations

Consider a solid capable of generating heat energy (source) throughout its volume and supplying that heat through a process of conduction to a passive medium (sink), which is assumed to be unheated by any internal Joule effects. The entire volume is considered to be enclosed in an adiabatic environment where no heat is removed from any surface through any process other than conduction (i.e., no radiation or convection). We will now examine closely the thermal behavior of a one-dimensional geometry (see Figures 1 and 2).

The flow of heat in a one-dimensional linear medium containing a continuous source of heat ($t > 0$) is described by the following equation[1]

$$\frac{\partial^2 V}{\partial x^2} - \frac{1}{\alpha_1} \frac{\partial V}{\partial t} = -\frac{A(x,t)}{K_1}, \; 0 < x < l, \; t > 0 \qquad (35)$$

where $\alpha_1 = K_1/\rho_1 c_1$ is the thermal diffusivity, (K, ρ, and c are thermal conductivity, density, and specific heat, respectively), $A(x, t)$ is the rate of heat generation per unit time per unit volume, and the temperature $V = T(x, t)$. It has been assumed that the thermal properties, K, α, c, and A are all independent of the temperature as well as the position within a given medium. In the more general case,[2,5] where $K, c = f(V)$, the solution of a nonlinear partial differential equation is required. For the general case, an exact analytic transient solution for the present model would not be possible; however, a steady-state solution may be found (see Appendix II) and compared with asymptotic limit in the following linearized analysis.

Equation (35) may be transformed into Laplace s-space using

$$L\{V(x,t)\} = V(s) = \int_0^\infty e^{-st} V(x,t)\, dt. \qquad (36)$$

Then,

$$\frac{\partial^2 V_1(s)}{\partial x^2} - q_1^2 V_1(s) = -\frac{A(s)}{K_1} - \frac{V_o}{\alpha_1}, \; 0 < x < l, \qquad (37)$$

and, similarly,

$$\frac{\partial^2 V_n(s)}{\partial x^2} - q_n{}^2 V_n(s) = -\frac{V_o}{\alpha_n}, l < x. \tag{38}$$

Here, V_o is the initial temperature of the solid at $t = 0$. Also,

$$q_n{}^2 = s/\alpha_n, \ (n = 1, 2, 3, \text{etc.}) \tag{39}$$

Substituting $V = U + \bar{A}$ into (37), where the particular source solution is

$$\bar{A} = \frac{A(s)\alpha_1}{sK_1} + \frac{V_o}{s}, \tag{40}$$

we obtain an ordinary homogeneous equation in $U(s)$. This equation is readily solved to give

$$U(x) = U(l) \cosh q_1 (l - x) + f(l) Z_1 \sinh q_1 (l - x), \tag{41}$$

where l is the length of the source (Region I) and

$$U(l) = V(l) - \bar{A}, \tag{42}$$

and where

$$Z_1 = \frac{1}{K_1 q_1}, Z_n = \frac{1}{K_n q_n} \tag{43}$$

B. Boundary Conditions

The appropriate flux boundary conditions for the source in Region I are

$$f(x) = -K_1 \frac{\partial V}{\partial x} = -K_1 \frac{\partial U}{\partial x} = \begin{cases} 0 \text{ at } x = 0 \\ f(l) \text{ at } x = l \end{cases} \tag{44}$$

corresponding to unilateral heat flow in the positive x direction. At all remaining interfaces (between each subregion), both the fluxes and temperatures are continuous (i.e., V and dV/dx are continuous).

For the purposes of this analysis, V_o, the initial temperature of the solid at $t = 0$, is selected to be equal to zero. The resulting solution of Equation (38) provides the temperature relationship between the sink regions and the source (all initially at zero temperature) after consideration of all the necessary boundary conditions, particularly at the interface where $x = l$ in Equation (44). In most practical cases, the device can assumed to be terminated by an infinite half-space resulting in a finite spreading resistance. During thermal transients, the device response time is considerably faster than that for the half-space sink. The transient solution shall ignore the spreading resistance and refer the device to an isotherm. A constant temperature is achieved by introduring a second source (i.e., V_A/s) into this analysis by using the superposition principle. This method provides the transient response resulting from the sink termination.

The only other case of interest is when $V_A = 0$ at $x_n \to \infty$ and the sink is thereby considered to be infinite in its length. Without the mechanism of heat removal by convection or radiation, this geometry would result in a continued increase in temperature due to $A(s)$.

C. *Source Condition (Uniform Heat Generation)*

In Laplace notation, we may represent a uniform power density A_o (watts/cm³) as

$$A(s) = \frac{A_o}{s}. \tag{45}$$

The primary source driving function $A(s)$ may consist of a periodic waveform of period T and pulse "on" width T_1. In Laplace notation,

$$A(s) = \frac{A_o(1 - e^{-sT_1})}{s(1 - e^{-sT})}, \quad \begin{array}{l} A(t) = A_o,\ nT < t < nT + T_1 \\ A(t) = 0,\ nT + T_1 < t < (n+1)T \end{array} \tag{46}$$

In many cases, it is of significant interest to study the "heating" behavior of our sample during and after a *single* pulse. This simplified case would provide useful rise and decay transient information for a specific geometry. In this case a delayed negative unit step is superimposed upon a unit step at $t = 0$, so that Equation (46) reduces to

$$A_2(s) = \frac{A_o}{s}(1 - e^{-sT_1}) \tag{47}$$

D. *Linear Thermal Network*

The solution to Equation (37), for the temperature distribution in the source region of the geometry shown in Figure 3 arising from $A(s)$ and with the necessary boundary conditions of Equation (44), is

$$\overline{V_1}(x) = \bar{A}\left[1 - \frac{C_1(x)}{C_1(l) + \sigma S_1(l)}\right], 0 < x < l, \tag{48}$$

where

$$\sigma = \frac{Z(l)}{Z_1} \tag{49}$$

$$C_1(x) = \cosh q_1 x$$
$$S_1(x) = \sinh q_1 x \tag{50}$$

and a driving point thermal impedance has been defined as

$$Z(l) = \frac{V(l)}{f(l)}, \tag{51}$$

where $Z(l)$ provides a direct algebraic means of introducing the passive $(V_A = 0)$ heat-sink parameters into the solution of Equation (48). Equation (38) may be solved for each individual sink region in a similar way. The terminal relationship for each subregion may be written in matrix form as

$$\begin{bmatrix} V_n(l_n) \\ \\ f_n(l_n) \end{bmatrix} = \begin{bmatrix} C_n(l_n) & -Z_n S_n(l_n) \\ \\ \frac{-1}{Z_n} S_n(l_n) & C_n(l_n) \end{bmatrix} \begin{bmatrix} V_{n-1}(l_{n-1}) \\ \\ f_{n-1}(l_{n-1}) \end{bmatrix} \tag{52}$$

where $l_n = x_n - x_{n-1}$ corresponds to the length of the nth subregion, and $q_n^2 = s/\alpha_n$. The form of solution in Equation (52) is readily extended to a cascaded series of subregions, each characterized by a set of constants, K_n, ρ_n, C_n, α_n, etc.

The function $Z(l)$ may be found by multiplication of the sink matrices of Equation (52), and substitution into Equation (48) with appropriate boundary conditions. For the case where a finite length of sink is terminated at a constant temperature V_A, we may use superposition. Initially, $V_A = 0$, so that the multiplication of the subregion matrices of Equation (52) gives,

$$\begin{pmatrix} 0 \\ f_A \end{pmatrix} = \begin{bmatrix} A & B \\ C & D \end{bmatrix} \begin{pmatrix} V(l) \\ f(l) \end{pmatrix} \tag{53}$$

where it is readily found that AD-CB = 1. The primary source driving point impedance of Equation (51) for $V_A = 0$ becomes,

$$Z(l) = \frac{V(l)}{f(l)} = -B/A. \tag{54}$$

It is now necessary to let $A(s) = 0$ and solve for the temperature distribution in Region I due to a constant source V_A at $x = x_n$. Reversing the flow of heat at $x = l$ and using Equations (52) and (44),

$$V_A(l) = Z_1 \frac{C_1(l)}{S_1(l)} f_A(l). \tag{55}$$

The subscript A is used to denote dependance on source V_A with flow in the negative x direction. Introducing a finite source V_A (i.e., $0 \to V_A$, $f_A \to -f_A$, $f(l) \to -f_A(l)$) into Equation (53) we have,

$$\begin{bmatrix} A & B \\ C & D \end{bmatrix} \begin{bmatrix} 1 \\ -\frac{1}{Z_1} \frac{S_1(l)}{C_1(l)} \end{bmatrix} = \begin{bmatrix} \frac{V_A}{V_A(l)} \\ \frac{-f_A}{V_A(l)} \end{bmatrix} \tag{56}$$

or,

$$V_A(l) = \frac{V_A}{A - \frac{BS_1(l)}{Z_1 C_1(l)}}. \tag{57}$$

Then, using Equations (41), (55), and (57) and noting the direction of $f_A(x)$,

$$V_{1A}(x) = V_A \left[\frac{C_1(x)}{AC_1(l) - \frac{B}{Z_1} S_1(l)} \right]. \tag{58}$$

Combining the results of Equations (47), (49), (54), and (58),

$$V_1(x) = \overline{V_1(x)} + V_{1A}(x)$$
$$= \bar{A}\left[1 - \frac{C_1(x)}{C_1(l) + \sigma S_1(l)}\right] + \frac{V_A}{s}\left[\frac{C_1(x)}{AC_1(l) + \sigma S_1(l)}\right]. \tag{59}$$

It may often be necessary to know the variation of the average temperature in the active layer (Region I). For this case, using Equation (59),

$$V_{AV}(x) = \frac{1}{l}\int^{l} V_1(x)\,dx,$$

$$= \bar{A}\left(1 - \frac{S_1(l)}{lq_1[C_1(l) + \sigma S_1(l)]}\right) + \frac{V_A}{s}\left(\frac{S_1(l)}{lq_1A[C_1(l) + \sigma S_1(l)]}\right)$$
$$\text{for } 0 < x < l$$

E. *Inverse Transformation*

The resulting temperature distribution, $V_1(x, t)$ is found from the Laplace transform by the use of an inversion theorem, and contour integration. Thus,

$$V_1(x, t) = \frac{1}{2\pi i}\int_{\gamma - i\infty}^{\gamma + i\infty} e^{st} V_1(x, s)\,ds, \; \gamma > 0. \tag{50}$$

For the present problem, where we are dealing with finite subregions, the transforms are free of branch points and usually have poles at the origin and along the negative real axis. The inverse transforms of Equation (59) reduce to a sum of residues of $V_1(x, s)$. In many cases, the transforms are readily obtained as tabulated functions.

Appendix II—The Effect of an Inverse Temperature Dependence of K on the Steady-State Solution

If the inverse temperature variation of the thermal conductivity K

is included in the thermal diffusion equation with $\partial T/\partial t = 0$ in the steady state, it is readily found[1] that for uniform heat generation throughout the active layer $(0 \leq x \leq l)$,

$$\frac{d^2}{dx^2}(\ln T) = \frac{-A_o}{K_1 T_o}, \; 0 < x < l \tag{63}$$

$$\frac{1}{T}\frac{dT}{dx} = -\frac{(A_o l)}{K_2 T_o}, \; l \leq x \leq l_2. \tag{64}$$

T represents the variable temperature and T_o is the reference ambient sink temperature (i.e., 300°K) at which the *constant* values of K have been calculated in the preceeding analysis.

Equations (63) and (64) are readily solved[2] and may be used for comparison with the two-layer case discussed in this paper. After introducing similar boundary conditions and neglecting sink resistance[15] for $x > l_2$,

$$\frac{T_s}{T_o} = \exp\left\{\frac{V_s}{T_o}\right\} \tag{65}$$

where T_s is the steady-state asymptotic temperature characteristic of the inverse temperature variation of K, and V_s is the steady-state asymptotic temperature relative to T_o for a constant value of K (Equation (22)). Substituting Equation (22) into Equation (65) and choosing a reference temperature $T_o \simeq 300$°K, we may calculate the actual device temperatures T_s and V_{so}. These results are plotted in Figure 9 versus the layer thickness l.

4.2 technology

For low average power devices it is sufficient to solder the device into the encapsulation with the substrate adjacent to the heat sink; the fabrication is then completed with a wire bond to the epitaxial layer contact. The structure is limited to power dissipations around the one watt level. Higher powers necessitate that the device be inverted; there are two possible alternatives in the construction.

Direct Bonding

The chip is bonded directly to the heat sink by thermocompression or ultrasonic means. Solder bonding is not preferred because all solders have a low thermal conductivity. Also, the device could be shorted by solder moving up the sides of the chip.

Gold to gold bonding is undoubtedly the best combination; the device must have a gold layer deposited onto the metal semiconductor contact. The thickness must be a minimum of 2-5 μm; thus, electroplated films are more convenient than evaporated. The copper heat sink also is plated with gold to the same thickness.

Thermocompression bonding requires a temperature of 250-300°C and pressures up to 2000 kg cm^{-2} to produce satisfactory bonds. The energy needed for ultrasonic bonding is yet to be determined experimentally.

Direct bonding has two disadvantages. First, the device chip must be cut to size and handled; often this step poses major problems with chip sizes as small as 100 μm square. Second, the bonding conditions are extremely severe; devices are easily damaged or degraded during bonding. The larger the chip, the easier direct bonding becomes, making the procedure ideal for high average power and high peak power pulsed devices. For continuous operation at high power levels, the integrated heat sink approach is favoured, although multi-mesa devices can be direct bonded.

Integrated Heat Sink

The integrated heat sink process, shown in Figure 4-3 and described in the key paper by Zettler and Cowley, involves several critical steps. The heat sink metal is electroplated (copper, silver, and gold have all been used successfully); good adhesion of the plated film to the contact metal is needed, so diligent control of the plating conditions is necessary. The etch thinning of the substrate must be uniform and parallel to the semiconductor heat sink interface to ensure that etch occurs simultaneously through all of the mesas. Also, the mesa etching of the individual devices has to be carefully controlled to achieve a high yield — once the etch reaches the contact metal it can undercut the device very quickly.

Final separation of the devices is achieved using a razor blade or sharp scalpel; the units are subsequently soldered or ultrasonically bonded into an encapsulation.

The IHS process can produce any size device, but generally the final heat sink diameter is approximately four times that of the device. The total device yield per wafer therefore is considerably less than for direct bonding. For device diameters $>$ 150 μm the direct bonding procedure is more economical, with a possible yield 16 times greater than for IHS.

A problem encountered with InP has been that the plating of the thick heat sink has changed the I-V characteristics of the device. The cause of this alteration is probably due to strain between the metal and the semiconductor. It is important, therefore, to use high purity plating solutions and to aim for soft and ductile films.

Comparison of Techniques

Apart from the relative advantages of each process that have been discussed, it must be considered that the technology of both IHS and direct bonding is difficult. The former requires critical wafer processing; the latter, critical device bonding. The spreading resistance can be made small by multimesa design[76], which obviates the need for expensive diamond heat sinks.

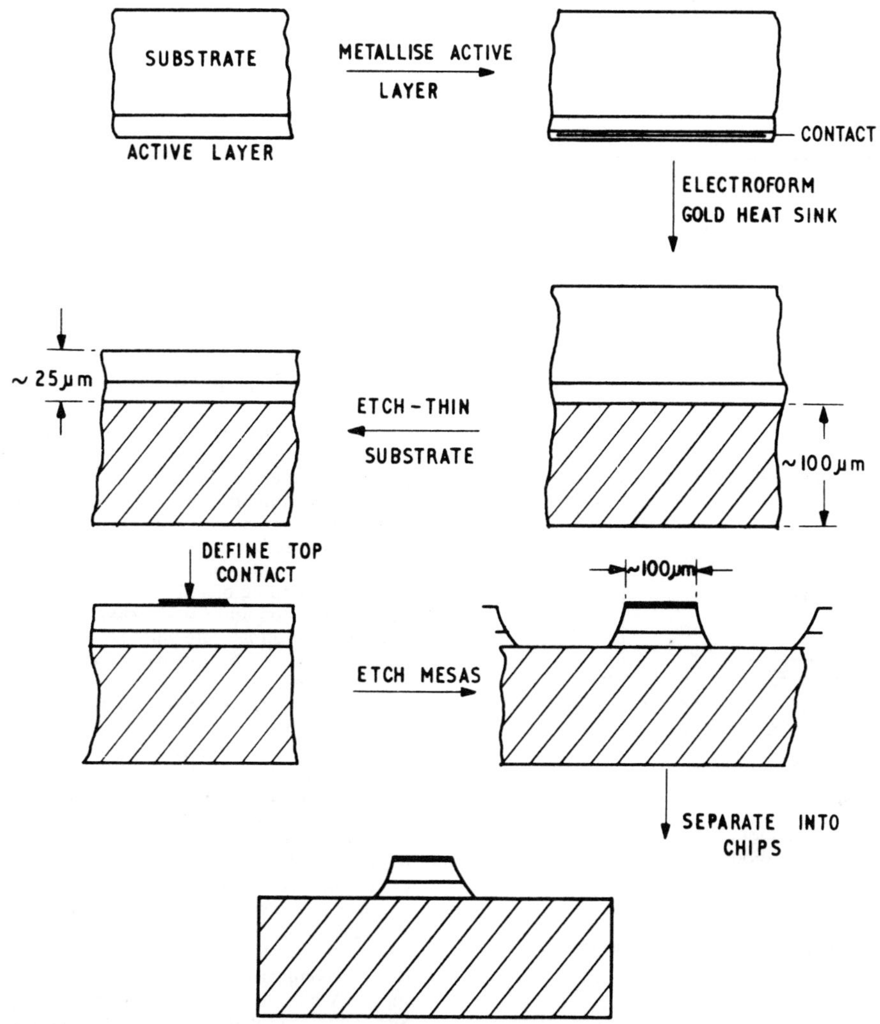

Figure 4-3/Integral Heat Sink Process.[79]

InP transferred electron devices have to work at higher power densities than comparable GaAs devices, and this necessity introduces special problems in the heat sink which can only be overcome by working with small area devices. Hence the IHS becomes essential for CW operation[77]. The only other possibility is to use the direct bonded multimesa approach[78] which, in many ways, combines all of the disadvantages of IHS and direct bonding.

Few attempts to heat sink from both sides of the device have been made, apart from thick LSA diodes where the active region is often 100 μm. The difficulty has been in the design and assembly of an adequate encapsulation incorporating the double heat sink. All the devices discussed have thermocompression bonded gold wire or ribbon contacts to the substrate side.

Batch Fabrication Of Integral Heat Sink IMPATT Diodes

R.A. Zettler and A.M. Cowley
(Hewlett-Packard Co., Palo Alto, California, US)

Electronics Letters, Vol. 5, No. 26 27 (December 1969)
Pages 693-694

key paper

A new technique for IMPATT-diode fabrication is described, in which the critical portion of the diode heat-flow path is formed as an integral part of the diode. The technique has economic advantages over conventional methods in that all processing steps are performed while diodes are still in wafer form. In addition, the technique permits passivation using standard dielectric deposition techniques, and the diode geometrical shape is such as to suppress premature edge breakdown. Typical experimental results are shown.

The fabrication of IMPATT-oscillator devices has as one of its primary problems the heat sinking of the diode chip. For example, it is quite common for device manufacturers to individually thermocompression-bond an inverted mesa chip to a gold-plated copper heat sink, as in Fig. 1A. Although this technique can sometimes produce nearly ideal thermal interfaces between the diode and the heat sink, it is a relatively difficult and slow operation, and is not very reproducible.

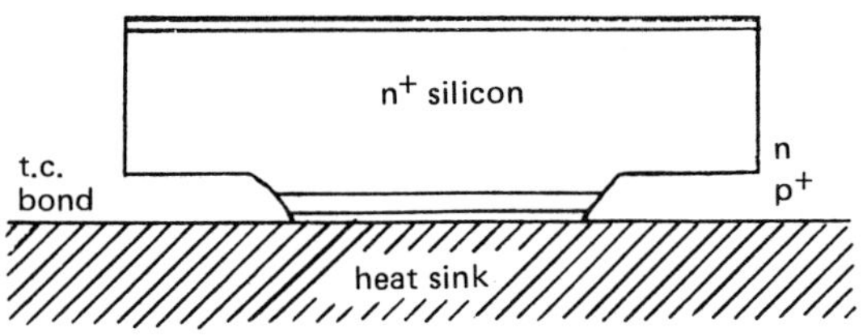

Fig. 1A *Conventional 'flip-chip' technique for mounting IMPATT diodes*

This letter describes a simple technique for the batch fabrication of integral heat-sink IMPATT diodes which offer several distinct advantages over the most common approaches in the areas of reproducibility and packaging ease, without sacrifice of device performance.*

The basic fabrication technique is illustrated in Fig. 1B and consists of the following steps:

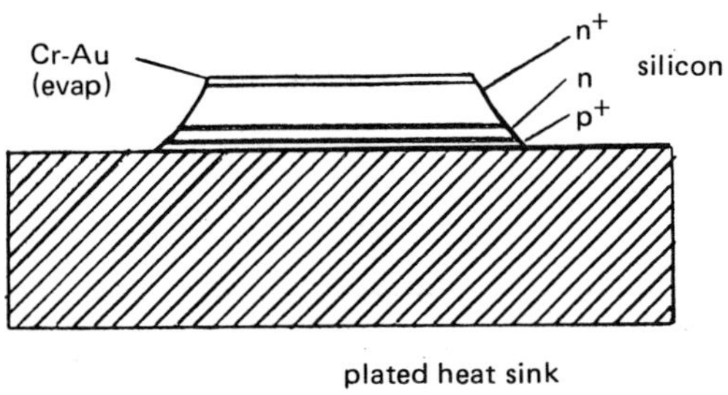

Fig. 1B *New approach for IMPATT-diode fabrication, in which diode and critical portion of heat sink are fabricated in wafer form*

(*a*) A $p^+ n$–n^+ double epitaxial (or diffused) silicon starting wafer is metallised on the junction (p^+) side of the wafer.

(*b*) The wafer is then electroplated over the entire p^+ surface with gold, copper or silver to a thickness of several thousandths of an inch.

(*c*) The substrate or n^+ side of the wafer is chemically or mechanically thinned to a thickness of 1·0–2·0 10^{-3} in.

(*d*) The wafer is metallised on the n^+ side, standard photoresist techniques are applied, and the silicon wafer is etched through to its plated metal substrate leaving mesas on the metal substrate having the general shape as shown in Fig. 1B.

After separation, the integral-heat-sink chips can be handled in the same manner as right-side-up mesa devices and can be soldered or bonded into packages or stripline circuits with relative ease, since the heat generated at the diode junction has been permitted to spread in the high-thermal-conductivity metal substrate, thus making the introduction of a solder interface or nonuniform bond much less critical.

In addition to the cost advantages of batch fabrication, the integral-heat-sink technique offers several other advantages over the conventional approach of thermocompression or ultrasonic bonding of individual inverted mesa chips. First, with both silicon and GaAs IMPATT diodes, the conventional profiles use n on n^+ substrates so that the p^+ side is nearest the heat sink. With a normal mesa etched from the p^+ side, the radius of curvature of the junction is such that the electric field at the periphery of the diode is increased over its value in the bulk. With a mesa etched from the n^+ side, the radius of curvature is opposite that of the right-side-up mesa, and the field at the periphery is reduced from that in the bulk. This tends to suppress any edge leakage or premature breakdown in the device. Secondly, the individual bonding of inverted mesas with its need for a very uniform thermal bond at the top of the mesa does not permit the use of straightforward passivation techniques, particularly since the device is often damaged during the bonding and must be etched *in situ*. The configuration of the devices in our new approach is such that batch passivation of the diodes can be easily accomplished by depositing or evaporating dielectric films. We have demonstrated this in our laboratory with SiO_2 films deposited from silane at temperatures in the region of 325 C. The microwave performance of devices passivated in this manner was neither degraded nor improved compared with unpassivated devices, but handling of the chips during soldering was further simplified.

Results: Diodes fabricated in this manner have been packaged in standard varactor-type ceramic pill packages and operated in simple singly tuned coaxial cavities.[2]

With diodes designed primarily for X band operation, output powers in excess of 1 W c.w. have been obtained at frequencies from 6 to 13 GHz. These results were obtained without specific optimisation of the circuit impedance at the second harmonic of the operating frequencies.[3] Efficiencies as high as 10% have been achieved, with 8% being very common.

Thermal resistance was measured in the avalanche-breakdown mode using a variation of the technique described by Hartz.[4] Fig. 2 shows data accumulated for a number of diodes from different fabrication runs. The solid line was calculated for the configuration shown in the inset, using the commonly accepted values for thermal conductivity of gold, copper, silicon, and solder (tin). The broken line represents

* The technique described has also been independently proposed by Hoeflinger[1]

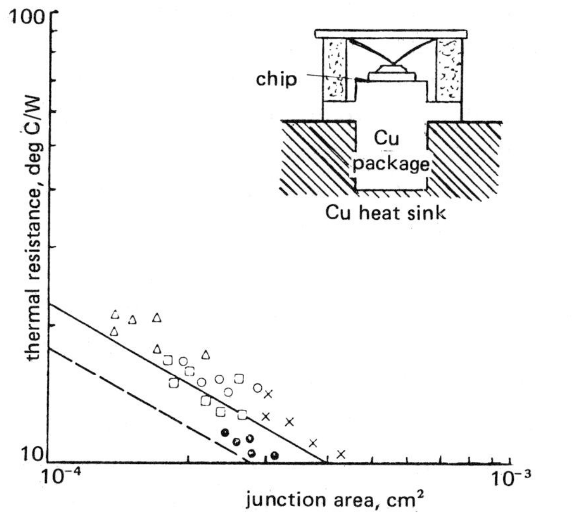

Fig. **2** *Measured thermal resistance of experimental diodes compared with calculated values for a circular diode.*

Shown for comparison is thermal resistance of conventional silicon mesa on 'infinite' copper heat sink, assuming no thermal resistance in bond

△ ○ □ × } circular mesa

● elongated mesa

the thermal resistance of a conventional silicon mesa on an infinite copper heat sink, assuming no thermal resistance in the bond to the heat sink, usually not a good assumption in practice. Our diodes designed for X band operation are generally fabricated with $A \simeq 2 \times 10^{-4}$ cm²; the typical value of θ is about 16–17 degC/W. We can easily achieve substantial (700–800 mW) c.w. microwave powers with junction temperatures below 200°C, and are occasionally able to extract 1·0 W from a diode with less than 200°C junction temperature. With higher junction temperatures, 1·0 W is easily obtainable. Burnout of the diodes occurs for junction temperatures in the vicinity of 350°C.

In addition, devices using a long thin mesa (approximately 5 : 1 length/width ratio) have given 15–20% improvement in thermal resistance, for junction areas of about 2×10^{-4} cm² (Fig. 2).

With the obvious desire of the microwave industry for more solid-state power, it is quite likely that IMPATT-diode arrays will come into increasing use. The simple fabrication technique described above will permit direct fabrication of parallel arrays, will simplify the construction of series-parallel arrays, and should considerably ease the task of component manufacturers who want to install IMPATT-diode chips in microwave integrated circuits.

The authors wish to acknowledge the skilled assistance of Mrs. M. Bryan, Mrs. E. Petherbridge, and R. C. Patterson and M. R. Gavette in fabricating and testing the diodes described in this letter.

References

1 HOEFFLINGER, B.: 'Recent developments on avalanche diode oscillators', *Microwave J.*, 1969, **12**, pp. 101–115

2 COWLEY, A. M., *et al.*: 'The singly-tuned IMPATT oscillator', to be presented at 1970 IEEE Solid-State Circuits Conference, Philadelphia, Pa.

3 SWAN, C. B.: 'IMPATT oscillator performance improvement with second-harmonic tuning', *Proc. Inst. Elect. Electron. Engrs.*, 1968, **56**, pp. 1616–1617

4 HAITZ, R. H., STOVER, H. L., and TOLAR, N. J.: 'A method for heat flow resistance measurements in avalanche diodes', *IEEE Trans.*, 1969, **ED-16**, pp. 438–444

4.3 diamond heat sinks

For many years, diamond has been investigated as a heat sink material; encouraging results have been obtained[80], although the process has never gone into production. Improvements in device technology so far have managed to keep abreast of higher thermal requirements without resorting to special materials. But as demand increases for higher output powers, the eventual inclusion of diamond in a commercial device is inevitable.

The diamond with the best thermal conductivity is type IIA[81], with a room temperature conductivity of 25 W/°K/cm (6 times that of copper). The thermal conductivity of diamond decreases as a function of 1/T; thus, some advantage is lost at elevated temperatures. In a device structure, a portion of the thermal resistance is in the semiconductor, with some in the metallisation – whether a diamond heat sink is used or not. An improvement in overall thermal resistance of perhaps two is therefore a realistic design goal for a diamond heat sink.

The diamond industry presently supplies selected and processed diamond blocks and inserts suitable for direct incorporation into the device/package assembly, however, they are costly.

4.4 measurement of thermal resistance

The thermal resistance of devices can be calculated with acceptable accuracy from the theory discussed; but only assumptions can be made about the departures from the ideal during the device processing, particularly with respect to the various interfaces between the active layer and the semi-infinite heat sink. For this reason, a simple on-line measurement method is invaluable for quality control.

Transferred electron devices have a bulk resistance which is temperature dependent. It is therefore possible to compare the electrical resistance measured with a short pulse to that measured with dc[82]. Again, allowances must be made for contact resistances and doping fluctuations. Meyer[82] has developed analytical expressions for these considerations.

IMPATT's are characterised by the variation of breakdown (avalanche) voltage with temperature[83], which gives a good first approximation to operating temperatures.

Since heat dissipation is a serious problem only with the power generating devices, the heat sinking of mixers and varactors is not critical. Field effect transistors pose a particular problem in that they are planar devices and integral heat sink techniques are not applicable.

chapter five

the encapsulation

Device packages at microwave frequencies are extremely expensive, and often impose limitations on the electrical or thermal characteristics of the unit. A selection of diode packages is shown in Figure 5-1; all of these components have an alumina ceramic insulating sleeve with OFHC copper pedestals (on which the chip is mounted) and kovar flanges and covers.

Figure 5-1/Microwave Diode Encapsulation

The package is designed to mechanically protect and hermetically seal the device, to provide a good thermal path, and to have parasitic reactances which do not affect the performance of the device at high frequencies. Commercially available packages meet all these conditions (at least up to 20 GHz); it is unlikely that there will be any significant advances in the technology in the near future. Plastic encapsulation is unlikely to offer any cost advantage, as the final assembly would be difficult. An immediate alternative is to mount the chip directly into the microwave circuit; this procedure has many advantages, although the hermetic seals are difficult and the circuit/device becomes an expensive item.

As the applications of microstrip and stripline microwave integrated circuits increase, more devices are mounted directly into a circuit to make a complete circuit function block.

At frequencies above 20 GHz, commercial packages present serious problems through parasitic reactances, which can be only by moving to smaller dimensions and using low dielectric constant materials. One technique is to provide a small quartz post which is metallised on the top and fixed to the diode mount; the bond wire electrically connects the top of the quartz and the diode, while the quartz post provides the mechanical strength for pressure contact. An integrated process has been developed[84] to produce a similar structure using glass-ceramic as the insulator material. The technique is described in the key paper by Colliver. The glass ceramic has good dielectric properties; it can be screen printed directly on copper to form a good chemical bond; and it can be shaped easily to any desired form. No packages have been developed to allow heat removal from both faces of the device, although this feature would be desirable and practicable.

Microwave Performance Of Glass-Ceramic/Copper Encapsulations

key paper

D.J. Colliver
(Royal Radar Establishment, Great Malvern, Worcs., UK)
Electronics Letters, Vol. 8, No. 23 (November 1972)
Pages 569-570

Indexing terms: Encapsulation, Glass, Ceramics, Cavity resonators, Microwave measurements

A previous letter by Davies *et al.* has described the methods of preparing microwave encapsulations using a glass–ceramic/copper encapsulation. Various configurations have been evaluated over the frequency band 26–40 GHz with c.w. Gunn oscillators. Results show that frequency saturation effects due to encapsulation parasitics can be avoided by the use of glass–ceramic encapsulations of suitable geometry.

When microwave devices are operated in waveguide cavities, the equivalent circuit comprises the device parameters, the package reactances and the transformation through the mounting structure to the external waveguide circuit. This type of cavity has often yielded unexpected results[2] in the tuning characteristic, which can be relatively insensitive to short-circuit position, often with abrupt changes in frequency and power output.

Theoretical and experimental studies have been carried out[3,4] to explain these departures from the ideal circuit control that is required; however, these have so far been restricted to circuits operating at frequencies up to about 20 GHz. At higher frequencies, the device and mounting post become even more significant, because these occupy a larger fraction of the volume of the cavity. Taylor *et al.*[2] showed experimentally that the saturation frequency exhibited by devices operating in the band 26–40 GHz was dominated by the parasitic reactances of the encapsulation and the diode impedance. A number of laboratories have overcome this limitation by the use of quartz standoff posts in place of conventional alumina encapsulations, but this technique involves a number of extra processing steps in the fabrication of the device and is somewhat limited in its application.

Fig. 1 *Device mounted alongside glass–ceramic standoff post*
Glass–ceramic dimensions are 0·3 mm diameter × 0·15 mm high

To produce devices in this frequency range with good circuit control, a number of commercially available alumina encapsulations have been investigated, namely those described by Taylor *et al.*[2] and Frank and Day,[5] in comparison to encapsulations developed from glass–ceramic dielectric.

The glass–ceramic encapsulations were fabricated in two basic designs,[1] as an annulus and as a simple ceramic post of variable diameter/height ratios. This enabled a range of inductive and capacitive parasitics to be investigated. A mounted device is shown in Fig. 1. Throughout the measurements, devices of identical dimensions from the same gallium-arsenide wafer were used, and these were inserted into both full-height and reduced-height (1·5 mm) waveguide circuits. The bias post was mounted centrally in the broad face of the guide, with facilities for variation of bias-post diameter.

The device tuning curves were measured at a fixed bias voltage as a function of cavity length, measured from the centre of the post to the adjustable short-circuit plane.

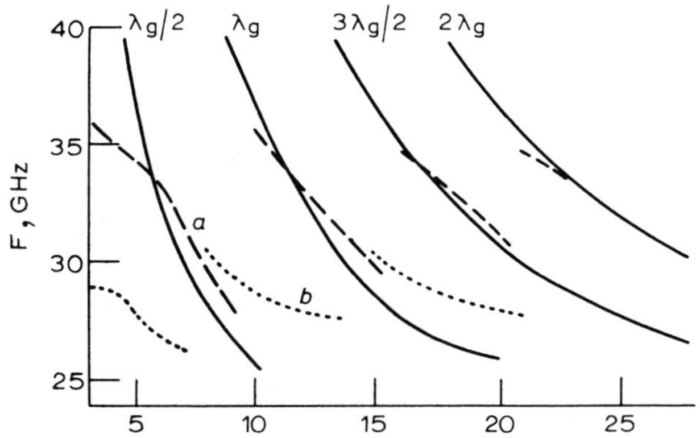

Fig. 2 *Comparison of tuning characteristics of low-impedance devices*

a Mounted alongside glass–ceramic post
b Mounted in conventional encapsulation, So[5]

The following observations were found to be valid for all the experiments carried out:

(*a*) The saturation frequency was independent of the waveguide height.

(*b*) The tuning characteristic was relatively independent of the bias-post diameter in the range 0·6–2·5 mm.

(*c*) Conventional alumina encapsulations were all limited to the frequency range 26–30 GHz with these low-impedance devices.

(*d*) The glass–ceramic annulus behaved exactly as for the conventional alumina encapsulation of the same lumped-circuit inductance and capacitance.

(*e*) Considerable improvement was achieved with the glass–ceramic standoff post arrangement, as can be seen in the comparison curves shown in Fig. 2.

The encapsulation reactances were calculated for each configuration used, and the appropriate parallel resonance frequency compared to that measured experimentally. The measured saturation frequency was always less than that calculated from the simple package reactance, but the frequency increased as the package reactance decreased, as expected. Package capacitances were in the range 0·025–0·3 pF and lead inductances in the range 50–200 pH, these low values being responsible for the improvements in the observed tuning range.

The glass–ceramic encapsulation offers good dielectric properties, together with design flexibility and low cost, compared with a conventional encapsulation. At high frequencies, the effect of encapsulation parasitics can be minimised, although other local-circuit effects, notably those due to the nonlinear device itself, will eventually limit the maximum frequency of operation.

The author would like to thank G. Partridge and S. E. Davies, of the General Electric Co., Nelson Research Centre, Stafford, for the supply of the glass–ceramic encapsulation, and S. E. Gibbs of the Royal Radar Establishment for valuable discussions throughout. Acknowledgment is made to the Director, Royal Radar Establishment, and to the Controller of HM Stationery Office for permission to publish this letter.

References

1 DAVIES, S. E., PARTRIDGE, G., and MCMILLAN, P. W.: 'Glass-ceramic–copper microwave encapsulations and mountings', *Electron. Lett.*, 1972, **8**, pp. 483–484

2 TAYLOR, B. C., FRAY, S. J., and GIBBS, S. E.: 'Frequency saturation effects in transferred electron oscillators', *IEEE Trans.*, 1970, **MTT-18**, pp. 799–807

3 EISENHART, R. L., and KHAN, P. J.: 'Theoretical and experimental analysis of a waveguide mounting structure', *ibid.*, 1971, **MTT-19**, pp. 706–719

4 EISENHART, R. L., and KHAN, P. J.: 'Some tuning characteristics and oscillation conditions of a waveguide-mounted transferred electron oscillator', *ibid.*, 1972, **ED-19**, pp. 1050–1055

5 FANK, F. B., and DAY, G. F.: 'High CW power K-band Gunn oscillators', *Proc. Inst. Elec. Electron. Eng.*, 1969, **57**, pp. 339–340

chapter six

failure analysis and reliability

All of the devices mentioned in this book are still under development. Reliability data is therefore limited, and often applies to devices superceded by ones of superior performance.

All of the power generating devices have high operating power densities; failure is catastrophic and its cause is consequently difficult to establish. Considerable data, though, has been acquired to identify some failure modes and form a statistical picture of reliability.

Transferred electron devices have operated for over 10,000 hours without failure; estimates of MTBF have been made.[85] Device failure is often associated with poor processing technology; the bond between chip and heat sink is an area particularly susceptible to thermal runaway and eventual breakdown. Contacting technology also plays a crucial role in the device reliability, and can affect the performance of the device over environmental operating conditions. Contact resistance can prevent the device from oscillating when switched on at low temperatures.

Life data with IMPATT's is also incomplete, although again the contact/heatsink region dominates the failure mechanisms[89]; Schottky barrier IMPATT's are a particular example, having high electric fields and temperature gradients at the metal semiconductor interface. Much of the reliability work on IMPATT's has concentrated on the contact metallurgy.

Mixer and varactor diodes have been researched more thoroughly; however, progress continues to be made in understanding failure mechanisms, particularly with regard to RF burnout.

appendices

appendix 1 properties of GaAs and InP

Property	Unit	GaAs	InP
Melting point	°C	1237	1070
Structure		Zinc blende	Zinc blende
Lattice parameter	Å	5.654	5.87
Band Gap E_g (opt)	eV (300°K)	1.43	1.26
Density	g cm^{-3}	5.32	4.78
Dielectric constant		12.5	12.35
Refractive index		3.3	3.26
Coefficient of expansion	°C^{-1}	5.9.10^{-6}	4.5.10^{-6}
Thermal conductivity	W°C^{-1}cm^{-1}	150/T for n type 120/T for n^{+}	~ 170/T
Atomic Number		31/33	49/15

appendix 2 properties of metals, alloys, and dielectrics

Material	Melting Point (°C)	Thermal Conductivity ($W\ cm^{-1}\ {}^\circ K^{-1}$)	Diffusivity ($cm^2 s^{-1}$)	Expansion Coefficient
Gold	1063	3.16	1.18	$13.9.10^{-6}$
Copper OFHC	1083	4.0	1.14	17.10^{-6}
Silver	960	4.27	1.71	$18.8.10^{-6}$
Aluminum	648	2.36	0.86	25.10^{-6}
Molybdenum	2650	1.46	0.58	
Tin	232	0.625	0.38	$21.4.10^{-6}$
Lead	328	0.34	0.25	$29.1.10^{-6}$
Diamond (IIA)	—	22.0	1.01	$1.2.10^{-6}$
(IIB)	—	13.0	0.59	
Alumina	—	0.505		$8.7.10^{-6}$
Quartz	—	—	—	$0.5.10^{-6}$
Gold — 12% Germanium	356	—	—	—
Gold — 6% Silicon	370	—	—	—
Gold — 20% Tin	280			

appendix 3 constants

Boltzmann's Constant
$K = 8.63.10^{-5}\ eV/{}^\circ K$
$= 1.38.10^{-16}\ ergs/{}^\circ K$

Electron charge
$q = 1.6.10^{-19}$ coulomb

Permittivity of free space
$\epsilon_o = 8.85.10^{-14}\ F/cm$

$KT/q = 0.026$ volt (300° K)

Planck's Constant
$h = 6.62.10^{-27}$ erg sec

references

1/ Mullin, J.B.; Straughan, B.W.; and Brickell, W.S.: "Liquid Encapsulation Techniques," *Journal of Physics and Chemistry of Solids, Vol. 26,* p. 782 (1965)

2/ Mullin, J.B.; Royle, A.; and Straughan, B.W.: "The Preparation and Electrical Properties of InP Crystals Grown by Liquid Encapsulation," *Institute of Physics and Physical Society Symposium on GaAs* (1970)

3/ Gilman, J.J. (ed.): *The Art and Science of Growing Crystals*, Wiley (1963)

4/ Cairns, B.; and Fairman, R.: "Effect of the $AsCl_3$ Concentration on Impurity Concentration in Vapour Grown Epitaxial GaAs," *Journal of Electrochemical Society, Vol. 115,* 327C (1968)
Vol. 117, 197C (1970)

5/ Joyce, B.D.; and Williams, E.W.: "The Preparation and Photoluminescent Properties of High Purity Vapour Grown InP Layers," *Institue of Physics and Physical Society Symposium on GaAs* (1970)

6/ Clarke, R.C.: *Private Communication*

7/ Knight, J.R.: *Private Communication*

8/ Rosztoczy, R.E.; Antypas, G.A.; and Casau, C.J.: "Distribution Coefficients of Ge, Sn, and Te InP Grown by Liquid Phase Epitaxy," *Institue of Physics and Physical Society Symposium on GaAs* (1970)

9/ Wood, C.E.C.; Tree, R.J.; and Paxman, D.H.: "Solution Grown Epitaxial InP for High Efficiency Circuit Controlled Microwave Oscillators," *Electronics Letters, Vol. 8,* pp. 171-173 (1972)

10/ Clark, G.M.; and Griffiths, I.: "An Investigation of X- and J-Band Transverse Gunn Oscillators," *Eighth MOGA Conference* (1970)

11/ Galginaitis, S.V.; Barnett, A.M.; and Henmann, F.K.: "Selective Liquid Epitaxy of GaP," *Institute of Physics and Physical Society Symposium on GaAs* (1970)

12/ Chang, L.L.; and Casey, H.C.: "Diffusion and Solubility of Zinc in InP,"

Solid State Electronics, Vol. 7, pp. 481-485 (1964)

13/ Cardwell, M.: *Private Communication* (1973)

14/ Colliver, D.J.; and Shaftoe, K.: *Unpublished RRE work*

15/ Asai, S.; Okazaki, S.; and Koders, H.: "Electrical Properties of n-type GaAs Layers Prepared by Solid-Solid Sulphur Diffusion," *Institute of Physics and Physical Society Symposium on GaAs* (1972)

16/ Murphy, R.A.; et al.: "Proton Guarded GaAs Impatt Diodes," *Institute of Physics and Physical Society Symposium on GaAs* (1972)

17/ Cho, A.Y.; and Hayashi, I.: "Epitaxy of Si Doped GaAs by Molecular Beam Method," *Metallurgical Transactions, Vol. 2,* pp. 777-780 (1971)

18/ Sedgwick, T.O.; Aboaf, J.A.; and Krongeib, S.: "Dielectric Films for Ge Planar Devices," *IBM Journal of Research and Development,* pp. 1-11 (1970)

19/ Irving, L.D.; and Colliver, D.J.: *Unpublished RRE work* (1973)

20/ Mullin, J.B.; et al.: "The Electrical and Photoluminescence Behaviour of Impurities in InP," *Institute of Physics and Physical Society on GaAs* (1972)

21/ Evans, C.A.: "Secondary Ion Mass Analysis: A Technique for 3-Dimensional Characterisation," *Analytical Chemistry, Vol. 44,* (Nov 1972)

22/ White, A.M.; et al.: "The Photoluminescence Spectrum of Bound Excitons in InP and GaAs," *Journal of Physics C, Vol. 5,* p. 1727 (1972)

23/ White, A.M.; et al.: "The Photoluminescence Spectra of Excitons Bound to Group II Acceptors in InP," *Solid State Communications, Vol. 11,* pp. 1099-1103 (1972)

24/ Morgan, D.V.; and Wood, D.R.: "The Analysis of Thin Epitaxial Layers of GaAs Using MeV α-Particles," *Physica Status Solidi, Vol. 17,* pp. K143-K147 (1973)

25/ Tietjen, J.J.; et al.: "The Origin of Macroscopic Surface Imperfections in Vapour Grown GaAs," *Institute of Physics and Physical Society Symposium on GaAs* (1968)

26/ Huber, A.M.; and Champier, G.: "Characterisation of GaAs Crystals," *Institute of Physics and Physical Society Symposium on GaAs* (1970)

27/ Halliwell, M.A.G.; Childs, J.B.; and O'Hara, S.: "Reflection on X-Ray Topography of Compound Semiconductors," *Institute of Physics and Physical Society Symposium on GaAs* (1972)

28/ *Surface Science, Vol. 25,* pp. 1-223 (1971)

29/ Smith, P.H.: "A Theoretical and Experimental Analysis of the Ellipsometer," *Surface Science, Vol. 16,* pp. 34-66 (1969)

30/ Raif, J.; et al.: "Rapid Non-Destructive Method for Measuring the Refractive Index and Thickness of Thin Dielectric Films," *Journal of Physics E, Vol. 6,* pp. 48-50 (1973)

31/ Van der Pauw, L.J.: A Method of Measuring the Resistivity and Hall Coefficient on Lamellae of Arbitrary Shape," *Philips Research Reports, Vol. 13,* pp. 1-9 (1958)

32/ Smith, R.A.: *Semiconductors*, Cambridge (1964)

33/ *Electronic Engineering* (June 1968)

34/ Stearn, J.W.; and Colliver, D.J.: *Unpublished RRE work* (1972)

35/ Copeland, J.A.: "A Technique for Directly Plotting the Inverse Doping Profile of Semiconductor Wafers," *IEEE Transactions on Electron Devices, Vol. ED-16, No. 5,* pp. 445-449 (1969)

36/ Baxandall, P.J.; Colliver, D.J.; and Fray, A.F.: "An Instrument for the Rapid Determination of Semiconductor Impurity Profiles," *Journal of Physics E. Vol. 4,* pp. 213-221 (1970)

37/ Hughes, F.D.: "The Characterisation of Epitaxial GaAs Using Schottky Barriers," *Acta Electronica, Vol. 15,* pp. 43-53 (1972)

38/ Copeland, J.A.: "Semiconductor Impurity Analysis from L.F. Noise Spectra," *IEEE Transactions on Electron Devices, Vol. ED-17, No. 5,* pp. 404-407 (1970)

39/ Smith, B.L.; and Rhoderick, G.A.: "Possible Sources of Error in the Deduction of Semiconductor Impurity Concenteation from Schottky Barrier (GV) Characteristics," *Journal of Physics D, Vol. 2,* pp. 465-467 (1969)

40/ Copeland, J.A.: "Switching and Low Field Breakdown in GaAs Bulk Diodes," *Applied Physics Letters, Vol. 9,* pp. 140-142 (1966)

41/ Sah, C.T.; et al.: "Thermal and Optical Emission and Capture Rates and Cross Sections from Photo and Dark Junction Current and Capacitance Experiments," *Solid State Electronics, Vol. 13,* pp. 759-788 (1970)

42/ Ikoma, T.; and Jeppson, B.: *IEEE Symposium on GaAs* (1972) (Unpublished)

43/ Weisberg, L.R.; and Schade, L.: "A Technique for Trap Determinations in Low Resistivity Semiconductors," *Journal of Applied Physics, Vol. 39,* p. 5149 (1968)

44/ Bube, R.H.: *Photoconductivity of Solids,* p. 292

45/ Forbes, L.; and Sah, C.T.: "On the Determination of Deep Level Centre Energy and Concentration by TSC Measurements Using Reverse Biased p-n Junctions," *Solid State Electronics, Vol. 14,* p. 182 (1971)

46/ Shockley, W.: "Electronics, Holes, and Traps," *Proceedings of the IRE,* pp. 973-990 (1958)

47/ Van der Ziel, A.: *Fluctuation Phenomena in Semiconductors* Butterworth (1959)

48/ Mircea, A.; et al.: "LF Noise in n type GaAs and its Correlation with HF Noise in Gunn Oscillators," *Acta Electronica, Vol. 15,* pp. 233-243 (1972)

49/ Nicollian, E.H.; and Goetzberger, A.: "The Si-SiO_2 Interface – Electrical Properties as Determined by the MIS Technique," *Bell System Technical Journal, Vol. 46,* pp. 1055-1133 (1967)

50/ Quast, W.: "Small Signal Admittance of the Insulator-n GaAs Interface Region," *Electronics Letters, Vol. 8,* pp. 419-421 (1972)

51/ Faktor, N.M.; and Ambridge, R.: Paper presented at the *Conference on Analytical Techniques for Crystal Cahracterisation,* Oxford (1973)

52/ Clarke, G.M.; and Griffiths, I.: "An Investigation of X- and J-Band Transverse Gunn Oscillators," *Eighth MOGA Conference* (1970)

53/ *Integrated Circuit Engineering* Boston Technical Publishers

54/ *Solid State Technology* (various issues)

55/ Colliver, D.J.; et al.: "Cathode Contact Effects in InP Transferred Electron Oscillators," *Institute of Physics and Physical Society Symposium on GaAs* (1972)

56/ Eimers, G.W.; and Stevens, E.H.: "A Composite Model for Schottky Diode Barrier Height," *IEEE Transactions on Electron Devices, Vol. ED-18, No. 12* p. 1185 (1971) 646

57/ Padovani, F.A.: *Semiconductors and Semimetals, Vol. 7A* (Willardson and Beer, eds.), Academic Press (1972)

58/ Mead, C.A.: *Ohmic Contacts to Semiconductors* (B. Schwartz, ed.), Electrochemical Society (1969)

59/ Panish, M.B.: "The Ga-As-Sn and Ga-As-Ge Ternary Systems," *Journal of Less Common Metals, Vol. 10,* pp. 416-424 (1966)

60/ Cox, R.H.; and Hasty, T.E.: *Ohmic Contact to Semiconductors* (B. Schwartz, ed.), Electrochemical Society (1969)

61/ Guetin, P.; and Schreder, G.: "Effects of Ion Bombardment on the Characteristics of Metal/n-GaAs Tunnel Contact," *Journal of Applied Physics, Vol. 43,* pp. 549-560 (1972)

62/ Furukawa, Y.; and Ishibashi, Y.: "Vapor Plating of Tin on to GaAs," *Japanese Journal of Applied Physics, Vol. 6,* pp. 787-793 (1967)

63/ Hilsden, F.J.: *Private Communication* (1970)

64/ Rhoderick, E.H.: "The Physics of Schottky Barriers," *Journal of Physics D, Vol. 3,* pp. 1153-1188 (1970)

65/ Mead, C.A.: "Metal Semiconductor Surface Barriers," *Solid State Electronics, Vol. 9,* p. 1023 (1966)

66/ Moroney, W.; and Anand, Y.: "Low Barrier Height GaAs Microwave Schottky Diodes Using Au-Ge Alloy," *Institute of Physics and Physical Society Symposium on GaAs* (1970) (also Ogawa, M.; et al., same publication)

67/ Upadhyayula, L.C.; Jolly, S.T.; and Huang, H.C.: "High Efficiency GaAs IMPATT Structures," *RCA Review, Vol. 35,* pp. 567-578 (1974)

68/ Smith, B.L.: *Private Communication* (1971)

69/ Cardwell, M.J.; and Peart, R.F.: "Measurement of Carrier Concentration Profiles in Epitaxial InP," *Electronics Letters, Vol. 9,* p. 88 (1973)

70/ Gibbons, G.K.; and Colliver, D.J.: *Unpublished RRE work* (1973)

71/ Nuese, C.J.; and Cannon, J.J.: "Silver-Manganese Evaporated Ohmic Contacts to p-type GaAs," *Journal of Electrochemical Society, Vol. 115,* pp. 327-328 (1968)

72/ Colliver, D.J.; et al.: "Material Selection For Efficient Transferred Electron Devices at Q-Band," *Electronics Letters, Vol. 6,* pp. 353-355 (1970)

73/ Fowler, R.H.: "Photoelectric Sensitivity Curves for Clean Metals at Various Temperatures," *Physics Review, Vol. 38,* p. 45 (1931)

74/ Gray, K.W.: *Private Communication* (1973)

75/ White, P.M.; and Gibbons, G.: "High Efficiency CW Operation of Anomalous InP Microwave Oscillators," *Electronics Letters, Vol. 8,* p. 166 (1972)

76/ Frey, J.: "Multimesa versus Annular Construction for High Average Power in Semiconductor Devices," *IEEE Transactions on Electron Devices, Vol. ED-19, No. 8,* pp. 981-985 (1972)

77/ Gibbons, G.; and White, P.M.: "CW and Pulsed InP Transferred Electron Oscillators," *European Microwave Conference Stockholm* (1971)

78/ Riley, T.J.: "Design Calculation for CW Millimetre Wave LSA Oscillators," *Institute of Physics and Physical Society Symposium on GaAs* (1968)

79/ Narayan, S.Y.; and Packowski, J.P.: "Integral Heat Sink Transferred Electron Oscillators," *RCA Review, Vol. 33,* pp. 753-765 (1972)

80/ Swan, C.B.: "Improved Performance of Si Avalanche Oscillators Mounted on Diamond Heat Sinks," *Proceedings of the IEEE, Vol. 55,* pp. 1617-1618 (1967)

81/ Seal, M.: "Diamond Heat Sinks," *Proceedings of the Industrial Diamond Conference,* pp. 191-198 (1969)

82/ Meyer, M.: *Thermal Resistance of Gunn Diodes: Analysis and Measurement, Vol. 1,* Siemmens Forsch, pp. 302-306 (1972)

83/ Haitz, R.H.; Stover, H.L.; and Tolar, N.J.: "A Method for Heat Flow Resistance Measurements in Avalanche Devices," *IEEE Transactions on Electron Devices, Vol. ED-16, No. 5,* pp. 436-444 (1969)

84/ Davies, S.E.; et al.: "Glass Ceramic-Copper Microwave Encapsulation and Mountings," *Electronics Letters, Vol. 8,* pp. 483-484 (1972)

85/ Caldwell, J.F.; and Rosztoczy, F.E.: "GaAs Gunn Diodes for Millimetre Wave and Microwave Frequencies," *Institute of Physics and Physical Society Symposium on GaAs* (1972)

86/ *Compound Semiconductors, Vol. 1,* (R.K. Willardson and H.L. Goering, eds.) pp. 207-221

87/ Yarborough, D.W.: "Status of Diffusion Data in Binary Compound Semiconductors," *Solid State Technology,* pp. 23-31 (November 1968)

88/ Gyulai, J.; et al.: "Alloying Behaviour of Au and Au-Ge on GaAs," *Journal of Applied Physics, Vol. 42,* pp. 3578-3585 (1971)

89/ Murphy, R.A.; et al.: "Performance and Reliability of Ka-Band GaAs IMPATT Diodes," *IEEE S-MTT Symposium Digest,* pp. 315-317 (1974)